Observation, Experiment, and Hypothesis in Modern Physical Science

Studies from the Johns Hopkins Center for the History and Philosophy of Science

Peter Achinstein and Owen Hannaway, editors, *Observation, Experiment, and Hypothesis in Modern Physical Science*, 1985

Observation, Experiment, and Hypothesis in Modern Physical Science

edited by
Peter Achinstein
Owen Hannaway

a Bradford book

The MIT Press
Cambridge, Massachusetts
London, England

5242408 X

This book was set in Palatino
by The MIT Press Computergraphics Department
and printed and bound by The Murray Printing Co.
in the United States of America.

Library of Congress Cataloging in Publication Data

Main entry under title:

Observation, experiment, and hypothesis in modern physical science.

(Studies from the Johns Hopkins Center for the History and Philosophy of Science)
   "A Bradford book."
   Bibliography: p.
   Includes index.
   1. Physics—Philosophy—History. 2. Physics—Experiments—History. 3. Science—
Philosophy—History. 4. Science—Methodology—History. I. Achinstein, Peter. II.
Hannaway, Owen. III. Series.
QC6.026 1985   530'.01   84-20723
ISBN 0-262-01083-6

Contents

QC6
826
1985
PHYS

Preface

The Johns Hopkins Center for the History and Philosophy of Science was founded in 1969 for the purpose of sponsoring scholarly activities of interest to both disciplines. For the 1982–83 academic year a special theme was chosen—the testing of hypotheses in modern physics by observation and experiment—and several philosophers of science and historians of science were invited to contribute papers to a volume devoted to this theme. The result is the present work, the first in a projected series. Each volume in the series will address an issue that both philosophers of science and historians of science can usefully study.

Four of the chapters in this volume are devoted to particular experiments, the history of the instruments and techniques employed, and the hypotheses they are intended to test. Peter Galison describes the origin and development of the bubble chamber from its invention in 1952 to its employment in the early 1960s, Roger Stuewer provides a detailed account of a sharp dispute between Rutherford and Chadwick in Cambridge and Pettersson and Kirsch in Vienna over the interpretation of experiments in artificial disintegration, John Rigden gives the history of the magnetic-resonance method first demonstrated by I. I. Rabi, and Geoffrey Joseph suggests a statistical interpretation of quantum mechanics that can be employed to interpret the Stern-Gerlach and double-slit experiments.

The remaining chapters treat more general philosophical themes, usually with applications to modern physics. These chapters might best be understood, at least in part, as responses to two powerful claims frequently made about science. The first is that a scientific theory is a hypothetico-deductive system consisting of a set of hypotheses and of theorems that follow deductively and mathematically. The system is tested empirically by deriving conclusions, at least some of which can be verified by experiment and observation. If these conclusions are found to be true, support is thereby secured for the hypotheses. The second claim is that in the scientific vocabulary used to formulate the theory there is a distinction between "theoretical" terms (which purport to denote unobservable entities postulated by the theory, such as protons and quarks) and "observational" terms (which purport to denote objects and properties that are directly observable, such as lines on photographic plates and tracks in a bubble chamber). The theoretical terms have meaning only in the context of the theory in which they are used; their meaning can change from one theory to another. In contrast, the observational terms are theory-neutral. They can be applied and understood independent of any theory in which they appear. Hypothetico-deductivism and the "theoretical/observational" distinction are independent ideas, but together they form the basis for the logical positivist or empiricist philosophy of science that was so influential during parts of the twentieth century.

Each of the philosophical chapters has something interesting to say either in criticism or in defense of one of these claims. Dudley Shapere provides a general schema for the concept of observation that appeals to the idea that information is received directly by an appropriate receptor. But to flesh out this schema—to decide what counts as information, as directly received, and as an appropriate receptor—will (contrary to the tenets of those defending the usual theoretical/observational distinction) depend on prior theoretical knowledge. In contrast, Lawrence Sklar argues that considerations from Einstein's special and general theories of relativity dictate the need for an "absolute" (non-theory-dependent, noncontextual) distinction between the observable and the unobservable.

One problem with hypothetico-deductivism is that it provides either an incomplete or a fallacious picture of the testing of hypotheses by observation and experiment. Peter Achinstein criticizes the usual rules of inference from observations to hypotheses that hypothetico-deductive (h-d) theorists employ. He proposes a rule that takes into account both the plausibility of the hypothesis in light of the observations and the probability that there exists an explanatory connection between the hypothesis and the observations. Richard Jeffrey suggests an alternative definition of confirmation in terms of probabilities and proves two rules (Huygens's rule and Popper's rule) that show what probability basis can be given for the h-d method. One feature of this method, at least on most versions, is that only observations provide experimental evidence for a theory. Richard Boyd, however, asserts that this view is too limited. He argues that, in a perfectly reasonable sense, explanatory power and simplicity are "experimental" criteria that provide genuine confirmation for a theory. Furthermore, if one takes a realist interpretation of theories, as he does, then a distinction between "direct" and "indirect" evidence is of no fundamental significance. Ronald Laymon demonstrates that idealizations and approximations, which are often used in the physical sciences but rarely discussed in the philosophy of science, raise serious difficulties for the h-d method. Laymon develops a novel theory of confirmation in order to meet these difficulties.

Complementing the chapters that concentrate on broad philosophical issues, the historical contributions of Rigden, Stuewer, and Galison offer a close, even intimate picture of aspects of the practice of experimental physics in the twentieth century. All three of these chapters have strong narrative lines sustained by frequent references to such unpublished sources as laboratory notebooks, private correspondence, and personal interviews, and each author is sensitive to the structure and the dynamics of the research community in which the search for knowledge and fame is either fulfilled or frustrated. Stuewer and Galison give a vivid picture of the explosive growth in complexity and cost of the apparatus developed to explore the structure of the atom in the course of 40 years.

These historical chapters amply demonstrate that in many areas of modern physics the experiment is no longer the province of the individual but that of the team. This development is due, in part, to the technological sophistication of the experimental procedures and to the professional function of the modern laboratory in initiating a new generation of investigators into the theoretical and technical skills of the research frontier. That the scientific community and not the individual scientist is the critical unit defining the norms of scientific theory and practice is an insight we owe to T. S. Kuhn. The chapters by Rigden, Stuewer, and Galison suggest that, for twentieth-century physics at least, we must refine our notion of the experimental community. It is clear that this community is not, as some have assumed, a homogeneous entity composed of practitioners of equivalent status and commitment to the outcome of the experiment. As Galison points out, the community of experimenters in particle physics is integrated not only horizontally among the physicists but also vertically among physicists, administrators, graduate students, technicians, and even lay "observers" of experimental events. An experiment in modern physics is frequently a collective endeavor of distinctive groups of differing status, each with specialized functions in the design and performance of the experiment and sometimes with different investments in its outcome.

That the social stratification of the experimental community appeared even before the development of the industrial-scale laboratory is clearly shown in Stuewer's discussion of the Cambridge-Vienna controversy. This is essentially the story of how an experimental challenge to a theory of atomic disintegration advanced by the eminent Cambridge physicist Ernest Rutherford was resolved informally when Rutherford's assistant Chadwick successfully challenged the performance of the women employed as scintillation counters by the rival Viennese group. Such an attempt to defuse a dispute over an experiment was possible only because a division of labor had separated the eyes of the observers from the minds of the experimenters, so that the former could be evaluated in the light of technical performance without calling into question the scientific integrity of the latter.

With the transition from the "shop" culture of the laboratory of the 1920s (with its characteristic condescending exploitation of female employees) to the quasi-corporate enterprise of the 1960s (which, as Galison describes, was attendant upon the development of the bubble chamber), the division of labor has become even more pronounced and the scale has become so large that management techniques and practices are required to integrate the totality of the experimental activity. Efforts have been made—not without controversy—to replace human observers and scanners with automated data-processing techniques.

It is not surprising that one effect of this industrial culture on scientific practice has been to engender in some physicists classic sentiments of alienation from their experimental work and from nature itself. Galison provides two eloquent testimonies to this. The first is from the young Donald Glaser, who was repelled by the "factory environment" of accelerator physics; the second is from Luis Alvarez, who, shortly before winning the Nobel Prize in 1968 for the results of the huge bubble-chamber program at Berkeley, lamented the dullness and routine of nuclear physics and deplored how it had divorced the physicist from his experiments. It would appear that, for many scientists, Rigden's depiction of I. I. Rabi's ideal of an experiment where one knows one's results "at the end of the day" has passed from history into nostalgia.

Peter Achinstein and Owen Hannaway

Observation, Experiment,
and Hypothesis in
Modern Physical Science

1

Modestly Radical Empiricism

Lawrence Sklar

The observable/nonobservable distinction bears a heavy burden in empiricist approaches to theories. That such a distinction exists and that its existence is not excessively context-relative seem essential if observationality is to play as distinguished a role in our account of the sources of knowledge and of meaning as the empiricist demands. Beyond this, the distinction must be "hard" enough to persuade us that the differential attitude toward meaning, epistemic warrant, and (for some empiricist theories of theories) ontological commitment, which differentiates the observable from the nonobservable in empiricist accounts, can be rationalized relative to the distinction's nature. Too relative a distinction, or one that is too "soft" in other ways, will vitiate, with its modest nature, any plausibility attributable to the empiricist account.

In the realm of epistemic warrant, observations are, for the empiricist, foundational. All knowledge or reasonable belief (mathematics, logic, and the like to the side) begins with our knowledge of the contents of observation. For some empiricists, observational facts can be known with certainty. For the more modest, they are the only facts of the world that bear intrinsic, noninferential warrant for belief. Perhaps there are more modest empiricisms still; however, for any doctrine to be called empiricist at all there must at least be a hierarchical structure to our knowledge, with observational knowledge "closer to the base" of the pyramid of knowledge than the theoretical reasonable belief founded upon the observational basis.

Perhaps even more profoundly, observationality is, for the empiricist, the source of meaning. However meaning accrues to the theoretical vocabulary characterizing nonobservable entities and properties, it must first accrue to language describing the observational features of the world; only then can it percolate upward to the remaining portions of discourse. However modestly and delicately construed, some form of intending to use a word, accomplished by means of ostensive definition, must be at the root of our possibility of meaningful discourse. Whatever empiricism is, it entails the view that there is a world external to and independent of the language we use to describe it and that the junction point between language and world is located in the presence of the world to the language user in observationality.

But, of course, all this has come under merciless assault in recent years, having been subjected to arguments (some new and some revisions and modernizations of the old) intended to show that the whole empiricist program rests on a collection of mistakes, howlers, and non sequiturs. We are told that any possible distinction between observable and nonobservable is context-relative, merely pragmatic, soft, variable, and in any case nothing like the kind of distinction needed to support the empiricists' radically different construals of claims about the observable and the unobservable.

Further, we are informed, even if such a significant distinction between the observable and the nonobservable were possible, the observable could in no way play the significant roles reserved for it by the empiricists. Epistemically, foundationalism as a whole is simply wrongheaded. Knowledge is a web, a network of belief. There is no bottom, no top, and indeed no hierarchical direction at all. So there is no need for observational beliefs to provide a foundation on which the whole structure of knowledge rests—which is a good thing, as they could not possibly play such a role were a candidate for the job needed.

Semantically, too, we are told, the empiricists' claim of a special role for observations in the grounding of meaning is a myth. Language is public, and meaning is use. If there is anything left for meaning to be at all, it is something to be approached from the viewpoint of a holistic, functionalistic account of communication. There is no place left in any serious meaning theory for a special role of observations, functioning through ostensive definitions, to provide the basis of meaning accrual— again, a good thing, since Wittgenstein has shown us that observations in the empiricist sense, even if they existed, simply could not play the role demanded of them by the empiricist account of meaning.

At least two related but quite distinct accounts of knowledge and meaning are offered to take the place of the discredited empiricist account: naturalism and pragmatism. I intend to sketch, in the broadest and crudest possible strokes, some of the aspects of these two alternative accounts of knowledge and meaning. Then I plan to look at the epistemic and semantic role that observationality seems to play in a number of apparently special theoretical contexts. I will argue that neither the naturalistic nor the pragmatic account of the role of observationality does justice to the special role the observational/nonobservational distinction plays in these special theoretical contexts, and that these two alternative accounts also fail to do justice to the apparently very special place of the observational in epistemic and semantic critique of theories (again in these special contexts). I will argue that the empiricist account, whatever its ultimate faults, seems to do a better job of reconstructing and rationalizing our actual scientific practice in these special cases. Then I will try to explore a little the question of just what makes these cases special and therefore supportive of empiricism against its rivals. In particular I will ask whether the support given to the empiricist account in these cases can be generalized to argue that, overall, empiricism just seems to do a better job than its rivals of accounting for facts and intuitions.

The naturalist tells us to abandon "armchair a priorism" in philosophy. If there is anything we can reasonably say about the world and our place in it, it is only that which is a consequence of our best currently available theory of the world. This theory, which summarizes our best established scientific knowledge of the world, is the repository of our knowledge of the world and tells us all there is to know. Nothing comes before the theory and nothing stands outside it. If epistemic critique and semantic analysis are to be done at all, they are to be done as merely specialized components of the general scientific attempt to grasp the nature of the world.

The theory, according to the naturalist, determines what is observable and what is not. Observers and their observations are as much in the world as tables, clouds, and quarks, and if there is anything to say about them our theory of the world says it. This, of course, could be compatible with many positions on the observable/nonobservable distinction. But, in fact, the naturalist camp has a fairly unified view. The theory of the world is physicalist. Observers are more or less reliable measuring instruments, reponding to stimuli impinging on them from the remainder of the world in causally correlated ways so that we can infer, more or less reliably, from the observers' functionally characterized states to states of the impinging environment. Of course, from this point of view the observational/nonobservational distinction is highly context-dependent and relative. Functionally characterized brain states are "observations" relative to the psychologist's interest in experiments on perception; however, relative to the astronomer's interest in stellar constitution, spots on photographic plates or the output of image-amplifier tubes function just as well as the "observations" correlative to the particular phenomena being observed.

Knowledge, from this naturalist perspective, is just true belief acquired by a reasonably reliable process. The only grain of truth remaining of the empiricist view that observation is at the ground or root of all knowledge is the empirically established truth that, as a reliable indicator of what the world is like, what we take to be observations (in the naturalistic sense of how observations actually function in science) are more reliable indicators of the true state of the world that are, for example, wild guesses or the intuitive states of the seer. This fact of reliability can, of course, be explained by the causal origin of observations correlating them to the states of the impinging world. Epistemic critique, from this point of view, can only amount to demanding of a theory of the world internal self-consistency. Something is wrong if the theory tells one, in its own terms, that one's grounds for believing it are unreliable. Beyond this one cannot go.

About meaning we have much less in the way of firm agreement from the naturalists, but all naturalists agree that an account of meaning is to be in terms of some sort of functionalist account superimposed on an underlying physicalist world view. The role of language in public communication, and meaning as some sort of "coarse grid over use," are common elements of the sketchy naturalistic semantics we have been offered. Although constraints of rationality may leave a semantic account of meaning distinct in kind from the physicalist lawlike basis that describes the world of things and events from a naturalistic point of view, one thing is clear: There is no place in a theory of meaning for some sort of private first-person intending of words to things by the mediation of ostensive association of words with observational content in any kind of respectable, naturalistic meaning theory.

It is harder to pin down the pragmatic approach than the naturalist, even in the crude way we must be satisfied with here. Essential, surely, is the denial of any kind of viewpoint external to our immersion in our present schema of belief from which the schema as a whole could be epistemically or semantically examined and criticized. There is only the ongoing process of piecemeal revision, guided by precepts of rejection and innovation, themselves the product rather than the presupposition of our accepted ways of acting in the world, where action includes as but one component the activities of speaking and describing.

Whatever the distinction between observable and nonobservable may be for the pragmatist, it is most assuredly context-dependent and relative. We speak correctly of persons observing the red patch before them (or even appearing to them in hallucinatory experience), but equally correctly of the physicist observing the passage of the K meson in the Wilson cloud chamber. Of course, our distinction between what we observe and what we infer may sometimes have a point, say in offering an appropriate in-context critique of someone's illegitimate assurance. (A defense attorney asks a prosecution witness: "But you didn't actually observe the murder, did you? You were on the other side of the door." However, hearing the screams of the victim from the other side of the door certainly is "observing the murder" when contrasted with the experience of the murder being gained by the jurors in the courtroom.)

What is the epistemic special place of observation for the pragmatist? For the pragmatist, what is desirable is parasitic on what is desired. Similarly, what is believable is parasitic on what is believed. Our rules for rational believability are simply the reflective synopses of the principles by which we do in fact accept beliefs as warranted. From this point of view observations do have a kind of epistemic priority, but it is nothing like that given to them by the empiricist. Since observations do, in general, constitute the locus of firm, shared agreement, and since agreed belief is the only ground on which agreed rational believability can be based, the fact that beliefs accrued by observation are usually beliefs shared with some degree of certainty places them in a more or less central point in our structure of accepted truth. But that is all there is to their centrality.

Semantically, with the pragmatist as with the naturalist, observations are reduced to a much lower role than that granted them by the empiricist. Meaning, if it is anything at all, is use. A notion of meaning may play a role in some discourse used to talk (more or less theoretically) about discourse, but no special place will be played in that account by any mythical act of intending a word to designate an item of "pure experience." A *reductio ad absurdum* of the empiricist view is provided, from this standpoint, by mere reflection upon the fact that before experience can enter into our cognitive structure of belief it must be experience "always already" conceptualized and linguified. All terms function according to their total role in our language scheme for finding our way about in the world; all discourse is "always already" theoretical discourse, and all experience is "always already" framed in terms of our conceptual apparatus with its embodied theoretical presuppositions.

Of course, the answer to the empiricist's question as to how then language connects up with a preconceptualized world is that the very idea of such a world is a well-lost myth, part of the naively dualist scheme of world and representing perceiver.

As I noted earlier, my aim is not to pursue in extensive and careful detail just what the empiricist, naturalist, and pragmatist options are, but rather to look at their suitability, crudely characterized as they may be, to give a rational reconstruction of our actual scientific practice in a couple of apparently very special cases of scientific theorizing. Later we shall consider whether these cases are really as special as they first appear to be, or whether instead they ought to be taken as paradigmatic of the case of rational scientific decision making, differing from the more usual cases only in having presented in the special cases in a bold and surface way certain features that are only more subtly discernible in the more standard cases.

The special cases I have in mind are Einstein's two "purifications" of physics: the purification by elimination of the aether frame, which is the fundamental accomplishment of special relativity, and the purification by the elimination of global inertial reference frames, which is the fundamental accomplishment of general relativity.

In the first case there was an antecedent theory, Maxwellian electromagnetism, which, as Einstein clearly saw, possessed an element usually otiose for the prediction of observable phenomena: the absolute velocity of the system relative to the aether frame of electromagnetism. The null results of the famous round-trip experiments on light eliminated what could plausibly be construed as the last possibility for finding some observational consequence of uniform motion relative to the aether frame. Einstein's genius allowed us to see that, with a sufficiently radical revision of the most fundamental ideas of space and time, a new electromagnetic theory could be constructed that would give the usual observational results in the standard experimental cases and the null results in the case of the new round-trip experiments, without the peculiar "putting absolute velocity in only to take it out again" that infected the earlier compensatory theories. Of course he then went on to invent a novel mechanics properly invariant relative to the new spacetime picture, but this, important as it is to physics, is for us only a derivative consequence of the fundamental move.

The origin of general relativity can be viewed from a similar perspective. The principle of equivalence shows us that the traditional joint theory of mechanics plus gravitation posits a distinction without an observational difference. A world with a uniform gravitational field everywhere is indistinguishable by any observational means from one without that field. The theoretical otioseness can be cashed out by moving from a theory that treats the global inertial frames of spacetime and the gravitational field separately to a new pseudo-Riemannian spacetime in which nothing but the geodesic structure of the spacetime exists, determining the local inertial reference frames and simultaneously fixing the inertial and gravitational properties of the world everywhere. Once again there is much left out in this picture (such as the fact that it is now relativistic gravitation that is assimilated to spacetime structure and not Newtonian gravitation, and the importance of the generation of the new field equations connecting geodesic structure to mass-energy source), but for our purposes these are again not nearly so interesting as the fundamental move of purifying an antecedent theory of elements that are totally otiose from an observational point of view.

What is special about these cases of the construction and the acceptance of scientific theories? For one thing, the distinction between the observable and the nonobservable is crucial to the whole scientific enterprise in these cases. Unless we are assured that in our notion of the observable consequences of a theory we have genuinely captured all possible observables, the reasoning just does not go through. In both special relativity and general relativity we are told to delimit ourselves to coincidences and continuity along paths traversible by causal signals in trying to map out the spacetime structure of the world. If we allow ourselves to go beyond this (countenancing, say, simultaneity for spatially separated events, or the absolute magnitude of the gravitational force field, or the geodesic structure of spacetime itself among the class of things that could, in principle, be observable), then the whole Einsteinian program of demonstrating to us the theoretical otioseness of aspects of established theory, and then revising theory in such a way as to eliminate from it those elements totally irrelevant to observational prediction, becomes an impossibility. If we could ever, in principle, determine simultaneity for distant events in a noninferential, observational way, then all of the arguments designed to show us that the aether frame is an in principle otiose element of physics fail. And the argument goes through in the same way regarding the allegedly otiose notion of a global inertial frame dispensed with in general relativity.

This is not to say that the features of the world called "observable" in presentations of relativity are what the radical empiricist philosopher would count as really, truly, genuinely observable; only that what is counted in these theories as forever unobservable must be, in principle, in that class. For Einstein's arguments to be the least bit persuasive we must be assured, in a manner independent of the theories proposed and ultimately accepted, that there are nonrelative, noncontextual limits on the domain of what is to be counted as observable. This holds true independent of one's views of the status of the theories one ultimately accepts in an Einsteinian way. Whether one believes that the theories are to be interpreted realistically or instrumentalistically, and whether one believes that the choice of alternative theory is a matter of convention, of *a priori* plausibility, or of projection from past accepted theory, the very structuring of the scientific decision process requires that one not be prepared to challenge Einstein's hard, context-independent, and irrevocable limitation on the domain of the observable.

Other aspects of this problem of theory choice have a degree of "specialness" as well. The theories presented (Minkowski spacetime, curved spacetime) are novel and radical. Whereas the older theories they are to replace have a certain familiarity and built-in-ness as components of our ordinary way of looking at the world, the new alternatives—the ones any rational person would accept according to most of us with intuitions about this—present us with a view of the world sufficiently startling that it take us a lot of effort and use of analogy to feel that we really even understand exactly what the picture of the world presented to us by the new theory amounts to.

Furthermore, we have in these cases a presentation in a fully worked out form of something philosophers frequently allege to exist in other cases but have a hard time demonstrating. We have alternative, fully characterized and developed theories of the world, all of which agree with each other on what has been presupposed in the context to exhaust all possible observational data. It is one thing to be told that life might be a dream or that one might be a brain in a vat and yet all one's experiences might be the same; it is another thing again to have two fully developed theories before one, and a fully characterized notion of what constitutes all their consequences plausibly characterized as observational, along with a demonstration in the formally fullest manner that, as incompatible on the surface as the theories might be at the nonobservational level, their observational consequences are probably the same.

Nor, in this special case, can any of the observationally equivalent alternatives be glibly dismissed as so intrinsically implausible as to be unworthy of serious scientific consideration. Perhaps we can dismiss dream-life or brain-in-a-vat hypotheses as beyond plausibility—mere philosopher's alternatives that no rational man in the course of framing a theory about the world would even seriously consider. But in the special cases we have been considering such instant dismissal of alternatives would plainly be out of order. The alternative accounts to those we do accept (special and general relativity) not only are viable scientific hypotheses; they are the hypotheses all reasonable people in the scientific community did hold to (sometimes only implicitly and without even noticing that they were hypotheses) before Einstein pointed out to them that new and better alternative accounts of the world were available.

How do the claims of the naturalist fare in an attempt to understand what is going on in the special scientific situations on which we have focused our attention?

First there is the naturalist's claim that any relevant observational/ nonobservational distinction must be merely part of an established naturalistic picture of the world, and that, seen as such, it can be clearly seen to be at best context-relative. The theory tells us what is observable. But does it in these cases? I doubt it. Notice the independence of the assumption that the observational facts grounding spacetime theories must needs be local facts and facts about material systems. Coincidence of events and continuity of the path of a particle are legitimate components of observationality. Global facts such as simultaneity for distant events or facts about the structure of the spacetime itself (whether its geodesics are straight lines) are not even considered as possible components of the observable world. Where do these assumptions that must be made for the whole critique to get underway come from? From a naturalistic investigation of the actual constitution of observers as components of the naturalistic world? From a physicalist examination of their structure as more or less reliable recording instruments responding causally to the data? I doubt it. There is built into this scientific practice an "a prioricity" of what is observable that, whatever its origin, just does not seem to fit the naturalist's model. This can be seen clearly when we reflect on the fact that observationality as it is meant in these critiques just cannot be taken to be the context-relative thing the naturalist demands it to be. To be sure, once we have adopted general relativity we can speak of the astronomer "observing the null geodesic structure of the world" by watching the course of unimpeded light rays in outer space. That notion of observing is plainly not the same one we needed to get general relativity in the first place, though, for if we could have, all along, observed the geodesic structure of spacetime, we could have told from the beginning whether we were inertial or, instead, noninertial but falling along with everything else in a uniform gravitational field, and we would have never dropped the global inertial frames and gravity in favor of local inertiality (that is, in favor of general relativity) in the first place. The critiques Einstein gave us, however they are to be read, seem to require an absolute notion of at least nonobservability to get them underway, a distinction between the really nonobservable and the observable quite unlike anything compassable within the purely naturalistic scheme.

Epistemically, the naturalist tells us that all we can hope for is an indication that a method of inference is, according to the theory of the world that we do accept, a reliable method for coming to the truth. We know *a priori*, that such a notion of evaluating a method will be of no help to us when we are puzzled about just which of a number of seemingly equally good theories of the world we ought to believe, where the theories are not minor hypotheses about things whose general nature we already understand but radically distinguished general accounts of the structure of the world.

Given the data behind special relativity, ought I to believe that theory or perhaps one of the aether-theory alternatives? The naturalist says "Believe the theory that is determined to be believable by the methodology that will, according to your best theory of the world, most often lead to the truth." But how is this any help at all? If the world really is as Einstein tells us, then the usual procedure that leads us to special relativity is the one that is reliable. If the world is, rather, as the aether theory tells us, then the usual method is not reliable, for it misled us in one of the fundamental cases in which it was applied. But, of course, what we want to know is which alternative we ought to believe, sitting here with the data and with general inductive good practice and dubious about the fundamental facts of spacetime.

Lots of options do occur to us that have the virtue of being relevant to this situation. We can opt for skepticism and withhold judgment. We can opt for simplicity or conservatism or *a priori* plausibility to resolve the dilemma. We can develop a confirmation theory that gives the alternatives differing degrees of credence, even relative to exactly the same observational confirmations. Or we can try to undercut the skeptical problem in the traditional positivistic manner of arguing that there is no theory choice to be made at all, since observationally equivalent theories are fully the same theory differently expressed.

None of these may be satisfying solutions, but they are at least relevant options to be proposed here. That we, given our theory of the world, externally evaluate others as more or less reliable in their means and methods for discerning what is or is not the case is true enough, but it is totally irrelevant to the internalist problem of trying to extrapolate from what we can observationally discern to what the world is like. All of this is clear enough in an appraisal of reliabilism from a general perspective. However, faced with the problem of real, fully formulated alternatives, all of which coincide on the presupposed and generally unquestioned basis of what is taken in the scientific context to exhaust the observable, the criticism of reliabilism becomes at least more persuasive and more dramatic.

The naturalist does, of course, have a reply to this empiricist critique of his position as incapable of rationalizing the inference from observable data to inferred theory. The line we would expect would go something like this: If we can establish, on the basis of our best accepted theory to date, the reliability of some principle allowing us to infer from data to generalization, then we are justified in using that rule to project from new data to new generalizations. For example, relying on our accepted paleontological-geological theory of the world, we are justified in inferring from a newly discovered fossil type to the existence of a previously unsuspected extinct species rather than to some curious creationist alternative. But how would such a reliabilist rationalization of projective inference work in the cases in question? Only, I think, at a rather high-powered meta level. If our best accepted theory to date could tell us, for example, that simple theories of the world were overwhelmingly more often true than their more complex alternatives, then we could, on this naturalistic ground, rationalize the choice, say, of special relativity against the aether theories. I am extremely dubious, however, that any kind of reliabilist account of grounding the usual principles that allow us in the broadest cases to make our leap beyond data to theory can be made plausible. Where we are making limited inferences from only somewhat novel data to only somewhat novel new theories that are small components of our overall world view (as in the paleontological case noted above), we can have some hope that a broader component part of our accepted theory of the world can serve to rationalize our theory choice. Where, however, we are making global inferences from broad reaches of fundamental data to theories that themselves define our overall world picture, and where we have little to go on other than the traditional notions of simplicity or *a priori* plausibility in making our choice, I doubt that any attempt to characterize the rationalization of this kind of theory choice as being of the reliabilist type can succeed.

What about the naturalistic approach to semantics? How well does it fit in with what we seem to be impelled to say about meaning when dealing with the cases to which our attention has turned?

It is harder to say anything definite here than it was in the case of naturalistic epistemics, but I think that the general naturalistic approach tends to leave out something of crucial importance when it deals with cases of this special kind.

To say that meaning is derivative from use is vague and harmless enough. Where I doubt that the usual naturalistic approach to semantics functions adequately in our problem cases is where it disavows any special role in a meaning theory for ostensive association of words with directly presented situations of the world.

We need meaning theory in the problem cases primarily to resolve the question of the equivalence or the nonequivalence of observationally equivalent theories. I need not go into detail concerning all the options available here (Ramseyite approaches to theoretical terms, semantic analogy approaches, Craigian equivalence doctrines about theories, and so on) to say at least this: Any resolution of the problem of the standards for equivalence of theories seems, at some point, to require a radically asymmetric treatment of terms that are taken as referring to nonobservable entities and properties and terms that function by referring to the observable.

Not only the positivist, but even the realist, who wishes to have some standard for determining the equivalence or the nonequivalence of theories, seems to require that a certain part of the vocabulary of science be distinguished from the remainder in an adequate treatment of the accrual of meaning. Whether one believes the aether theories and special relativity to be equivalent (because of their observational equivalence) or to be nonequivalent (because of their nonisomorphism on the theoretical level) despite their observational equivalence, one will find oneself, when one takes the problem of equivalence seriously, treating the terms of the presupposed observational language radically differently from those of the nonobservational discourse. However it is done, sooner or later one finds oneself treating the theoretical discourse as parasitic on the observational. For some radical positivists the former is simply termwise definable by the latter. For some realists the theoretical terms are holistically fixed by their Ramsey sentence role in the total theory. Both approaches, however, require that the observational discourse come fully equipped in a meaningful way prior to its functioning in the theories in question.

Perhaps a relativity is implicit here. Vocabulary that is observational relative to one problematic will function as theoretical in another context. Perhaps there is no ground level of irreducibly observational discourse. Nonetheless, I think that ultimately one will have to allow for some discourse a distinguished role, in that its meaning is fixed not by role-in-theory but by some version of the ostensive intending of word to world mediated by presentation of the world in awareness so familiar from the empiricist tradition and so fervently eschewed by most naturalists.

How well does the pragmatist approach to the notion of observationality and to the epistemic and semantic role of observations fare when brought to bear on the special cases to which we are attending? Once again, I think that careful examination will show that empiricism, whatever its faults, gets us closer to what is going on than does pragmatism.

The pragmatist, perhaps even more fervently than the naturalist, wishes to deny the existence of any firm, context-independent distinction between the observable and the nonobservable. As before, I do not think this does justice to what goes on in the Einsteinian critiques. Once again, the claim is not that what is taken as observable in these critiques constitutes the ultimate level of pure, noninferential observationality, or even that such a level exists. Rather, the claim is that what is counted as nonobservational in the Einsteinian critiques must, for these critiques to be even remotely plausible, be nonobservational in some strong, in principle nonrelativized way. Were it even conceivable to us that we could, in the real sense of observability, observe whether distant events were simultaneous or observe the geodesic structure of spacetime itself, then the demonstration on the in-principle otioseness of the earlier theories would not go through. I do not believe that one does full justice to Einstein by viewing his alternatives, special and general relativity, as simply more plausible alternatives than the aether and spacetime-plus-gravitation theories they replace. What is crucial to Einstein's accomplishment is his demonstration of the methodological deficiency of the earlier theories. They simply are not the right kind of doctrines to be even considered legitimate candidates for the correct explanation of the world. Einstein's argument requires a conviction that some aspects of the world in these earlier theories' ontologies are unobservable in a rigorous, non-context-dependent way. This is not to deny, for example, that we are then permitted to talk, in the loose sense, of astronomers observing the curvature of spacetime.

Consider, in addition, the pragmatist claim that all observations are theory-laden. Perhaps they are, but the whole point of the Einsteinian critique, which emphasizes the restrictive observational basis we can truly count on in constructing spacetime theories, is to show us that with careful reflection we can, in a specific context, purge our characterization of the observational data of those theoretical presuppositions that stand in the way of our discerning what is truly empirically relevant in our theories and what is conceptually otiose. The whole point of the critiques, with their emphasis on coincidence and continuity along traversable spacetime paths as the limited body of truly observational data available to us, is to get us to see that, once purged of theoretical preconceptions, our data can be captured by a theoretical apparatus that is far more parsimonious in its ontology than were the earlier theories and is superior to them in that what has been discarded ought never to have been tolerated by a respectable methodology anyway.

Does this mean that we can eventually discover a basis of observationality purged of all theoretical preconceptions? Perhaps not. It is not so radical a radical empiricism that I am espousing. The important thing about the empiricist claim, which the pragmatist denies and which must be invoked to make sense of the Einsteinian critiques, is the hierarchical nature of our science, in which, lower in the hierarchy, are observations that in a given theoretical decision-making context have been detheorized in such a way as to purge them, as observations, of theoretical presuppositionality that would get in the way of a fair epistemic evaluation of the contending theories. Once again, since we have opted for one theoretical approach, our language of observationality may be reinfected with theoretical presupposition. Even our immediate data of private awareness may become structured, in the familiar Gestaltist way, with our theoretical bias. But it is the possibility of a rigorous purging of the relevant theorization of observational data that is relevant, and that is just the sort of thing Einstein shows us how to carry out in these particular cases.

Does the pragmatist do justice to the epistemic place of observations in these cases? I think not. Remember that for the pragmatist the only thing that observation statements have in favor of their special epistemic role is that they are a class of usually agreed upon, uncontroversial truths. I think that misrepresents the case, however. I think it is more plausible to claim that our naive spacetime picture garnered far more instinctive agreement among the community than would the truth of the null results of the Michelson-Morley experiment. Nonetheless, the latter takes dominance over the former when theoretical decisions are made. Theories, as well entrenched as we can image a theory to be, must make way in the face of even a few novel bits of contradictory observational data.   .

Of course there will be cases where we are so convinced of the truth of a theory that we will not take alleged data contradicting it very seriously. The response of the physics community to Miller's later alleged positive results of round-trip experiments shows us that. Yet there is still a fundamental sense in which Popper is right that, when the issue is between what a theory predicts and what the observational results tells us is (contrary to the theory's prediction) really the case, the observations must take epistemic precedence if there is to be anything like a coherently rational science at all. Does this mean that there are incorrigible, indubitable, or irrefutable reports of observation, or anything like that? No. Once again the fundamental point of empiricism is not that "foundations" exist but that there is a hierarchical order to our epistemic structure in which observation is foundational relative to theory. This is what the thoroughgoing pragmatist coherentist denies.

When we look at how we get from data to theory, what these special cases tell us about pragmatism (as opposed to empiricism) must be construed more subtly.

Some basically empiricistically minded philosophers, looking at the underdetermination of theory by data that is clearly present in the cases we are considering, opt for a theory of theory choice that invokes such notions as ontological simplicity and methodological conservatism to allow the selection as "best" of one of the many alternatives, all of which are equally compatible with all possible observational results. Surely, if the choice is done that way we are well on the road to some kind of pragmatist epistemology. How else, other than by the reduction of truth to ultimate warranted assertability and the reduction of assertability to the systematization of what we in fact take to be warrantedly assertible, could the connection between simplicity or conservatism and truth ever be motivated?

However, deeper reflection on semantic issues will show us, I believe, that what these special cases really reveal to us is the necessary absorption into pragmatist accounts of meaning of very significant portions of the empiricist account. A standard complaint lodged against pragmatism, with its coherentist notions of warrant and truth, is that too many incompatible worlds could all be the case on the pragmatist view. Recently we have been assured, on the basis of arguments stemming from the problem of radical translation and the inevitable role in it of the principle of charity which attributes to others a general agreement with us as to the facts of the world, that we need not worry about such "alternative coherent worlds" ever arising. Given a coherent account of the world, we will perforce interpret it as, in general, the same account of the world that we offer in our home world picture.

It has been claimed that this pragmatist solution to the problem of too many coherent world views is basically verificationist. Indeed, I think it is. Rather than go into that, all I want to do here is show how reflection on the cases at hand will indicate how a careful construal of pragmatist attitudes toward semantics may allow us to see that at least some pragmatists are more empiricist than they would like to think themselves.

Just as the recent pragmatists evade the issue of multiple coherent worlds, some have tried to avoid the problem of theory choice among spacetime theories with common observational consequences by arguing that there is no choice to be made. All the observationally equivalent alternatives are said to be mere alternative expressions of one and the same theory. I think it is clear that the only way that this line could be made to work, in general, would be on the basis of a positivist view that what a theory says is just what is said by its totality of observational consequences. Surely that requires the full empiricist panoply of hard observational/nonobservational distinction and the very special role of observations in grounding meaning, which are the hallmarks of empiricist semantics.

Actually, I think much more is true. Even someone who wishes to claim that aether theory and special relativity are not equivalent theories, and then to be a skeptic, perhaps, or else to invoke some principle above and beyond conformity with the data that rationalizes theory choice, must still offer a coherent account of just what makes theory expressions expressions of one and the same theory (when they are) and inequivalent expressions of distinct theories (when they are not). I believe that one really cannot give such an account in any coherent way without presupposing a good deal of the empiricist's apparatus. In particular, I do not think that, without a rather rigid distinction between the observable and the nonobservable (again, a distinction at least rigid enough to assure us in any context that a body of propositions contains all that could be plausibly construed as observable, assuring us that all in the remainder is nonobservable in a context-independent way), we can construe any plausible general principles of theoretical equivalence, whether these principles (as in positivism) construe aether theory and special relativity as equivalent or whether (as in some Ramsey sentence approaches) they construe them as nonequivalent.

Generalizing, I believe the following to be true: Any pragmatist who wishes to avoid the familiar accusation that his principles of epistemic warrant will allow as warranted (and, from a pragmatist viewpoint, as "true") too many alternative internally coherent worlds will have to come up with a notion of equivalence of world pictures that parallels the positivist picture of equivalence for theories applied in the cases we have been discussing. To this extent, such a pragmatist will be presupposing, in a manner very distasteful to him, the empiricist notion of genuinely nonobservable aspects of the world and the empiricist idea that observational facts play an important and distinguished role as the "ground" of the accrual of meaning to propositions about the world.

In summary, I am arguing that neither naturalism nor pragmatism does justice to the nature or role of observations in the context of Einstein's critiques.

Both pragmatism and naturalism advocate a merely relative and contextual distinction between the observable and the nonobservable. However, Einstein's critical results do not fully make sense unless at least some consequences of theories can be taken by us to be, in principle, nonobservational in a nonrelative and non-context-dependent way. In addition, naturalism, with its doctrine that what is observable is determined by the theory, neglects *a priori* aspects of the observable/nonobservable distinction that allow us to make that distinction prior to accepting a theory to cover the data and independent of the theory we will ultimately adopt. Pragmatism, with its doctrine that observations are all intrinsically theory-laden and that facts are "soft all the way down," fails to do justice to the way in which someone like Einstein deliberately shows us how to at least partially (and relevantly for his purpose) detheorize our observations so that they can be conceptualized in a manner independent of the theories from among which we are making our choice of which to believe.

Neither naturalism nor pragmatism does justice to the epistemic role of observations in the theoretical contexts in question. Naturalism restricts us to asking questions about the reliability of methods, evaluated externally and relative to an adopted theory of the world. What we need to attain in the scientific case in question, however, is an understanding of the internal role observations play in rationalizing for the undecided decision maker one of his possibilities as the rational theory to believe. Pragmatism, which can deal with the special epistemic role of observations only by characterizing them as the sorts of things commonly agreed upon by the members of the community without doubt or disagreement, fails to do justice to the quite asymmetric role played by observations (no matter how dubious) and theories (no matter how entrenched and agreed-upon) in the process of scientific decision making.

Nor can naturalism or pragmatism do justice to the fundamental semantic role played by observationality in the context in question. Both approaches, each locked into one or another use theory of meaning, fail to provide a radically distinctive role for observables as opposed to nonobservables in an account of meaning. Resolution of the scientific problems we have been discussing requires an analysis of the notion of equivalence of theories, however. Whether one adopts a positivistic account in which any observationally equivalent theories are taken to be fully equivalent or instead an account that allows for observational equivalence without full equivalence (say, by demanding structural isomorphism at the theoretical level for full equivalence), it can still be shown that observational identity is crucial for theoretical equivalence and that, as a consequence, radically distinct accounts must be given in one's theory of meaning to the accrual of meaning by observational terms and the accrual of meaning for the terms referring to the nonobservable entities and properties. Although a naturalistic-pragmatic use account of the latter, relative to an assumed meaningfulness already accruing to the former, may be plausible, such an account will not do for the observational part of the language.

The fundamental claim being made here is this: The first-person, internalist perspective is an essential component of any thoroughgoing epistemic and semantic critique of theories. It is indispensable and cannot be replaced without loss by either a physicalistic-naturalistic epistemology and semantics that has been naturalized or a pragmatist "always-alreadyism" or internal coherentism with respect to epistemic grounding or semantic comprehensibility.

The perspective I have been emphasizing is a hierarchical one. With regard to epistemic assurance and to semantic comprehension, some things are more foundational than others. The existence of some ultimate foundation, some bottom to the hierarchy, is not of the most fundamental importance. What is essential is that there is a "closer to" and a "further from" the foundation, and that this directionality is provided by a measure of the degree to which we have moved away from the observational basis of theory. What is important is not whether we can ever find a kind of assertion that, being purely a report of the immediately observable, is immune to epistemic doubt and has its meaning given by some infallible ostensive process. What is crucial is that we can develop an epistemic and semantic critique of theory that proceeds by moving to a relatively purer epistemic and semantic level. In addition, it is crucial that this procedure is the process of extracting from the theoretical context that which is to count as more given and less theory-laden—i.e., that which is to count, relative to the context, as the observational basis of theoretical decision making as opposed to theoretical presumption.

Now, it might be thought that such a line of reasoning appears plausible only because of the very special nature of the theorizing context from which our examples of scientific practice have been chosen. These contexts are indeed special in some ways. The theory choices (aether vs. special relativity, gravity vs. spacetime curvature) are totally formulable, unlike the vague and unformalizable options we talk about in the more philosophical context (such as material world vs. brain in a vat). We are in a situation where a genuine theory choice must be made. We cannot rely on everyone simply agreeing *a priori* that all but one of the alternative theories before us may simply be dismissed as philosophers' nightmares that are not worthy of serious consideration in practice. The epistemic-semantic critique of the theories is, in the context in which we are working, an essential part of the very formulation of the theories in question. In the development of special and general relativity, the investigation into what is and what is not observable and the role of that distinction in characterizing the epistemic and semantic bases of the theories were essential to getting Einstein's practicing science underway. Here the epistemic-semantic critique is integral to the ongoing scientific program, not a philosophical after-thought. In the cases in question, the distinction between what is to count as observational (coincidence, continuity along timelike paths) and what is nonobservational (nonlocal spatio-temporal relations, the structure of the spacetime itself) is clear and uniformly presupposed as a given by the scientific community.

Granted, all of this is rare in the scientific situation. Usually science goes on without such an epistemic-semantic critique, leaving that for philosophical "moppers up." But I think the ways in which the particular problem situation focused on here are special make it clear that, in principle, all our commonsense and scientific beliefs ought to be susceptible to the same sort of epistemic-semantic critique. That this is so, that in any such critique reliance on a first-person, internalist perspective is essential, and that from that perspective both epistemological grounding and semantic comprehensibility require a special privileged role for the observable are, I think, at the core of what empiricism is all about.

2

Observation and the Scientific Enterprise

Dudley Shapere

Two concepts, or rather the problems associated with them, have lain at the heart of the philosophy of science for the past half century: the concept of observation and that of reason. Positivism took the former for granted as the basis of its interpretation of science and collapsed when it found that observation could not be sharply separated, as that view required, from any infusion of theory. Its critics of the 1950s and the 1960s, in effect rejecting the role of reason in science, failed thereby to account for that of observation. Their successors, hoping to do justice to the role of reasoning through a two-tiered picture of science, each tier having its own standards of acceptability (a line already taken by positivism), simply repeated the errors of the past, and in any case failed to confront the problem of observation effectively or, indeed, at all. The centrality of the two concepts and their associated problems long antedates our century. Classical empiricism, identifying observation with perception and the latter with immediate, uninterpreted awareness, not only failed to explain why observation constitutes good reason—evidence—for or against beliefs, but also, through the very poverty it demanded of what was to count as the observational base of knowledge, divorced itself from any possibility of accounting for the knowledge we do have. The classical opponents of that empiricism, focusing on the concept of reason, failed to deal with that concept successfully, partly because they failed to acknowledge or at least to make clear the role that observation does have and partly because of their insistence— an insistence held in common with classical empiricism and with much though not all of twentieth-century philosophy—that a good reason be an absolute guarantee or at least be based ultimately on absolute guarantees.

Observation: The Solar Neutrino Experiment

In a recent article,[1] I offered a new view of what counts as an "observation" in a range of cases in contemporary science. According to the analysis given in that paper, x is observed (observable) if information is received (can be received) by an appropriate receptor and if that information is (can be) transmitted directly (i.e., without interference) to the receptor from the entity x (which is the source of the information). But this statement alone, I argued, is merely schematic:

> ... specification of what counts as directly observed (observable), and therefore of what counts as an observation, is a function of the current state of physical knowledge, and can change with changes in that knowledge. ... More explicitly, current physical knowledge specifies what counts as an "appropriate receptor," what counts as "information," the types of information there are, the ways in which information of the various types is transmitted and received, and the character and types of interference and the circumstances under which and the frequency with which it occurs.

(In other words, the two conditions by themselves, without any further specification in terms of the current state of knowledge with regard to each of their key terms, cannot be taken as sufficient conditions for determining that something counts as an "observation.") These assertions were detailed through a close examination of the solar neutrino experiment, which has been in continuous operation since 1967 and which was designed to test our theory about how stars produce their energy. According to that extremely successful theory, the energy is produced in the central core of a star by thermonuclear reactions, the most important of which is the conversion of hydrogen into helium. The small excess of mass of four hydrogen atoms over one helium atom is converted into energy according to the familiar $E = mc^2$ relation. For a star of the sun's mass, the basic hydrogen-converting process is believed to be the so-called proton-proton sequence of reactions, which begins with the interaction of two hydrogen nuclei (protons). In the reactions that follow this one, three alternate subchains, each having a calculable probability of occurring, are possible. One of these involves the production of the radioactive isotope boron 8, which decays and releases a highly energetic neutrino.

Thus far, we have in the type of observation situation exemplified by the solar neutrino experiment what I have called the "theory of the source": For an observation of the source (or, more specifically, of the processes occurring there) to be possible, there must also be a "theory of the detector (or receptor)." In this case, the possibility of observation rests on the fact that the boron 8 neutrino is highly energetic. Neutrinos are notoriously very hard to observe. They react only extremely weakly (i.e., rarely) with other matter; that is, of the three types of fundamental interactions relevant in current elementary-particle physics, they participate only in the weak interaction. But a highly energetic neutrino is easier to capture than a less energetic one. It proves possible, both theoretically and technologically, to build apparatus capable of capturing a few of them—enough to make possible significant scientific conclusions—and this has been done. As is abundantly illustrated in the paper cited above, the experiment is universally said by astrophysicists (with a few variations that are fully explained in terms of the view developed in that paper) to be a case of "direct observation" of the center of the sun—this in spite of the fact that that core region is buried underneath 400,000 miles of material under conditions of temperature, pressure, and opacity that would seem to rule out the possibility of any sort of direct access to it, especially "observational."

That the capture process is spoken of as constituting "direct observation" of the processes occurring in the center of the sun, and not, for example, as a basis for "inferences" from observational data to "hypotheses" about those processes, is justified by the "theory of the transmission": The captured neutrinos having had an extremely low probability of interacting (i.e., of having been interfered with) between their production point in the core of the sun and their capture by the "appropriate receptor," the information obtained from these neutrinos is said to constitute "direct observation." (If, however, reasons arise for doubt as to whether the neutrinos received are from the center of the sun, the description of the experiment will "retreat" from the statement that the center of the sun has been observed to the more cautious statement that neutrinos have been observed. Further such "retreats" are always possible, although, according to the argument of the earlier paper, they will be made only if there arise specific reasons to doubt the propriety of a given level of description.) The propriety of the description of the experiment as a "direct observation" is brought out partly by contrast with paradigmatic examples of talk about inference from observational data to hypotheses as to what goes on "behind the scenes" of what is "observed." Such an example is provided by information received via photons (electromagnetic processes). Consider the visible light received by us from the sun. Like all solar energy, it is originally produced, in the form of photons, in nuclear reactions in the solar core, and from there it is transmitted to the surface, from whence it is sent on to our "appropriate receptors" (telescopes, cameras, spectroscopes, eyes, and so on). But there is a great difference between the neutrinos and the photons produced at the solar core. The latter are highly energetic and of short wavelength. Their passage outward to the surface is long and tortuous. Their mean free path under the conditions prevailing there being well under one centimeter, they undergo frequent and drastic interference through processes specified by current physics, so that the photons finally emitted at the surface differ radically in character from those produced at the core. The energy has been degraded from the high-energy, short-wavelength photons originally produced to the relatively low-energy, long-wavelength ones finally emitted at the surface. The latter, however, once released at the surface, can be expected to travel without further serious interference to our receptors of electromagnetic energy. Hence we speak of *observing* the surface of the sun by means of those receptors (including eyes), but of the necessity (because of the uncertainties involved) of *inferring*, on the basis of such observation, what goes on in the solar interior.

It is clear that a great deal of prior knowledge enters into specifying the "observation situation" in cases such as the solar neutrino experiment; among much else, note the following prior knowledge, without which the experiment would be, in the most literal sense, inconceivable. In the theory of the source, various aspects of the theories of the strong, electromagnetic, and weak interactions, particularly in regard to nuclear processes, enter in; also playing central roles are the theory of stellar structure, the theory of stellar evolution, and various specific data about the sun. In the theory of the transmission, the theory of weak interactions plays the important role in the case of neutrinos; however, the contrast of observational status between neutrino and photon reception requires also information about factors contributing to the opacity of the sun to the passage of photons there (i.e., interference with the photons). Those factors are specified in modern physics as bound-bound, bound-free, and free-free absorptions and electron scattering; the difficulties of estimating each—that is, definite and specific ways in which our knowledge of those interferences falls short—are also laid out by current physics. Finally, a great deal of prior knowledge is required in the "theory of the receptor," among which may be mentioned once again the theory of the weak interaction and knowledge about nuclear reaction rates, both theoretical and experimental, but also a great deal of "practical" knowledge. For example, there are requirements having to do with the proper sort of detecting material: Large quantities of it will have to be available; it should, in this sort of experiment, contain in plentiful supply a substance that in interaction with a neutrino will undergo inverse beta decay to produce a radioactive atom of a sort (namely, a noble gas) that, despite the almost infinitesimal amount produced in the enormous quantity of detector material, can be demonstrated to be easily and thoroughly recoverable therefrom; the radioactive decay period of those atoms should be of a specifiable convenient length; and the decays should be easily registered by existing radioactive-decay counters. We must know the kinds of circumstances required for setting up the apparatus: It must be deep underground to be shielded from cosmic rays (whose effects, it is known, might in certain specifiable sorts of cases mimic those sought in the experiment), and once the apparatus is underground precautions must be taken against similar mimicking effects due to radioactive processes in the cave walls or in the content of the apparatus. To summarize (and add a few more items of knowledge prerequisite for the conception of the experiment): We must bring to bear theories of weak interactions and of nuclear reactions, experimental determinations of reaction rates, cosmic-ray physics, the chemistry of noble gases, the properties of perchloroethylene (the detector material), knowledge of the composition (particularly the radioactive content) of the cave walls, technological information as to how to airproof the apparatus (and theoretical information as to why this must be done), technological information about the capabilities of radioactive-decay counters (both in general and in reference to idiosyncracies of the individual counters employed), information as to how to clean the apparatus in such a way as not to interfere with its fulfilling its experimental function, and much else. In all three components of the observation situation, moreover, the kinds of errors and inaccuracies to which the information is or may be subject is also given by prior knowledge—for example, the range of uncertainties in reaction rates, the types of background interference that might lead to the production of unwanted radioactive elements in the apparatus, and how to overcome the dangers thus posed.

Even from this brief summary of the argument of the above-mentioned paper, it is apparent that at least in a wide variety of cases in science (such as the solar neutrino experiment)—cases of telescopic, microscopic, and spectroscopic "observation," for example—determination of what is to count as "observational" rests on a great deal of prior information. In such contexts, what is referred to as "observational" is far from

being the pure uninterpreted "given," distinguishable from all theoretical infusions, that much of classical and also much of modern empiricism claimed it must be.[2] Note, too, that what is referred to as
"observational" in these modern scientific contexts, though it diverges
in this crucially important way from what classical empiricism thought
of as "observational," nevertheless plays the very same primary epistemic roles assigned to observation by the empiricist tradition and at
least some aspects of ordinary usage, namely of being the basis of
testing beliefs and of acquiring new knowledge about nature. In what
follows, I want to extend this analysis of "observation" in certain ways
that will provide the basis for placing it within a larger picture of science
and scientific change, a picture in which that analysis of observation
is an integral part.[3]

The Role of Background Information

The pervasive role of prior knowledge in shaping the conception of
an experiment becomes even clearer when we consider the following
additional ways in which it functions in the theory of the source.

• That there is a specifiable region of the volume of a star that can be
marked off as a distinct region that can be studied in at least relative
isolation from its environment, namely the core of the star, is determined
by the prior knowledge that the pressures in that region will produce
temperatures leading to the occurrence there of processes that do not
(with only rare and, for our purposes, trivial exceptions) occur elsewhere
in the star, namely thermonuclear reactions. That the prior information
itself deserves to be relied on—that it deserves to be considered prior
knowledge—rests on a vast body of highly successful other information
about the behavior of gases and about theoretical and experimental
nuclear physics. That all this prior knowledge is applicable and relevant
to the sun (and to stars generally) rests on spectral analysis, which
reveals the stars to be composed of the same materials to which that
knowledge has been successfully applied in earthly laboratories, and
on the enormous success achieved in understanding stellar structure
and evolution in those terms.[4] Not only has all this prior information
been vastly successful; in almost all relevant respects, there is no viable
alternative information that could be brought to bear, and there is no
specific and compelling reason to doubt it, either in other applications
or in its application to this case.

• That this region, the core of the star, is one of importance for investigation is likewise determined by prior knowledge—in this case by the theory of the source of stellar energy, which says that that source lies in the nuclear fusions that take place (again with what are for present purposes trivial exceptions) in the cores of main-sequence stars. That theory has proved remarkably successful relative to initially plausible rivals and to the objects and processes with which it is concerned as an explanation. Indeed, there are no longer any serious or even any proposed alternatives to that theory of the sources of stellar energy. In particular, before the solar neutrino experiment there was no specific reason to doubt that theory, the experiment having been set up not because there was any difficulty in or any alternative to the theory, but simply because, in the case of so fundamental a theory, an experimental test "just to make sure" has been found to be generally advisable if the experiment is feasible. Before the experiment, no one doubted that it would bear out the theory fully.[5] The diverse body of general claims and experimental data constituting what is loosely called the "theory" of nuclear reactions has also proved trustworthy in general, even though there is considerable variation among those claims with regard to the degree of specific doubt to which they are open. Nonetheless, those doubts, and their relative importance, are in general clearly specifiable. Indeed, the failure of the solar neutrino experiment to bear out the predictions of the body of theory and other information making up the "background information" of the experiment led to a spate of investigations, both theoretical and experimental, to improve the status of that background information.

• Within the class of objects and processes being investigated (nuclear processes in stellar cores), the particular object to be selected for study (in this case, those processes in the core of the sun) and the particular processes to be examined (in this case, the third sub-branch of the proton-proton chain, and in particular the production of highly energetic neutrinos in the decay of the isotope boron 8 in that subchain) are also determined by prior knowledge. Among this complex body of background information are the following: our knowledge of the relative nearness of the sun as compared with other stars (and of course our knowledge that the sun is a star, and in particular a main-sequence star), our knowledge of the current technological impossibility of studying the relevant processes in any other star, and the rather last-minute realization, through an application of relatively recent discoveries, that the specific processes to be studied were within current technological capability.

- That that object (the solar core) can be taken as representative of the class of objects under study, or, more exactly, the respects in which and the extent to which it can be so taken, is also determined by prior knowledge. The sun is known to be a star, in particular a main-sequence star, typical of main-sequence stars in utilizing the conversion of hydrogen into helium as the source of its energy. But the relevant background knowledge about nuclear reactions and the masses and internal conditions of main-sequence stars stipulates that there are two alternative elaborate processes by which this conversion can take place: the proton-proton chain, mentioned earlier, and the carbon-nitrogen chain (the former dominating in less massive stars, the latter in more massive stars). It is possible to calculate on theoretical grounds the relative contributions of each of these processes in a star of rather intermediate mass like the sun; it turns out that the sun's energy production is heavily dominated by the proton-proton chain. With this knowledge, it becomes possible to extract well-grounded expectations that and how information obtained by studying these processes in the solar core will be applicable to the cores of main-sequence stars in general. Again, all this "background information" has proved highly successful in accounting for stellar characteristics and stellar evolution; indeed it has been extended to provide a generally excellent account of the origins and abundances of the chemical elements beyond hydrogen and some helium. There is no specific, compelling reason to mistrust it, either in general or in its present application.

These examples of some of the ways in which "background information" shapes scientific research bring out several points that provide a basis for putting the entire preceding discussion of observation into the context of a broader view of science and scientific change. In the first place, we see as an important element in these examples the clear delineation of a subject matter to be investigated, together with a rationale for the importance of investigating that subject matter;[6] we see also a rationale for selecting a particular item of the type of that subject matter as the item to be investigated. The rationale in all these instances is established in the light of certain background information. We see, further, that it is not just any beliefs that function as "background information." In our contemporary case of the solar neutrino experiment, we see that there are severe constraints governing what can be so used;

what is employed in this case are, wherever possible, prior beliefs that have proved successful, are free from doubt, and are clearly shown to be relevant to the subject matter under investigation. Where beliefs falling short of these conditions are utilized, attempts are made to make them more satisfactory in that regard.[7] (More will be said about these ideas of "success," "freedom from doubt," and "relevance" in what follows.) Further, the above examples bring out the fact that there is a background history of the immediately employed background information, a history in which that success and that freedom from doubt have been established. With these points as a basis, let me now proceed to put my discussion of the solar neutrino experiment, and in particular of the concept of observation operative in that experiment, into the context of a broader view of science and of the knowledge-seeking enterprise more generally.

Reasons, Domains, and Background Information

Since this broader picture of science will have to do with an explanation of the "rationality" of science, I will begin with a general discussion of what is involved in the concept (or at least one aspect of the concept) of "a reason." In everyday arguments about factual subject matter, one typical kind of move is to charge that what an opponent has said in defense of his claim is irrelevant to the subject matter under discussion. In general, such a charge is meant to allege not that the opponent has put forward a bad reason in favor of his thesis but that he has put forth no reason at all. This everyday sort of argumentative move thus provides an important clue to what is involved in "giving a reason," as opposed to failing to do so, in the case of everyday disputes and everyday processes of deciding about factual matters. In such everyday disputes or decisions, whether a consideration counts as a reason for accepting one side or another depends crucially on there being a subject matter which that argument is about and on there being other information relevant to that subject matter. Furthermore, since the characteristics and boundaries of that subject matter and the relevance thereto of proposed considerations can themselves be subject to debate, it follows that whether a given consideration counts as a "reason" depends crucially on how definitively the subject matter can be formulated and how clearly the relevance of proposed considerations can be established.

However far it has departed from the characteristics of its ancestor, science is, in a sense that will be deepened as we proceed, a rational descendant of everyday concept and beliefs. It is therefore not surprising that these features of the everyday notion of "a reason" in the case of factual disputes or decisions should find their descendants in modern science—that that line of descent should provide an important key to understanding the sense in which and the extent to which modern science is "rational" in its procedures and its beliefs. For a central feature of science consists in the formulation and revision, over its history, of areas or fields of investigation that (as a first approximation, which I shall shortly make more accurate) may be viewed as consisting of two distinguishable aspects. First, there is a body of information to be investigated (elsewhere I have called this the "domain" of the area or field[8]); second, there is a body of "background information," that is, a body of beliefs that have been found to be relevant to the domain of investigation. These ideas of "domain" and "background information" correspond respectively to the notions of "subject matter" and "other information relevant thereto" that appear in the everyday notion of a reason, which I have just discussed; the former are the descendents, respectively, of the latter, and thus together they constitute the descendents of that everyday idea (or that aspect of the everyday idea) of what enters into "being a reason" for or against a belief.

Thus, the formation of domains and of background information relevant thereto constitutes at least an important part of the development of the rationality of science. More specifically, the development of science consists (in part only, of course) in a gradual discovery, sharpening, and organization of relevance relations, and hence in a gradual separation of the objects of its investigations and what is directly relevant thereto from what is irrelevant to those investigations—a gradual demarcation, that is, of the scientific from the nonscientific. In that development, science aims at becoming, as far as possible, autonomous in its organization, description, and treatment of its subject matter: at becoming able to delineate its domains of investigation and the background information relevant thereto, to formulate its problems, to lay out methods of approaching those problems, to determine a range of possible solutions, and to establish criteria of what is to count as an acceptable solution, all in terms solely of the domain under consideration and the other beliefs that have been found to be relevant to that domain—that is, to make its reasoning in all respects wholly self-sufficient. Indeed, it is to the extent that science has come to be able to rely solely on the subject matter and relevant "background information" alone, without appeal to "outside" considerations—to the extent, that is, that science has been able to internalize the considerations on which it relies—that science has become rational. That, let me reemphasize, is (or is a part of, or more exactly is what we have come to view as) proceeding rationally in the search for scientific knowledge.

This process of "internalization," of course, captures only one aspect of the rationality of science; as we have seen in the examples given earlier, the background information employed in the delineation of domains and the examination of them must also be "successful" and "free from specific and compelling reasons for doubt." But those aspects of scientific rationality also develop, and indeed co-develop, with the organization of science into domains and information relevant thereto. As such organization becomes better defined, notions of "success" and "reasons for doubt" also sharpen: Clear delineation of subject matter and of information relevant thereto makes it possible to say more clearly what we are expected to account for, how to account for it, whether we have succeeded in doing so, and what specific sorts of doubts there may be that we have done so.

For all the splendid achievements of the Greek and medieval thinkers, the domains approach to the seeking of knowledge about nature was at best peripheral to their thought; in general, they attempted to explain nature as a whole—all at once, so to speak—rather than by piecemeal investigation. The widespread adoption of the latter method, of isolating specific, well-delimited areas for investigation, was an important step in the development of the modern approach to inquiry about nature.[9] With the full adoption in science of the piecemeal (domains) approach to the search for knowledge, the "success" of a theory has come to be judged, in large part, in terms of how completely and precisely it accounts for the items of its domain. However, this statement is subject to several important qualifications. In the first place, the domain for which the theory is responsible can alter over time, and can even be discarded as a domain (the responsibility can change or even disappear). Whether the theory is responsible for certain specific items—whether those items really belong within that domain—may also be debatable. Description of the items, and therefore the exact nature of the theory required, can also change. In other words, what the domain is—the items to be investigated, and what it is about them that needs to be accounted for—is specified by background information, and if that background information changes, so does the domain and therefore so does what a particular theory must do to be successful. Then, too, there are other conditions of success of a theory besides that of accounting for its domain; for example, it is often insisted, especially in sophisticated areas of science, that a theory must account for its domain in the same terms employed by other theories that have been successful in accounting for other but related domains.[10] Still more generally, what it is to "account for" a domain, being also a function of background knowledge, can change with changes in that knowledge. So, despite the importance of noting that theories are responsible for accounting for their domains, what this involves is a function of the state of science at the time. The criteria of success are determined by background information, and that fact calls attention to the dynamic element of science: the change to which its criteria are open. There is nothing necessarily final—at any stage, as far as we have reason to believe—about our divisions of nature into domains, and there is therefore nothing necessarily final about what we take to be successful accounts of nature.

The alterability of our ideas of successful approaches to nature is not, however, limited to possible changes in what we take to be criteria for successful accounts of domains. As was noted above, even the general requirement that a theory is responsible to its specific domain is not something that has always been a central ingredient of the knowledge-seeking enterprise. Nor is that piecemeal approach something that must be maintained always, as long as we seek knowledge. For example, we might learn someday that the hard-won conclusions of that day could have been arrived at in a holistic way had we only thought carefully enough, and then we might obtain further conclusions about nature by those means. But perhaps more important for understanding the present approach of science is the fact that science, having through the efforts of generations divided the world into distinct domains for investigation, attempts ultimately to obliterate those very distinctions by arriving at more and more unitary conceptions and treatments. There is of course no contradiction. Science has found that with understanding of particular domains it manages to gain through its piecemeal approach, it becomes possible to unify that understanding with the understanding of other domains. Having succeeded with the method of dividing the world into distinct areas of investigation, we have learned that it is possible to find elaborate links between those areas, in many cases even removing the very distinctness with which we began. The piecemeal approach has thus become a means to an end, the end of unification of knowledge (an end that may or may not be achievable). Therefore, the process of unification, which is so impressive that it sometimes distracts us from the importance of the piecemeal approach or leads us to disparage it as "specialization," does not invalidate the success we have been able to achieve in that direction by precise delineation of the subject matters of our investigations. On the contrary, the unifications we have achieved themselves contribute to the possibility of such delineations and make them more flexible, thereby contributing ultimately to further unification.[11] Thus, despite the complex intertwinings of relationships in astronomy (relationships that also encompass numerous areas of physics), we are able, in the solar neutrino experiment, to focus on the core of the star as the object of investigation, and to do so more effectively, bringing to bear much more background information than we could have had science not arrived at those complex relationships. That is, the requirement (found in both the ordinary concept of "a reason" spoken of above and the rational descendant of that concept in science) that a subject matter be specified if what counts as "a reason" is to be specified remains in force even in a highly unified science.

The existence of domains also provides a basis for classification of some of the major types of reasons for doubt that can arise in science: *domain problems*, which have to do with inadequacies about the domain under consideration, such as incompleteness or imprecision (when determined on bases other than a theory of the domain, i.e., in terms of the domain and background information other than a theory of that domain; *theoretical problems*, which have to do with the need to "account for" the domain and which therefore require appeal to entities or processes other than those covered by the domain description; and *theoretical inadequacies*, which concern failings of those explanatory theories themselves. There are two main classes of theoretical inadequacies: problems of *incompleteness* and problems of *incorrectness*; only the latter lead to rejection of the theory (though not necessarily to its uselessness). As I have discussed these types of problems or doubts elsewhere,[12] and will do so in future papers, I will not discuss them here. What must be noted here is that, as with the idea of "success," science is in principle subject to deep changes with regard to what is to be counted as a "reason for doubt." Because the above partial taxonomy of types of reasons for doubt in science depends crucially on the domains approach to the knowledge-seeking enterprise, it involves profoundly significant departures from holistic approaches and conceptions of "reasons for doubt."

It is perhaps not surprising that the ideas of "success" and "reason for doubt" (as those ideas function in science) should be tied so intimately to the concept of domain and to the domains approach to inquiry about nature, for we have seen that the idea of relevance to a subject matter is of central importance in at least one idea or aspect of "reasoning" in everyday life. That raises questions about the sense in which the Greeks (or any others who have not sought knowledge through the piecemeal "domains" approach) could be said to have been engaged in "reasoning"—something that no one would deny. Was theirs (and all other non-"domain" types of reasoning) a different sort of reasoning, departing perhaps from some other everyday intuition? Or was it perhaps based on the same intuition but interpreted very differently (perhaps less clearly)? Were they (and other types of nonscientific reasoners) reasoning in a way related somehow to scientific reasoning? More generally, one cannot help feeling that the ideas of "success" (at least explanatory success) and "reason for doubt" have always been, and always will be, necessary ingredients in the idea of a search for knowledge. I will say something more about this question (though I will show only the general direction in which an answer is to be found) at the end of this chapter. For the present, let us turn to a fuller examination of some of the ways in which science is subject to change.

Internalization, the Rationale of Scientific Change, and the Normative Aspect of Science

In the process of scientific development I have described—the marking off of domains and of information relevant thereto—problems and their solutions occasionally lead to the rejection or modification of background beliefs or to the addition of new background information relevant to a given domain; sometimes these changes lead, in turn, to profound alterations of the fabric of science. Areas for scientific investigation are reformulated, split, or unified, and fragments are reattached to other areas in the light of newly found or newly understood relevance relations. Even the descriptive language of domains may in some cases (as in the chemical revolution at the end of the eighteenth century) be radically revised to incorporate new beliefs that, wherever possible, have shown (or, as is necessarily the case in early science at least, come ultimately to show) themselves to be well founded.[13] The problems associated with particular domains become altered, as do the lines between recognized "scientific" problems and questions that are classed as "nonscientific." What counts as "observation" of a subject matter, too, may be altered; both the entities or processes to be studied and the methods of observing them can develop with the acquisition of new successful and doubt-free beliefs. New background information, or old information formerly considered irrelevant, is found to be relevant to a particular domain. Old methods are rejected or reinterpreted, new ones are introduced; new standards of possibility and acceptability arise.

In the light of this process of internalization, it is no wonder that the development of science involves not only changes in substantive views about nature but also changes in the descriptive language of science, changes in the body of problems regarding nature so described and classified and in the criteria of genuineness and importance of those problems, changes in the methods by which those problems are approached, changes in the standards or criteria of what can count as a possible solution (explanation) of those problems, and also as to what can count as an acceptable solution of them (including what can count as evidence for or against a proposed solution), and even changes in the conception of the goals of science. Such changes, insofar as they are unequivocally scientific, are aimed at increasing the autonomy, the self-sufficiency, of scientific argument, for it is in such autonomous argument that the rationality of science resides; it is through arriving at the possibility of such autonomy—such "reasoning"—that science has found itself able to build on what it has learned, in order to learn new things.

In short, the more science has attained in the way of well-formulated and clearly relevant background information, the more it has available as a basis for pushing forward, for building new ideas on the basis of the best beliefs it has available—those that have shown themselves to be successful, to be free of specific doubt, and to have clear relevance relations to one another. It is thus that science can attain, and to a considerable extent has attained, what is appropriately (as a rational descendant of an everyday concept) described as a rationale of development. On the basis of those well-founded (successful, doubt-free, and so on) beliefs, including its conceptions of its subject matter as well as information relevant thereto, science is able to develop new hypotheses, new problems, new methods, new standards, and even new goals for itself, or at least to modify or refine its earlier ones where such changes are called for.

The centrality of this process of "internalization" of scientific considerations—the clear conversion of its considerations into "reasons,"
which themselves are justified and tested in terms of what we find out
about nature by their means—has rarely been appreciated. It has, for
instance, been continually ignored by philosophers of science, who
have almost universally tried to explain science and its changes in terms
of distinct "levels" of thought and activity, each level having its own
distinctive methods of procedure and justification. Usually the "higher
level" (metaphysics, metascientific concepts and rules, paradigms, hard
cores, preferred preanalytic intuitions about rationality) is seen as in
some way governing the "lower" one, and in many such views the
"higher" level is held to be immune to revision in terms of what happens
on the "lower." Such interpretations of science are fundamentally mis-
guided, for it has become a central aim of science to remove, wherever
possible, such distinctions among "levels," if they exist, in favor of an
integrated approach, "internalizing" the separate levels to achieve an
interaction of ideas (methods, standards, and so on) in which all elements
are subject to revision or rejection or justification in the light of what
we learn about nature.

In no case do the rational procedures of science either rest on or
produce any guarantees. We may, as far as we know, always find
reason to doubt any aspect of scientific belief. Even what count as
"reasons for doubt" and criteria for deciding the degree of seriousness
of a doubt are themselves subject to the possibility of alteration or
rejection in the light of new beliefs arrived at through application of
those very criteria. Our very views of our goals and of what counts as
success in attaining those goals are, again as far as we know, subject
to the possibility of doubt, and in that sense they are "hypotheses."

Thus, for example, in the development of science we have learned
not to take as "reasons for doubt" any alleged "reasons" that apply
indiscriminately to any proposition whatever—to both a proposition
and its negation. "Philosophical doubts," such as the possibility that
a demon might be deceiving me or that I may be dreaming, play no
role in any specific scientific investigation; they are not reasons for
taking or refusing to take any particular direction in science. Such
skeptical doubts, we have found in the actual practice of seeking knowledge, are only misleading ways of reminding us that, as a matter of
learned fact, doubt may rise. However, the possibility of doubt arising
is not itself a reason for doubt of any particular proposition; more
exactly, it is no reason to abstain from using the best beliefs we have—

those that have been found to be successful and free from specific (and compelling) doubt—to build on.

What is to count as success, too—what is to count as a reason in favor of a belief—is something that can change with the accumulation of new well-founded beliefs. For example, one criterion of success may prove unsatisfiable or only poorly satisfiable (perhaps, but not necessarily, in the light of some further criterion of successful satisfaction), while another, even though initially considered less important, is satisfiable, so that the first is abandoned or demoted in importance while the second becomes primary.

Finally, even the aims of science, its goals, can undergo change in response to the development of successful and doubt-free beliefs. Not only can aims (the perfection of earthy materials) be surrendered with the abandonment of unsuccessful criteria of success, but new ones can be introduced as new criteria of success develop (searching for components rather than perfection of material substances). More generally, out of successful ideas and approaches spring normative principles, not only on such abstract levels as "Attend only to doubts that are directed at specific beliefs" or "Employ as background beliefs only those that have proved successful and free from specific doubt" or "Aim at the internalization of scientific reasoning," but also more specific principles such as "Try to explain all phenomena in terms of matter in motion" or "So construct explanations of elementary particles and forces that they take the form of renormalizable locally gauge-invariant field theories." Successful approaches become normative guiding principles as well as beliefs about the world; it is a mistake to separate sharply the descriptive and normative in science. And, as the last two examples show, such principles too can change. What ought to be done and sought for in science is, like all else in science, a function of the state of science: Ideas and methods that have proved successful achieve, at least temporarily, the status of guiding principles, determining the ways science should go. More accurately, this internalization of its normative principles is itself a normative ideal for which science has learned to strive.

If it is indeed true that science is able to increase its ability to define its subject matter, the information relevant thereto, and so forth, so as to be able to resolve its problems solely in terms of the subject matter with which the problem is concerned and the other information specifiably relevant thereto—if, that is, science is able to become more clearly "rational," and if it has (as a matter of contingent fact, not of necessity) actually done so, then it follows that much of the appeal to historical cases in the philosophy of science has been inconclusive or invalid. Historical cases can of course be used as counterevidence to general theses about science, if those theses are claimed to hold for all instances of science, including historical ones such as that being adduced as counterevidence. If, on the other hand, the general thesis is asserted to hold only for science that has reached a certain stage of maturity (earlier work being classed either as "primitive" or as not "science"), and if the case being adduced as counterevidence is one that precedes that stage, then the case is not valid counterevidence. (It might, of course, be further argued that the case cannot or ought not be excluded from relevance.) In any case, historical cases can be used as positive evidence for theses about science only with the utmost care. In assessing the relevance of a historical case or some aspect thereof to a general claim about science, one must always beware of the possibility that science may have changed since the period of that case, and changed precisely in the aspect of the case that is being adduced in favor of the general claim, and furthermore that that change has taken place for good reasons. None of this should be taken to imply a rejection of the relevance of the history of science to the understanding of science; the present view is deeply committed to that relevance. However, the relevance is dynamic, in that full understanding of the rationale of our modern scientific views is to be gained through understanding the rationale of their development—including the development of the idea of "rationale."

Science has, more and more, found it possible to achieve self-reliance in its reasoning, and, with the successes that have accompanied that possibility, to seek such self-reliance as a goal. But the ideal of complete autonomy of scientific investigation has not, even yet, been realized. Since the degree of such autonomy is a function of the available background information (the body of successful, doubt-free beliefs that have been found to be relevant to the domain under consideration), it is evident that that information will, in at least some cases, to some extent, be insufficient to permit the definitive formulation of domains, problems, methods, and standards of possible and acceptable solutions. Where our knowledge is incomplete, we must look elsewhere for guidance to supplement that knowledge: to less well-founded beliefs, or to beliefs that, however well founded, have not been shown to be unambiguously relevant to the domain under consideration. Reliance solely on clearly successful and doubt-free beliefs whose relevance is clearly established has come to be seen as an ideal of science—an ideal that is not always fulfillable. Science, even today, even in its most sophisticated areas, is unable always to rely solely on what are unambiguously "reasons." Even though background information in the 1920s had imposed many clear restraints on the formulation and range of possible solutions of the problem of the source or sources of stellar energy, that background information could not provide adequate guidance for the selection of any one of those possibilities. Again, theories of the strong and weak interactions were constructed in the 1930s and the following decades on a basis that was less than doubt-free, namely in terms of what deserves (though with qualifications) to be called "analogy" with the highly successful quantum electrodynamics, even though the field-theoretic exchange-particle approach of the latter could not be expected necessarily to be applicable to those other domains. In yet other cases, we are forced to rely on what I have elsewhere called "conceptual devices," such as idealizations or models—concepts or propositions or sets of propositions that, among other characteristics, are subject to specific doubt.[14] In other words, the further development of science in the light of its accumulated "background knowledge" does not provide a "logic of discovery." In the first place, its rational procedures depend wholly on the content of belief rather than on purely "formal" considerations, and they develop as that content develops. Second, they do not necessarily produce unique and unambiguous solutions of problems (for instance), though they sometimes may. Third, precisely because of their dependence on the content of scientific knowledge, which is incomplete, they are not always sufficient to provide clear (or in some cases any) indication of the direction in which to proceed. (Such is the present situation with the solar neutrino experiment: Being unable to explain the discrepancy, we can only look to a new experiment.) We must not, however, allow these points to delude us into supposing that "analogies" helpful in resolving scientific problems may come from anywhere; it is, to say the least, highly unlikely that new approaches in contemporary particle physics, if they do not come from within physics itself, will come from anywhere else except mathematics.

Concluding Remarks: Rational Descendance, Observation, and Reason

In this survey I have outlined only a few major aspects of the general view I have been discussing; many other aspects have gone unmentioned. One of the most obvious omissions is that I have not explained the basis of the intuition, discussed above, that a consideration counts as a "reason" only if it is relevant to a subject matter. The basis lies in an interlocked group of concepts in which the knowledge-seeking. enterprise has its roots (intellectually, psychologically, and perhaps— though this must be forever hypothetical—historically). A discussion of that "framework" (if I may use that almost universally misused expression) would have to be extensive, partly because its full treatment (and hence that of the general concept of "a reason") is inseparable from that of a large and different topic, namely the concept of truth.[15] An understanding of that framework is also necessary for resolving the question of whether the ideas of "success" and "reason for doubt" (and also of "relevance," of which I have said little explicitly here) must not have been always present in any approach to inquiry about nature. Although it is impossible to go into these two topics fully here, I will at least outline some of the major points concerning that "framework."

The concepts of "reason" and "truth" are (as I argue more fully elsewhere[16]) interlocked, and the reasons for that interlocking themselves run deep. The three major traditional theories of truth—on the one side the correspondence theory, and on the other the pragmatic and coherence theories—are complementary to one another, in the sense that each side captures an aspect of the concept of truth that the other side ignores, and also in the sense that each side faces devastating objections that the other side answers. When this is seen we are able to see how the correct insights of each side can be reconciled while avoiding the failings of the other, and when that is realized we are able to see that the concepts of "reason for doubt," "success," and "relevance," having to do with the general concept of "a reason," are tied, respectively, to the correct insights of (respectively) the correspondence theory, the pragmatic theory, and the coherence theory. We are also able to see why our "framework" regarding the concept of "a reason" involves these ideas, related in these ways: Together with certain other ideas, they form a "framework" from which the search for knowledge springs. That framework, a contingent and schematic one, is not necessarily the one in which the knowledge-seeking enterprise will eventuate; indeed, science has already deviated significantly from it. It can be shown, however, that those deviations are "rational descendants" of that framework.

This idea of a "rational descendant" is important in understanding scientific change, and in addition throws further light on the extent to which "analysis" of these framework ideas is possible. Scientific terms in general, whether ones like 'electron' or like 'observation' ("meta-scientific" ones) or like 'reason for doubt' or 'success' or 'relevance' (which are more general still) may, at a given stage of development in a particular area of science, have relatively precise criteria of application. (That is, "what counts as" an electron or an observation or a reason for doubt or as success in that particular context may be fairly clearly specified.) But those criteria can change ("what counts as" an x can change), and can change for reasons. (No vicious circle is present here when x is 'reason,' if we keep in mind that the development of criteria of rationality is itself a bootstrap process and that the criteria themselves are not necessary conditions.) The term in question—'electron,' 'observation,' 'reason for doubt,' 'success,' or whatever—may thus be applied to a succession of such criteria which are related to one another by a chain-of-reasoning connection in a relation of ancestor-and-descent (or cousinhood). There need be no features in common to all the particular uses in the chain; if there are, those common features have no special status as constituting the "meaning" of the term. In other words, the general concept of 'reason' is just a family of reason-related criteria; the same is true, of course, of such "concepts" as "success" and "relevance," and also of ones like 'electron' and the allegedly "meta-scientific" concepts of the logical empiricists. There are no general "meanings" in terms of which such concepts can be presented completely adequately in an encapsulated way. (Thus, it is no wonder that philosophers have failed to pin down those meanings.) Such "concepts" can be exhibited by examples from their families and through the connections between "members" of those families, and they can be discussed in a general schematic way, but "analysis" in the sense of provision of a set of necessary and (or even or) sufficient conditions, covering all possible cases of application of the term, is not in general possible. (I say "not in general" advisedly, because on the present view it is possible under certain circumstances to specify and even to discover such conditions.) Indeed, the very search for such conditions is itself a misunderstanding of the nature of scientific change. Concepts develop just as do all other aspects of our views of nature, and they are prone to the same sorts of deficiencies and openness that are characteristic of other aspects of our view of nature. It follows that full discussion of such ideas as 'reason for doubt,' 'success,' and 'relevance' requires an understanding of the general view outlined in this chapter rather than such discussion being a precondition of the clarity of that view. But more immediately, the schematic character of the "concepts" of "success" and "reason for doubt," and their roots in a general framework from which science has departed, may help account for the feeling (described above) that those "concepts" are necessary ingredients of the knowledge-seeking enterprise at all its stages.

For these reasons, I have been unable here to go into the question of how and why the concept of reason in modern science descends from, but also departs from, everyday notions and from the framework of ideas from which science itself is descended. To do that would be to do for that concept what was done for the concept of observation in "The Concept of Observation in Science and Philosophy," where it was shown that the concept of observation as it appears in the context of the solar neutrino experiment departs from ordinary and philosophical concepts that associate (indeed, except for minor qualifications, equate) observation with perception. Furthermore, I argued there that departure, in which the link between observation and perception is broken or at least severely attenuated, is made for good reasons—namely, in order to maximize the epistemic contribution of observation to the search for knowledge. This aim is achieved by tying (for good scientific reasons) what counts as an observation to the current state of knowledge, in essentially the way I sketched toward the beginning of this chapter. Thus, what used to be called a "metascientific term," with a "meaning" independent of the ongoing content of science, has undergone the process of "internalization" and has itself become an integral part of the body of science, for, as is shown in that earlier paper, it is precisely the assimilation of observation to the general category of 'interactions' that constitutes the important point in understanding the role of observation in the search for knowledge and the testing of beliefs. The concept of observation as it operates in those scientific contexts is partly a generalization of and partly a departure from ordinary and philosophical concepts of observation, and both the generalization and the departure can be understood as being brought about for reasons. It is in this sense of a "chain-of-reasoning connection" between the relevant ordinary and scientific concepts that the latter is a "rational descendant" of the former; thus, the case of the concept of observation explains by clear example the concept of rational descent, which has been a recurrent theme in this chapter and which plays a central role in the understanding of scientific change, assuming the part formerly and so poorly played by the philosophical concepts of meaning and reference.

In closing, let me return to another of the problems of the philosophy of science I outlined at the beginning of the chapter. One of my accusations against classical empiricism was that it failed to explain why observation constitutes good reason—evidence—for or against beliefs. Insofar as the view I have presented is "empiricist" (or, more exactly, a rational descendant of empiricism), it is so on empirical grounds and on the basis of considerations subsidiary to the concept of what it is to be a reason. In other words, that all our knowledge of nature comes through observation of nature is not an ultimate doctrine but a derivative one; observation plays an epistemic role in the search for knowledge *because* it provides reasons for or against beliefs. Furthermore, that doctrine is itself not a necessary one, but it expresses the approach to inquiry about nature that we have learned satisfies successfully what it is to be a "reason" for or against a belief about nature. (The empiricism that is being advocated may thus be described as "contingent empiricism.") However, it must be emphasized that the concept of what it is to be a "reason," too, is a product of experience; we have learned that things in nature have relevance relations to one another, and that, along with such ideas as "success" and "understanding," which have co-developed with our view of nature and of "reason," it is through employment and systematization of discovered relevance relations that understanding of and successful dealing with nature are achieved.

Notes

1. D. Shapere, "The Concept of Observation in Science and Philosophy," *Philosophy of Science* 69 (1982): 485–525.

2. It is clear that much about the "observation situation" is determined by background information; nevertheless, not everything about it is so determined. Although what we are to measure and how we are to measure it are determined by background information, the specific value of what is measured is not. More precisely, a value of that specific sought bit of information—described, given the status of information, and elicited in the observation made in light of the theory of the receptor—may or may not be predicted by the background theory in question; however, whether the value actually obtained by the receptor corresponds to the predicted value is independent of the background beliefs. It is "given"—not in the traditional empiricist sense that it is found as a result of pure perception, free of all prior belief, but rather in the sense that, having been marked out by our best available background ideas, having been appropriately described in terms of those background ideas, and having been accessed

in terms of the "theory of the receptor," its specific value is independent of, not given by, those background ideas.

3. Further aspects of this picture are discussed in the introduction to my book *Reason and the Search for Knowledge* (Dordrecht: Reidel, 1983).

4. Distance measurements, and therefore geometrical theory, are also involved as background information in these considerations about stars.

5. In the mid 1960s there were not a few who thought the experiment was not even worth performing, as the theories to be tested seemed so firm. The solar neutrino experiment has thus increased the status of the normative principle that all our theories, no matter how great our confidence in them, should be subjected to the test of direct observation.

6. There is a deep relationship between the existence of a certain type of problem (what will below be called "theoretical problems") regarding a domain and the descriptive vocabulary for the items of the domain. More explicitly, in mature science the descriptive vocabulary science has evolved as appropriate for a particular domain often itself (or rather, in virtue of the background information on which that vocabulary's interpretation and appropriateness rest) indicates clearly the central (theoretical) problems regarding that domain. In other words, there develops, in mature science, a degree of artificiality in the distinction between what might be loosely referred to as "facts" and "problems," though that artificiality is limited by the fact that the problems may ultimately be resolved while the description remains appropriate. These points cannot be discussed in this brief chapter, where I will generally speak in terms such as the above ("a subject matter to be investigated, together with a rationale for the importance of investigating that subject matter") or in similar locutions that suggest but do not explain the relationship in question. Full discussion will be given in my *Concept of Observation in Science and Philosophy*, forthcoming from Oxford University Press. (That book should not be confused with the article of the same title, referred to in note 1. The article forms part of one chapter of that book.)

7. In many cases, and for certain purposes, it is possible and sometimes even necessary to use "idealizations" (or, more generally, "conceptual devices") as background information, despite the knowledge that the ideas involved must be false in a precisely specifiable sense. See my "Notes Toward a Post-Positivistic Interpretation of Science," part II, in P. Achinstein and S. Barker, eds., *The Legacy of Logical Positivism* (Baltimore: Johns Hopkins University Press, 1969), reprinted in *Reason and the Search for Knowledge*. The role of conceptual devices in science will be treated more fully in a future paper, where the topic will be placed in the context of the broader view of scientific change outlined here.

8. Domains are discussed in my "Scientific Theories and Their Domains," in F. Suppe, ed., *The Structure of Scientific Theories* (Urbana: University of Illinois Press), reprinted in *Reason and the Scientific Enterprise*; see also chapters 14 and 15 of the latter book.

9. The adoption of this piecemeal approach was a gradual process, taking place primarily in the sixteenth through eighteenth centuries. It was of course not

the only important methodological innovation of that period. Others included the pervasive use of idealizations, experiment, and mathematics, all of which, though found in earlier studies, now gradually became ubiquitous features of the knowledge-seeking enterprise. (Isolation of subject matters was not what was needed in all cases, however; astronomy had to be reunited with physics after a millennium and a half of instrumentalism on the part of most of its professional workers.) A full elaboration of the present view of the knowledge-seeking enterprise would give due account to these further aspects of modern scientific reasoning.

The above comments about Greek and medieval thinkers and their ways of trying to understand nature should not be interpreted as denying that piecemeal approaches, or any of the other approaches mentioned in this footnote, are sometimes to be found in their work. However, their use of any one of these approaches tended to be sporadic or isolated from employment of the others.

10. As when we seek a local gauge theory for all the fundamental interactions. That sort of theory has been immensely successful with the electromagnetic interaction—an example that will appear again later in this chapter to fill out the present point. For further examples see "Scientific Theories and Their Domains."

11. For some aspects of the flexibility of domains in sophisticated science see *Reason and the Search for Knowledge,* chapter 14, "Remarks on the Concepts of Domain and Field."

12. "Scientific Theories and Their Domains" (note 8 above).

13. See my "Reason, Reference, and the Quest for Knowledge," *Philosophy of Science* (March 1982): 1–23, reprinted in *Reason and the Search for Knowledge;* also see, in that book, chapter 15, "Alteration of Goals and Language in the Development of Science."

14. "Notes Toward a Post-Positivistic Interpretation of Science" (note 7 above).

15. The impossibility of going into the concept of "truth" here also explains why I have not said more about the role of conceptual devices (idealizations, abstractions, models, etc.) in science; that topic, too, is inseparable from that of the concept of truth. Indeed, the question of "scientific realism" requires a full understanding of the role of conceptual devices.

16. See the introduction to *Reason and the Search for Knowledge.* Further details will be given in *The Concept of Observation in Science and Philosophy.*

3

Observations, Explanatory Power, and Simplicity: Toward a Non-Humean Account

Richard N. Boyd

Introduction

Two Puzzles about Confirmation

Truisms from empiricist philosophy of science often turn out to be false, but one such truism is certainly true: Scientific knowledge is experimental knowledge. It is characteristic of scientific research that observational evidence plays a decisive role in the resolution of the issue between contending hypotheses, and whatever sort of objectivity scientific inquiry has depends crucially on this feature of the scientific method. It may be disputed what the limits of experimental knowledge are, or how theory-dependent observations are, or how conventional or "constructive" scientific objectivity is, but it is not a matter for serious dispute that the remarkable and characteristic capacity scientific methodology has for the resolution of disputed issues and for the establishment of instrumental knowledge is strongly dependent upon the special role it assigns to observation. In some way, observations permit scientists to use the world as a kind of court to which issues can be submitted for resolution. However "biased" the court may be, the striking success of scientific methodology in identifying predictively reliable theories must be in significant measure a reflection of that court's role. Call a theory *instrumentally reliable* if, and to the extent that, it yields approximately accurate predictions about observable phenomena. Similarly, call methodological practices instrumentally reliable if, and to the extent that, they contribute to the discovery and acceptance of instrumentally reliable theories. It is unproblematical that the crucial prole of observation in science contributes profoundly to the instrumental reliability of scientific methodology.

Once this special epistemological role of observations is recognized, it is natural to investigate other features of scientific methodology by comparing or contrasting the role they play with the special role played by observation. In this chapter I apply this strategy to two features of scientific methodology. The first of these is the systematic preference that scientific methodology dictates in favor of explanatory theories. The second goes by several names; what I have in mind is the methodological preference for theories having the property or properties that philosophers typically call simplicity or parsimony and scientists often call elegance (or, perhaps, beauty) instead. The standards for theory assessment (call them the nonexperimental standards) required by these features of scientific methodology are, at least apparently, so different from those set by the requirement that the predictions of theories must be sustained by observational tests that it is, initially at least, puzzling what they have to do with the rational scientific assessment of theories or with scientific objectivity.

Simplicity, Explanatory Power, and Projectability: Why the Puzzles Are Serious

When we think of scientific objectivity, two importantly different features of scientific practice seem to be at issue: *intersubjectivity* (the capacity of scientists to reach a stable consensus about the issues they investigate and to agree about revisions in that consensus in the light of new data or new theoretical developments) and *epistemic reliability* (the capacity of scientists to get it (approximately) right about the things they study). If we focus exclusively on the first component of scientific objectivity, then the role of the preference for explanatory theories and for simple theories may not seem especially puzzling. Suppose that, for whatever reason, scientists prefer simple and explanatory theories. Perhaps the preference for simplicity reflects a basic psychological law and the preference for explanatory theories reflects a feature of graduate training in science; the source of the preferences does not matter. Suppose as well that, as a result of common indoctrination in their professional training (a common "paradigm" in Kuhn's sense), scientists share basically the same standards of explanatory power and relative simplicity. Under these conditions, the methodological preference for explanatory and simple theories could as readily contribute to the production of a stable scientific consensus as could scientists' common recourse to the results of observation. Indeed, the contribution to the establishment of consensus might be greater, since the consensus-making effects of appeals to observations sometimes depend upon considerable luck or ingenuity in the design of experiments or in the making of relevant observations in nature.

Similarly, even if we focus on the second component of scientific objectivity (the capacity of scientists to get it right in their views about the world), some features of the contribution of the nonexperimental standards of theory assessment to scientific objectivity may seem unpuzzling. Suppose that we follow Kuhn (as we should) in holding that judgments of explanatory power and simplicity are determined by standards embodied in the current research tradition or "paradigm" (Kuhn 1970). Suppose, further, that we follow Kuhn (as we should not; see Boyd 1979, 1982, 1983) in holding that the theoretical structure of the world that scientists study (its fundamental ontology, basic laws, and so on) is constituted or constructed by the adoption of the paradigm. In that case the contribution of nonexperimental standards to the epistemic reliability of scientific methodology with respect to theoretical knowledge will seem unproblematical. After all, it would be hardly surprising that paradigm-determined standards of the acceptability of theories should be a reliable guide to the truth about a paradigm-determined world.

When we turn to the question of the contribution of such standards to the epistemic reliability of scientific methods with respect to our general knowledge of observable phenomena—that is, their contribution to the instrumental reliability of those methods—the situation is quite different. In the first place, the instrumental reliability of scientific methodology cannot be plausibly explained solely on the basis of the supposed paradigmatic construction of reality postulated by Kuhn and others. The fact that anomalous experimental results (results that contradict the expectations dictated by the theoretical tradition or "paradigm" in theoretically intractable ways) occur repeatedly in the history of science and are important in initiating "scientific revolutions" (Kuhn 1970) is sufficient to show that the capacity of scientists to set it right in their predictions about observable phenomena cannot be explained by assuming that the observable world is "constituted by" or "constructed from" the paradigm that determines their methodology. The data from the history of science simply do not permit such an interpretation (Boyd 1983).

Moreover, nonexperimental criteria of theory acceptability are absolutely crucial to the methodology by which scientists achieve instrumental knowledge (Boyd 1973, 1979, 1982, 1983, forthcoming). Briefly, this is so for two reasons. In the first place, nonexperimental criteria determine which theories are taken to be "projectable" in Goodman's (1973) sense. Of the infinitely many generalizations about observables that are logically compatible with any body of observational evidence, only the (typically quite small) finite number of generalizations that correspond to theories that are simple, are explanatory, and otherwise satisfy nonexperimental criteria are candidates for even tentative confirmation by those observations. Thus, many possible and experimentally unrefuted generalizations about observables are simply ruled out by such criteria (Boyd 1972, 1973, 1979, 1980, 1982, 1983, forthcoming; van Fraassen 1980).

To make matters more puzzling, in the testing of hypotheses that have been identified in this way as projectable, scientific methodology requires that a theory be tested under circumstances that are identified by other projectable rival theories as circumstances in which its observational predictions are likely to prove false. From the extraordinarily large body of predictive consequences of a proposed theory we identify those few whose testing is adequate for its confirmation by pitting the proposed theory against its few rivals that satisfy the nonexperimental criteria. To a very good first approximation this is the fundamental methodological principle governing the assessment of experimental evidence in science (Boyd 1972, 1973, 1979, 1980, 1982, 1983, forthcoming). Both judgments of projectability and assessments of experimental evidence for claims about observables thus depend on nonexperimental criteria of the sort that I am discussing. They play a crucial epistemic role in scientific methodology, and thus, like the practice of subjecting theories to observational tests, they contribute to the epistemic reliability that characterizes scientific objectivity.

The same point may be put in another way. Van Fraassen (1980, p. 88) discusses the various nonexperimental theoretical "virtues" and concludes that they should be treated as pragmatic rather than epistemic constraints on theory acceptability: "In so far as they go beyond consistency, empirical adequacy, and empirical strength, they do not concern the relation between theory and the world, but rather the use and usefulness of the theory; they provide reasons to prefer the theory independently of questions of truth." What we have just seen is that this approach is not tenable. We cannot think of the nonexperimental virtues as additional purely pragmatic criteria of theory acceptability above and beyond the criterion of empirical adequacy, for they are essential components in the methodology we have for assessing empirical adequacy. They may also be desirable "independently of questions of truth" (although I doubt it); however, what is striking about their methodological role is precisely that they are central to the ways we assess observational evidence for the truth of generalizations about observables.

We really do have an epistemological puzzle, then. On the one hand, it seems pretty clear that scientific objectivity depends crucially upon the practice of deciding scientific issues by referring those issues to adjudication by the world via experimental or observational testing of proposed theories. That this practice should contribute to both components of scientific objectivity seems unproblematical. On the other hand, it appears that judgments of the aesthetic or cognitive merits of theories play a role in establishing the epistemic reliability of scientific practice comparable to that played by the criterion of experimental confirmation—indeed, such considerations seem to be part of the very methodology by which adequate experimental confirmation is defined. We need to ask how nonexperimental criteria of this sort can play a role so similar to that played by observations in sound scientific practice.

Traditional Empiricist Approaches to the Puzzle

Traditional logical empiricist philosophy of science treats the two nonexperimental criteria I am discussing quite differently. In the case of simplicity and related criteria, I think it would be fair to characterize the approach of logical empiricists as varying, depending upon whether they were doing abstract epistemology of science or applied philosophy of science. In the former case, simplicity was almost always treated as a purely pragmatic theoretical virtue. Often the rationality of preferring simple theories (all other things being equal) was explained in terms of rational allocation of time: It was more rational to investigate first the computationally less complex theories rather than those whose testing would require longer and more difficult computations. Variations on this theme of simplicity as a factor in intellectual economy are characteristic of the pragmatic treatment of the issue within twentieth-century logical empiricism.

In the context of applied philosophy of science—the examination of epistemological and logical issues surrounding particular issues in the various sciences—the situation was quite different. In general, logical empiricists treated issues of scientific methodology more descriptively when they undertook to do applied philosophy. That is, they identified methodologically important features of scientific practice, which they characterized in relatively nonanalytical terms (such as "simplicity," "parsimony," or "coherence"). They then cited the standards set by such features in offering solutions to philosophical problems in particular sciences. What they tended not to do, in such contexts, was emphasize the "rational reconstruction" of methodological principles in the light of the verificationist accounts of scientific knowledge and scientific language that formed the basis of their more abstract philosophical investigations. There is little doubt that this departure from verificationist strictures in applied philosophy of science was a reflection of the fact that the anti-realist perspective dictated by verificationism cannot serve as the basis for an adequate account of the epistemology or the semantics of actual science (Boyd 1972, 1973, 1979, 1980, 1982, 1983, forthcoming). In any event, the general pattern of departure from strict verificationism in applied philosophy of science was clearly manifested in many applications of the methodological principle of preference for simpler theories. In dealing with actual disputes in science, logical empiricist philosophers of science typically took the preference for simpler theories as a basic principle in the epistemology of science and cheerfully cited it as relevant to the determination of answers to questions that were plainly substantive rather than pragmatic. If this practice admits a coherent philosophical rationalization within the empiricist tradition, its rationalization probably lies in positions like that of Carnap (1950), according to which many substantive questions are held to be intelligible only when they are understood as arising within a theoretical perspective that is itself purely conventional and is chosen on essentially pragmatic grounds. Positions of this sort are anticipations of "constructivist" positions in the philosophy of science, such as those of Hanson (1958) and Kuhn (1970), and they are probably best thought of as intermediate between verificationist anti-realism and the anti-realism of these latter positions (Boyd 1983). In any event, no matter how their philosophical practice might be rationalized, logical empiricists routinely treated the methodological preference for simple theories as though it were on a par with more obviously epistemic norms of the scientific method when they were dealing with philosophical issues arising out of actual scientific theories or scientific practices.

In the case of explanatory power, standard logical-empiricist accounts have all been variations on a single basic account, the deductive-nomological (D-N) theory of explanation, which has been employed both in the abstract analysis of scientific methodology and in applications to particular scientific issues. The key idea is that what it is for a theory to explain an event is that it is possible to carry out an *ex post facto* prediction of the event from the theory together with suitable specifications of conditions antecedent to the event in question. The explanatory power of a theory consists in its capacity to serve as the basis for such "retrodictions." As logical empiricists knew, the adoption of this sort of analysis of explanatory power affords what appears to be a neat (indeed, elegant and even simple) solution to the puzzle of the relationship between scientific objectivity and the methodological principle of preference for explanatory theories. A successful explanation by a theory of some fact has just the same logical form as the confirmation by that fact of an experimental prediction of the theory. An explanation amounts to a demonstration that some event that has occurred previously can be retrospectively interpreted as an experimental test of the theory on which the explanation is based—a test which the theory passes. Thus, it is hardly surprising that the observational testing of theories and the practice of preferring explanatory theories should play similar roles in establishing scientific objectivity; they are the same practice, except for the largely irrelevant retrospective character of the latter. The methodological preference for explanatory theories is just a special case of the more general preference for theories that have survived experimental testing.

Three features of the D-N account make it especially attractive and plausible:

• It has the consequence—plausible in the light of the integrative nature of scientific understanding—that the explanatory power of a theory depends upon the theoretical setting in which it is applied. That it has this consequence is a reflection of the acknowledged role of previously established "auxiliary hypotheses" in the derivation of testable (or applicable) observational consequences from a given scientific theory.

• It is appropriate to the conception of causation prevailing in the philosophical tradition in which it arises. This is so because the D-N account is simply a verificationist "Humean" gloss on the "unreconstructed" preanalytic conception that to explain an event is to say how it was caused.

• It portrays the methodological preference for explanatory theories as a special case of a general epistemic principle, of which the principle dictating a preference for theories whose observational predictions have been confirmed is also a special case.

To these we should add a feature that almost all logical empiricists intended as a feature not only of the D-N account of explanation but also of their accounts of all other features of scientific methodology:

• Philosophical accounts of scientific methodology should all honor the distinction between the "context of discovery" and the "context of justification." In particular, they should invoke principles of deductive logic and statistical reasoning, but not principles of inductive logic of the sort that might be thought to provide rational principles for the invention or discovery of scientific theories. Accounts of the nature of theory confirmation should be entirely independent of contingent empirical claims about how theories are invented.

Despite its attractiveness, the D-N account of explanation proved vulnerable to a number of *prima facie* objections. These fall into three rough categories. In the first place, there seem to be clear-cut cases of scientific theories that explain events even though they do not yield deterministic predictions of their occurrence. Second, there are retrodictions from laws that fit the D-N account of explanations but do not seem to be genuinely explanatory. Finally, even where the laws in question appear to be deterministic, there are clear-cut cases of explanations in which it seems doubtful that the explanation is founded on information sufficient to allow the deduction of a retrodiction of the explained event.

It will not be my aim here to examine in detail any of these objections to the D-N account, or the rebuttals to them, since I hope to raise difficulties for the D-N account of quite a different sort. Suffice it to say that the first of the objections has typically been met by requiring only that there be a statistical prediction of the explained event deduced from the laws in question. (For criticisms of this approach and a defense of a related alternative see Salmon 1971.) The second objection has typically been met by holding that the apparently deficient D-N "explanations" are indeed explanations that their apparent deficiency reflects merely their failure to meet purely pragmatic standards of, for example, practical or current theoretical interest. Against the third sort of objection, the typical reply has been (depending on the case at issue) either to identify suitable "tacit" premises to make the deductive prediction of the event possible or to assimilate the case to the statistical version of the D-N account. I think it is a fair summary of the literature in the empiricist tradition to say that the first and the third of the objections we are considering have been seen as the more pressing and that the treatment of the second objection in terms of pragmatic considerations has typically been taken to have been largely successful. For the sake of argument, I will assume throughout the chapter that an adequate empiricist solution to the first of these problems exists. I will speak of "predictions" or of "retrodictions" deducible from scientific theories on the understanding that these terms cover the relevant sort of statistical prediction or retrodiction in cases where deterministic predictions or retrodictions are not possible.

The Aims of the Chapter

The D-N account of explanation and the "Humean" account of causation from which it derives are, in their numerous variants, the most durable legacy of the tradition of logical positivism within professional philosophy. (No doubt extreme noncognitivism in ethics is even more durable if we consider the thinking of those who are not professional philosophers.) What I intend to show here is, first, that these legacies of positivism are inadequate even as first approximations to the epistemic task of explaining how considerations of explanatory power are able to play a methodological role analogous to that played by observational testing in science. An adequate explanation of this phenomenon requires that we adopt an account of explanation appropriate to a scientific-realist conception of scientific theories and scientific knowledge (see Smart 1963; Putnam 1975b; Boyd 1972, 1973, 1979, 1980, 1982, 1983, forthcoming).

I shall argue that nevertheless logical empiricists were right in proposing an account of explanation having the first three features mentioned above: that it portray the explanatory power of a theory as depending upon the theoretical setting in which it is applied, that it be consonant with an appropriate account of causation, and that it treat explanatory power as epistemically relevant in the same way that success in making observational predictions is. I will offer a realist account of a wide class of scientific explanations that meet these criteria and that avoid the difficulties plaguing the D-N account and its variants.

I will indicate how the realist account of explanation can be extended to a closely analogous treatment of the other nonexperimental criterion of theory acceptability we are considering: simplicity. Indeed, I will argue that, in an extended but well-motivated sense of the term, both simplicity and explanatory power are "experimental" criteria of theory acceptability. They reflect indirect theory-mediated evidential considerations that can be accounted for only from the perspective of scientific realism.

Finally I shall argue that these realist treatments of the nonexperimental criteria show that the fourth feature of logical empiricist accounts of scientific methodology—the sharp distinction between context of invention and context of confirmation—cannot be sustained. An adequate account of the epistemic role of observations or of the nonexperimental criteria of theory acceptability requires that we countenance inductive inferences at the theoretical as well as the observational level. The epistemic reliability of such inferences depends both upon logically contingent facts about the particular theoretical tradition that human invention has produced and upon logically contingent psychological and social facts about the capacity of scientists to employ that tradition in the invention of future theories. No account of the epistemology of science that is independent of contingent claims about the social and psychological foundations of scientific practice can be adequate to the task of explaining how the epistemic evaluation of scientific theories works. The epistemology of science must be "naturalized" in a way that requires that the sharp distinction between theory invention and theory confirmation be rejected.

The Humean Conception of Explanation

The "Humean" Conception of Causation in Recent Empiricism

According to Hume's philosophical definition, a cause is "an object precedent and contiguous to another, and where all objects resembling the former are placed in like relations of precedency and contiguity to those objects that resemble the latter." Hume's reasons for adopting this definition are as close to twentieth-century verificationism as one can get in early empiricism. His account has the property (characteristic of later verificationist analyses of scientific notions) that, according to the analysis it provides, the cognitive content of a causal statement is a simple generalization of the cognitive content of the observation statements that are seen as providing evidence for it. No inference from observed regularities to natural necessities or causal powers is required for the confirmation of causal statements.

The version of Hume's account that prevails in twentieth-century empiricist philosophy is significantly different. Roughly, this account holds that an event e_1 causes an event e_2 just in case there are natural laws L and statements C describing conditions antecedent to e_2 such that from L and C, together with a statement reporting the occurrence of e_1, a statement describing the subsequent occurrence of e_2 can be deduced. This account, with variations intended to rule out "trick" cases and to accommodate statistical laws, has proved to be the most durable of the doctrines of logical positivism. The contemporary empiricist account is of course fundamentally verificationist in its content and its justification. "Metaphysical" commitment to such insensibilia as causal powers, underlying mechanisms, hidden essences, and natural necessity is eliminated in favor of the "rational reconstruction" of causal notions in terms of deductive subsumption under natural laws. As in the case of Hume's original formulation, the effect is to make the cognitive content of causal statements closely related to the cognitive content of the observation statements that support them. On an empiricist conception, the nonstipulative cognitive content of natural laws is exhausted by the observational predictions deducible from them, since scientific knowledge cannot extend to "unobservables." Moreover, confirmation of a body of laws consists solely in the experimental confirmation of just those predictions. Thus, the cognitive content of a body of laws consists in a predicted pattern in observations, and evidence for the laws consists in observations that instantiate the pattern in question. Just as in the case of Hume's analysis, events are causally related if they instantiate an appropriate pattern in observable phenomena and evidence for a causal claim consists of confirmation of instances of that pattern. What is different is the way the two "Humean" analyses of cause characterize the relevant patterns in observable phenomena.

The difference in formulation between Hume's account and the account that logical positivists adopted in his name reflects two important features of recent empiricist philosophy of science. In the first place, the contemporary formulations reflect the emphasis recent empiricists have placed on employing the results of modern logical theory in the "rational reconstruction" of scientific concepts. Where Hume's "natural" definition clarifies his philosophical definition by reference to the natural disposition of the mind to form associations of ideas, the contemporary definition refers instead to the logical integration of propositions into deductive systems. More important for our purposes is a special case of this sort of reconstruction: the syntactic conception of "lawlikeness." Hume's account of causation is incomplete without some answer to the question of what respects of resemblances are relevant in applying the definition he offers. It is rather plain that Hume's answer is provided by the "natural" definition of causation: The respects of resemblance that "count" are just those to which the mind naturally attends in forming general beliefs about property correlations. Logical positivists quite rightly rejected this particular form of philosophical naturalism. In its place they substituted an appeal to the notion of a natural law. Respects of resemblance "count" just in case they are the respects of similarity indicated as relevant by natural laws. Now, for any two nonsimultaneous events there will be some true general statement about events from which one can deduce a prediction of the occurrence of the subsequent event if one is given as an additional premise a statement reporting the occurrence of the antecedent event. If by a natural law one were to understand simply a true general statement about events, the contemporary "Humean" definition of causation would have the absurd consequence that causation amounts simply to temporal priority. The positivists' solution was to distinguish "lawlike" from non-"lawlike" generalizations and to understand the natural laws to be just the true lawlike generations. It was understood that lawlikeness should be a syntactic property of sentences—in particular, that it should be an *a priori* question which sentences were lawlike, although of course it would be an empirical question which of these were true (and thus laws). The problem of characterizing those generalizations that are lawlike is just the same problem as characterizing those generalizations that are "projectable" (Goodman 1973) or those kinds, relations, and categories that are "natural" (Quine 1969). In each case the question is which patterns in empirical data should be thought of *prima facie* as instantiations of causal regularities.

Within recent empiricist philosophy (see, e.g., Goodman 1973; Quine 1969) there have been proposed variations on the traditional positivist conception of lawlikeness according to which judgments of lawlikeness or projectability are not *a priori*. Successful inductive generalizations governed by particular judgments of projectability may be taken to provide empirical evidence in favor of the projectability judgments themselves, whereas unsuccessful inductive inferences may tend to disconfirm the projectability judgments upon which they depend. It will be important for us to establish just what variations on the traditional positivist conception of lawlikeness are compatible with the contemporary Humean conception of causation.

The Humean definition of causation, whether in its eighteenth-century or in its twentieth-century version, is essentially an eliminative definition. It is not an analysis of what we (as scientist or as laypersons) ordinarily take ourselves to mean when we talk about causal relations. No doubt we would ordinarily paraphrase causal statements in such terms as "makes happen," "brings about," or "necessitates," or in terms that refer to underlying mechanisms or processes. The Humean conception rejects definitions of causation in such terms not because they inadequately capture our preanalytic conceptions but rather because our preanalytic conceptions are held to be epistemologically defective. Neither natural necessitation nor most of the underlying mechanisms or processes to which we would ordinarily refer in paraphrasing causal statements are observable. Therefore, on the empiricist conception, knowledge of such phenomena—if there are any—is impossible. Our preanalytic conceptions of causation, if taken literally, would render knowledge of causal relations likewise impossible. The Humean definition of causation offers a remedy for this difficulty by "rationally reconstructing" our causal concepts in noncausal terms. Reference to suspect unobservable entities, powers, or necessitations is reduced to reference to patterns in observable data. This is the whole point of, and the sole justification for, the Humean definition. The appropriateness of various conceptions of lawlikeness must be assessed in the light of this essentially verificationist justification for the Humean definition. An analysis of lawlikeness—whatever its independent merits might be—is inappropriate for the formulation of a Humean definition of causation unless it is itself compatible with the verificationist project of reducing causal talk and other talk about insensibilia to talk about regularities in the behavior of observables.

This constraint is important because, just as our preanalytic inclination would be to paraphrase causal statements in terms of natural necessitation or underlying mechanisms, our preanalytic conception of the distinction between natural laws and accidental generalizations is probably equally infected with unreduced causal notions. We might, for example, propose to define as lawlike those generalizations that attribute the observable regularities they predict to the operation of a fixed set of underlying mechanisms, or perhaps to consider lawlike those generalizations that attribute the predicted observable regularities to underlying mechanisms that are relevantly similar to those already postulated in well-confirmed generalizations. Some such definition of lawlikeness might well be the correct one (indeed, I think that the latter proposal is very nearly right), but no such definition would be appropriate for the formulation of the contemporary version of the Humean definition of causation. If lawlikeness is already a causal notion, then the Humean definition fails to accomplish the desired eliminative reduction of causal notions to noncausal observational notions and is thus without any philosophical justification. It must be emphasized that any analysis of lawlikeness that referred to unobservable "theoretical entities" and "theoretical properties" or to unobservable underlying mechanisms or processes would be just as inappropriate for a formulation of the Humean definition as one that talked explicitly about "natural necessity." Such "secret powers" or hidden "inner constitutions" of matter have always been the paradigm cases of the sort of alleged causal phenomena reference to which the Humean definition of causation is designed to eliminate. To appeal to unobservable constituents of matter and their unobservable theoretical properties (such as mass, charge, and spin) is precisely to engage in a twentieth-century version of Locke's appeal to insensible corpuscles and their various "powers." Unreduced reference to, say, the charge of electrons *just is* reference to an unobservable causal power of one of the unobservable participants in the unobservable mechanisms underlying causal relations among observables. Reference to phenomena of this sort is precisely what the Humean definition must eliminate.

Similar considerations dictate a closely related additional constraint on definitions of lawlikeness suitable for formulations of the Humean definition. Suppose that a definition of lawlikeness were proposed that involved no unreduced reference to causal notions or to theoretical entities. Such a definition of lawlikeness might still prove inappropriate for the Humean definition of causation if in order to determine whether or not a statement fell under it one would have to rely on inferences from premises that themselves involve irreducible reference to causal notions or to theoretical entities. After all, the whole point of the Humean definition is to render causal statements confirmable even on the assumption that knowledge of unobservable phenomena is impossible. If judgments of lawlikeness can be made only on the basis of premises thus supposed to be unknowable, then the Humean project fails. As we shall see, this proves to be the case.

Explanation and the Humean Definition of Causation

For a wide class of cases, an explanation of an event is provided by a statement saying how the event was caused. On the Humean definition of causation, saying how an event was caused amounts to deductive subsumption of the event under natural laws together with specifications of antecedent conditions—in other words, deductive retrodiction of the event from initial condition statements and laws. The preanalytic conception of a wide class of explanations reduces to the deductive-nomological conception upon Humean rational reconstruction. This fact provides the only good reason there has ever been to accept the D-N account of explanation; to a good first approximation, the D-N account *just is* the Humean definition of causation.

As the recent empiricist conceptions of causation and lawlikeness depart significantly from our preanalytic conceptions, so the D-N account of explanation departs from our unreconstructed conception of explanation. Without doubt our preanalytic understanding of the central cases of scientific and everyday explanation would, if spelled out, invoke unreduced notions of causation and of causal processes and mechanisms. If unreconstructed causal talk were philosophically unobjectionable (as, I shall eventually argue, it should be), there would be no reason whatsoever to adopt the alternative D-N account. Indeed, the considerable difficulty defenders of the D-N account and its variants have had in accommodating paradigm cases of explanations (and of nonexplanations) to the definitions of explanation they have offered indicates just how far from compelling (or even plausible) the D-N account would be were it not for the verificationist objections to unreduced causal notions.

The Humean roots of the D-N account are evident in the literature, albeit in a somewhat unexpected way. A survey of the classical early papers defending and elaborating the D-N account and its variants (e.g., Hempel and Oppenheim 1948; Hempel 1965; Feigl 1945; Popper 1959) indicates that in the typical case Hume is never mentioned but it is taken for granted that the D-N account is appropriate for straightforward causal explanations. In Hempel and Oppenheim 1948 and in Hempel 1965 the Humean analysis (not so described) is very briefly appealed to in the case of causal explanations. Hempel and Oppenheim adduce the requirement that the explanans must have empirical content in support of the D-N account, and Feigl insists that it is possible to "retain the valuable anti-metaphysical point of view" in rival concep-

tions of explanation while adopting the D-N definition instead. None
of the early authors, however, spend much time elaborating these plainly
verificationist and Humean justifications for the D-N account. Instead,
insofar as the account is defended in detail, they defend its extension
to less clear-cut cases (teleological, motivational, or statistical explanations, for example). They take it for granted that, perhaps with the
help of a few verificationist "reminders," the reader will agree that the
D-N account of explanation is appropriate for ordinary causal explanations and will find controversial only its extension to other sorts of
explanation. The unself-conscious Humeanism in these early papers is
striking, but the fact that it is unself-conscious merely makes it clearer
that the philosophical justification for the D-N account lies in the fact
that it represents the Humean rational reconstruction of the notion of
causal explanation. At least for the central case of causal explanation,
no other philosophical justification is to be found.

   This situation persists in the more recent literature on the D-N account.
I think it is fair to say that, insofar as more recent philosophers have
defended the D-N account or its variants, their strategy has been to
offer rebuttals to a variety of putative counterexamples to the account.
These rebuttals have often been extraordinarily ingenious. It is nevertheless true that if we had no Humean and verificationist reasons for
accepting the D-N account in the first place these rebuttals to counterexamples would not by themselves constitute a good reason to accept
the account. The situation remains that the D-N account of explanation
is a preanalytically implausible analysis whose philosophical justification
lies in the presumed need to rationally reconstruct causal notions in
noncausal terms. Without this verificationist and anti-"metaphysical"
premise, the D-N account would be philosophically indefensible.

Humean Explanation and Evidence

I suggested above that the D-N account of explanation has three philosophical virtues: that it portrays the explanatory power of a theory as
depending upon the more general theoretical setting in which it is
applied, that it rests on a theory of causation appropriate to the philosophical tradition in which it arises, and that it portrays the methodological preference for explanatory theories as a special case of the
same methodological principle that dictates a preference for theories
whose predictions have been observationally confirmed. We have just
seen that the second of these claims is true, that at least for the central

case of causal explanations the D-N account *is* the Humean definition of explanation. The first is true just because the D-N account was understood in light of the principle of "unity of science" according to which a variety of different well-confirmed theories may legitimately be employed conjointly in making observational predictions. This principle is exactly the principle that entails the employment of "auxiliary hypotheses" in deducing the observational predictions that are to be tested in order to confirm or disconfirm a proposed theory. It is worth remarking that the "unity of science" principle is ineliminable if the D-N account of explanation is to be even remotely plausible. Even in the most typical and straightforward cases of causal explanation it is usually true that the event explained will not be retrodictable from the primary explanatory theory unless additional well-confirmed theories are also employed as premises. This point is as unchallenged in the empiricist literature as the corresponding point about the necessity of "auxiliary hypotheses" in the testing of theories.

Let us now turn to the third of these features. When an event is explained, the theories that are said to be explanatory on the D-N account are those that are employed in the retrodiction of the event. Thus, every successful explanation of an observable event has just the same logical form as a successful observational test of the relevant explanatory theories. This happy result is no surprise. It is characteristic of Humean conceptions of causation that the occurrence of a cause followed by its effect should be an instance of, and thus evidence for, the law or regularity whose existence is asserted by the appropriate "rationally reconstructed" causal statement. It would thus appear that the D-N account of explanation solves the epistemological puzzle about the evidential relevance of explanatory power as a nonexperimental criterion of theory acceptance. Really, according to the D-N account, the methodological preference for explanatory theories is not a non-experimental criterion. Instead it is the special case of the criterion that dictates preference for experimentally tested theories—the case that applies to experimental (or observational) evidence whose epistemic relevance is recognized only after the relevant observation has been made. For one of the nonexperimental criteria, at least, the puzzle appears to be resolved. The elegance of this proposed solution is surely one of the most attractive features of the D-N account.

Nevertheless, it is extremely important to recognize that—even by empiricist standards—the D-N solution to the puzzle of the evidential role of explanatory power is incomplete. Recall that explanatory power is only one of a number of apparently nonexperimental criteria of theory acceptability. Even if it should turn out that the explanatory power of a theory is just a matter of its experimental confirmation by belatedly recognized observational tests of its own predictions, the fact remains that there are some genuinely nonexperimental criteria that are central to scientific methodology. This must be the case, since, as Goodman demonstrated, judgments regarding the confirmation of theories require prior (though perhaps tentative and revisable) judgments of their projectability. The genuine nonexperimental criteria are just those that legitimately play a role in projectability judgments. It follows, of course, that no account of the epistemology of science that does not say something about the epistemic (as opposed to the purely pragmatic) role of the genuine nonexperimental criteria can be complete. What is striking is that the D-N account presupposes an appropriate solution to this problem because the notion of projectability or lawlikeness is appealed to in the very formulation of the D-N account. There are two possibilities: Either judgments of lawlikeness are simply judgments of explanatory potential (a plausible enough view in light of actual scientific practice) or there are additional or different components of such judgments. In the first case, the D-N account of explanation cannot be a complete account of the epistemic role of judgments of explanatory power, since it presupposes a nonexperimental role for such judgments. In the second case, the D-N account succeeds in the project of providing a Humean anti-metaphysical analysis of the epistemic role of such judgments only if a similarly Humean reconstruction is possible for the genuinely nonexperimental criteria. In either case, the view that the D-N account of explanation succeeds in offering an account of the epistemic role of judgments of explanatory power presupposes the possibility of providing a similarly Humean account of whatever nonexperimental criteria of theory acceptability there are.

Thus, providing Humean accounts of the evidential relevance of nonexperimental criteria and of explanatory power are not two independent tasks of empiricist philosophy of science; success in the former is a prerequisite for success in the latter. Once this fact is recognized, one can see that there is a significant *prima facie* difficulty in the traditional empiricist program. The traditional empiricist treatment of nonexperimental criteria other than explanatory power has been to treat such criteria either as purely pragmatic and thus epistemically irrelevant (as is typical when such criteria are described as "simplicity" or "parsimony") or as purely syntactic and thus conventional (as in the case of typical treatments of lawlikeness or projectability). As we have seen, there is no reason to believe that the epistemic contribution of nonexperimental criteria to the instrumental reliability of scientific methodology can be accounted for on solely pragmatic or conventionalistic grounds. The question is thus raised whether an adequate treatment of nonexperimental criteria is possible within the empiricist tradition, as the D-N account of explanation requires. In the next section I shall argue that the answer is no.

Toward a Non-Humean Alternative

Why a Non-Humean Alternative Is Needed

There is an extraordinarily rich and interesting literature in which various versions of the D-N account of explanation are criticized and defended with respect to their applicability to a wide range of kinds of explanation. What is characteristic of this literature is that philosophers have debated the applicability of the D-N account with respect to particular examples of scientific, historical, or psychological explanations that might be thought to resist subsumption under the D-N conception. Many of the criticisms of the D-N account represented in this literature are extremely important, as are many of the replies in its defense. Nevertheless, what I propose to do here is not to review this important literature but instead to argue directly against the D-N account on the grounds that the Humean definition of causation—which is its only philosophical basis—can now be seen to be wholly inadequate. I propose to adopt this strategy for two reasons. In the first place, it seems to me that the fact that for causal explanations the D-N account is an utterly straightforward application of the Humean definition of causation means that, unless a critique of the Humean definition is developed, the effectiveness of any criticisms of the D-N account for such cases will necessarily be reduced in the light of the support the D-N account receives from so well established a philosophical doctrine as the Humean definition. Moreover, it seems to me that recent criticisms of empiricist philosophy of science (including, of course, criticisms of the D-N account) have permitted us to develop enough anti-empiricist insights in the philosophy of science that a useful direct criticism of the Humean roots of the D-N account is now possible. What I propose to do in the remainder of this section is offer two sorts of criticisms of the Humean definition of causation (one more technical and the other more epistemological),

propose and defend alternative conceptions of causation and of explanation that are in the tradition of "scientific realism" rather than in
the tradition of logical empiricism, and indicate briefly how the proposed
alternative conception of explanation would apply to the problem of
explaining the epistemic role of explanatory power in scientific methodology. Since my aim is to indicate how recent critiques of logical
empiricism and its variants can be extended to a treatment of the issue
of explanation, I will rely heavily on recent work (including some of
my own) that is critical of empiricism. I will usually sketch the main
philosophical arguments involved, but I will not attempt to defend in
detail the anti-empiricist positions upon which my critique of the
Humean conceptions of causation and explanation depends. The present
work is intended as a contribution to a developing realist critique of
empiricism, not as an entirely self-contained refutation of the empiricist
conception of explanation.

A Technical Criticism

Suppose that L is a set of strictly deterministic natural laws that hold
in some possible world, C is a specification of initial conditions in that
world at some fixed time, and e is an event subsequent to that time
such that for systems governed by L an outcome just like e is necessary
whenever initial conditions satisfying the specification obtain. It is part
of any reasonable conception of causation—certainly it is part of any
typical empiricist conception—that the conditions satisfying C (or, perhaps, some proper subset of them) constitute the total cause of e. Thus,
on the Humean definition of causation, it should be true that the occurrence of e will be deductively retrodictable from L together with C.
But this need not be the case. If the determining function defined by
L is not general recursive in finitely many additional variables (representing physical constants), then it will certainly not be the case that
such deductive retrodictions are always possible, since if they were
they would provide a general recursive computation procedure for the
determining function. (I am here assuming that L is itself recursively
specifiable, and I am ignoring a more complicated possible case involving
infinitely many physical constants for which a similar result can be
obtained; see Boyd 1972 for the details.) Possible laws with this embarrassing property exists, and there is no general reason to suppose
that the fact that the determining functions they define are not effectively
computable for all possible initial conditions precludes their experi-

mental confirmation (again, see Boyd 1972). Thus, the Humean conception of causation is not universally applicable, even in those cases in which discoverable deterministic laws are governing. Exactly the same difficulties can arise in cases in which nondeterministic statistical laws govern the world in question. Thus, neither the contemporary Humean conception of causation nor any of the natural modifications of it that are appropriate for statistical laws can possibly represent a scientifically appropriate analysis of the concept of causation.

It might be thought that this difficulty could be remedied by taking the Humean analysis to rest upon the empirical assertion that the actual laws of nature define general recursive determining functions (or their statistical analogs). Of course this response would be entirely out of character with traditional empiricist philosophical methodology, which sought to provide rational reconstructions of scientific concepts that were justifiable *a priori*, but there is a long, if poorly developed, tradition of philosophical naturalism within empiricism, so we should certainly consider whether this particular appeal to the empirical might salvage the contemporary empiricist analysis of causation. It is very doubtful that it can. The reason is twofold. In the first place, even if it were demonstrated that the currently accepted laws of nature define determining functions that are general recursive (which may well be true), the well-established truism that all the laws we currently accept are likely to be only approximately true would prevent our immediately concluding that the true laws of nature have this property. It might seem that this difficulty could be overcome simply by inferring from the recursiveness of the determining functions defined by the currently accepted laws that the true natural laws probably also define recursive functions. This piece of inductive inference is, however, quite dubious methodologically. It is extremely doubtful that the hypothesis that the determining function defined by natural laws is recursive is itself a projectable hypothesis; it is doubtful, in other words, that one can reasonably infer this conclusion from the premise that various approximately correct "laws" define recursive determining functions. This is so because recursive determining functions are rather rare (if we treat physical constants as variables and thus count the number of basic forms of such laws, then there are only countably many of them) and each of them is approximated arbitrarily well by a continuum of extremely well-behaved (say infinitely differentiable) nonrecursive functions. Moreover, there is no reason to believe that the computability of determining functions has any physical significance. It is thus extremely risky to take the approximate truth of laws with recursive determining functions as even *prima facie* evidence that the true laws of nature specify such functions.

A somewhat more promising response might be to propose a natural revision in the formulation of the Humean definition itself. Suppose that, instead of requiring that the occurrence of the effect be deductively predictable from the laws together with relevant specifications of initial conditions, we require that statements completely specifying the effect hold in all of the intended models of the laws together with the specifications of initial conditions. At least for a great many cases (including all those discussed in Boyd 1972), a suitable notion of "intended interpretation" is available, and the Humean definition so modified will therefore identify causal relations in the appropriate way. The difficulties with this proposal are philosophical rather than mathematical. In order for the proposed definition of causation to be Humean, it would have to be the case that—on the conception of cause it advances—causal relations could be discerned in nature without recourse to knowledge of unobservable "theoretical entities" and their causal powers. There are extremely good reasons to believe that this is not so. The question of the confirmation of laws not all of whose observational consequences are deducible from them raises, in an especially clear way, a general problem in the experimental confirmation of theories. In general, when we accept the observational confirmation of finitely many of the infinitely many observational predictions of a theory as constituting sufficient evidence for its tentative confirmation, we are tacitly relying on some solution to what might be called the general problem of "sampling" in experimental design. By this I mean the problem of deciding, for any particular proposed theory, which reasonably small finite subsets from among the infinite set of observational predictions it makes are "representative samples" in the sense that observational confirmation of all their members would constitute good evidence for the approximate truth of the rest of the theory's observational consequences. The significance of this problem is especially easy to see when we consider the special case of theories not all of whose intended observational consequences are computationally available. In order to confirm such a theory, we would have to assure ourselves that from among the computationally available predictions of the theory a suitable representative sample can be formed.

I have argued (Boyd 1972, section 7) that for some theories with some computationally unavailable consequences it would be possible to reliably identify such representative samples by employing available theoretical knowledge of (typically unobservable) underlying mechanisms to determine under what various sorts of conditions the theory would be likely to fail, and by finding computationally available predictions of the theory regarding conditions of these various sorts. Such a strategy would permit confirmation of such theories even though the determining functions they define are not recursive, but it would not do so within the constraints required by a Humean conception of scientific knowledge. Prior theoretical knowledge of underlying unobservable causal mechanisms would be essential for the confirmation of such theories. Thus, the revised definition of causation we are considering would fail to be Humean, in that it would not portray causal knowledge as independent of knowledge of unobservable causal factors.

I have also argued (Boyd 1973, 1979, 1982, 1983, and especially Boyd forthcoming) that—for theories in general and not just for those with nonrecursive determining functions—no alternative to this procedure for solving the problem of sampling exists. I conclude, therefore, that the second response to the primarily technical objection to the Humean definition of causation also fails. The technical criticism is apparently successful.

It is, nevertheless, a good methodological practice in philosophy to be cautious in accepting primarily technical criticisms of broadly significant philosophical theses. Such theses often admit of unanticipated reformulations that are sufficient to avoid particular technical criticisms. The more important criticisms are those suggesting that the thesis in question rests upon a fundamental philosophical mistake. Such a criticism of the Humean definition is suggested by the epistemological rebuttal just offered to the possible revision we were considering. If the methods of actual scientific practice for resolving questions about sampling in experimental design rely upon prior (approximate) theoretical knowledge of unobservable causal factors, then, in particular, knowledge of such factors is actual and therefore possible. Thus, the empiricist conception that experimental knowledge cannot extend to unobservable causal powers and mechanisms must be mistaken and the philosophical justification of the Humean definition of causation rests upon a false epistemological premise. There is indeed considerable evidence that almost all the significant features of the methodology of recent science rest ultimately upon knowledge of unobservable causal powers and mechanisms (see Putnam 1975a, 1975b; Boyd 1973, 1979, 1982, 1983, forthcoming), and thus that the empiricist reservations about experimental knowledge of unobservable causal powers and mechanisms are profoundly mistaken. In the next section, I will explore in greater detail the consequences for the Humean conception of causation and explanation of the failure of the empiricist conception of experimental knowledge.

The Epistemological Inadequacy of the Humean Definition

The Humean definition of causation and the associated D-N account of explanation require acceptance of the "unity of science" principle and presuppose a Humean (that is, nonrealistic) but nevertheless epistemic account of the nonexperimental criteria of theory acceptance that determine judgments of lawlikeness and projectability. In fact, no satisfactory epistemological account of the "unity of science" principle is compatible with the empiricist's denial that we can have knowledge of unobservable causal powers and mechanisms, and no Humean account of lawlikeness and projectability can be epistemologically adequate.

Consider first the "unity of science" principle. Neither the Humean definition of causation nor the D-N account of explanation is even remotely plausible unless it is understood that the laws under which the caused or explained event is to be subsumed can be drawn from several different scientific disciplines or subdisciplines and conjointly applied in predicting an event. It must be possible for two laws that have been quite independently confirmed by specialists working in different areas to both be premises in the sort of deductive prediction to which the Humean definition and the D-N account refer. Moreover, it must of course be epistemically legitimate that independently confirmed laws be conjointly applied in this way to make observational predictions. After all, the point of the Humean conceptions we are considering is to reduce knowledge of causal relations and of explanations for events to knowledge of predictable regularities in the behavior of observable phenomena. Only if predictions obtained in accordance with the "unity of science" principle are epistemically justified would beliefs about the behavior of observables established by the sorts of deductive prediction we are considering constitute knowledge.

There is a further Humean requirement that applications of the "unity of science" principle must meet if the Humean conceptions of causation and explanation are to be justified. We have already seen that the

Humean conceptions are philosophically untenable if judgments of lawlikeness or projectability involve knowledge of unreduced causal factors. In a similar way, the Humean conceptions would be philosophically untenable if the applications of the "unity of science" principle upon which their plausibility depends themselves presupposed knowledge of unobservable causal factors. In fact this proves to be the case. The argument (see Putnam 1975b; Boyd 1982, 1983, forthcoming) can be summarized as follows: The principle that independently confirmed theories can legitimately be conjointly applied in making predictions about observables must presuppose some sort of judgments of univocality for the nonobservational (or "theoretical") terms occurring in the theories in question. Without the requirement that all the theoretical terms occurring in the conjointly applied theories occur univocally in their conjunction, the "unity of science" principle would dictate the absurd conclusion that one should expect approximately true observational predictions from the conjunction of well-confirmed physical theories of force together with well-confirmed theories of the role of force in international affairs even if one does not disambiguate the various occurrences of the lexical item "force" in the conjunction. Moreover, the principle for assessing univocality cannot be that theoretical terms are nonunivocal whenever they occur in different theories; such a principle would result in a very significant underestimation of the scope of the "unity of science" principle in actual scientific practice and therefore would be inappropriate for the defense of the Humean conceptions we are considering.

Once it is recognized that theoretical terms from quite different theories are sometimes to be counted as occurring univocally, it becomes clear that the "unity of science" principle makes a striking epistemological claim. Suppose that T_1 and T_2 are two theories from quite different scientific disciplines in whose conjunction no theoretical terms occur ambiguously. Suppose further that in the experimental confirmation of these theories neither was ever employed as an "auxiliary hypothesis" in the testing of the other. There will thus have been no *direct* experimental test of the conjunction of the two theories, except insofar as the predictive reliability of each of them taken independently has been tested by prior experiments. Nevertheless, the "unity of science" principle maintains, we are justified in expecting the conjunction of the two theories to be instrumentally reliable even in the absence of direct experimental tests, provided only that the univocality constraint on their constituent theoretical terms is satisfied. The univocality constraint is thus supposed to do real epistemic work in making possible what may be thought of as the *indirect* confirmation of the instrumental reliability of the conjunction of T_1 and T_2.

A good way of seeing what is going on is to consider what an empiricist might plausibly take the confirmation of a theory to amount to. Since no knowledge of theoretical entities is supposed to be possible, it would be initially natural for the empiricist to hold that when a theory is confirmed all that is confirmed is the approximate instrumental reliability of the theory itself. Recognition of the crucial role of auxiliary hypotheses in the testing of theories suggests replacing this instrumentalist conception with a broader one according to which the experimental confirmation of a theory amounts to the confirmation of the conjoint reliability of the theory together with the other theories that have been employed as auxiliary hypotheses in testing it. The "unity of science" principle requires a much broader conception. Experimental confirmation of a theory is supposed to constitute evidence for its instrumental reliability even when it is applied conjointly with other well-confirmed theories not even discovered at the time the evidence for the first theory was assessed! Something over and above the instrumental reliability of the conjunction of the theory with actually employed auxiliary hypotheses—something over and above even the instrumental reliability of the theory taken conjointly with currently established theories—is supposed to be confirmed when the theory is properly tested. That "something," the knowledge over and above the instrumental knowledge that has been directly confirmed, is represented in the theoretical structure of the theory, and the rule for extracting it is to make deductive predictions from the theoretical sentences in the theory in question together with the theoretical sentences that represent the similar "excess knowledge" in other well-confirmed theories. There is no plausible explanation of the instrumental reliability of this sort of instrumental knowledge-extraction procedure other than that provided by a realist conception of theory confirmation according to which confirmation of theories involves confirmation of the approximate truth of their theoretical claims as well as their observational ones. On such a conception, judgments of univocality for theoretical terms are judgments of co-referentiality, and what the "unity of science" principle licenses is deductive inferences from the partly theoretical knowledge embodied in independently tested theories to conclusions about the behavior of observables. Univocality judgments are crucial in establishing that the nonobservational subject matter of the two theories is really the same when it appears to be. (As a matter of fact, the situation is even more complicated: The unity of science also involves inductive inferences from theoretical knowledge. This additional consideration strengthens the case for a realist construal of both theory confirmation and univocality judgments; see Boyd forthcoming.)

It follows not only that knowledge of unobservable causal factors is possible but also that it is presupposed by the "unity of science" principle. The principle is tenable only on the assumption that knowledge of theoretical entities is possible, and it presupposes that the univocality judgments for theoretical terms that scientists actually make are reliable judgments about reference relations between theoretical terms and theoretical entities. This, in turn, requires that scientists have reliable knowledge of causal relations between unobservable causal factors and their own use of language. The "unity of science" principle thus presupposes just the sort of knowledge that the Humean conceptions are designed to "rationally reconstruct" away, and the Humean conceptions are thus philosophically indefensible.

Similar arguments show that judgments of lawlikeness and projectability are likewise infected with essentially non-Humean commitments to knowledge of unobservable causal factors. I have argued for this and related claims elsewhere (Boyd 1973, 1979, 1982, 1983, forthcoming); the basic argument can be summarized as follows.

We have seen that the solution to the problem of sampling in experimental design in mature sciences presupposes prior knowledge of unobservable "theoretical entities" or causal factors. In fact, the solution to this problem is intimately related to the solution to the problem of projectability. Roughly, theories are projectable just in case there is some *prima facie* reason to believe that they might be (approximately) true and thus some reason to treat them as live candidates for confirmation by observational evidence. The methodological rule for the solution to the sampling problem is this: Test a proposed theory under circumstances representative of those identified by other projectable theories about the same issues as those under which its predictions are most likely to be wrong. The theory-dependent judgments that go into solving the problem of sampling are just special cases of judgments of projectability.

In fact, as Kuhn (1970) has correctly maintained, this pattern of the dependence of scientific methodology on the ontological picture presented by the received theoretical tradition infects all the important principles of scientific methodology. For example, another important question in experimental design is what factors must be controlled for in setting up the experimental conditions. There are infinitely many factors about which it is logically possible that they could interfere with the intended functioning of experimental apparatus. We identify the relatively few factors that must be controlled for by applying our existing theories of underlying (and typically unobservable) causal mechanisms to identify those sorts of interference that it is reasonable to believe might operate in the relevant experimental conditions. Here too the methodological principle we employ is very intimately connected with judgments of projectability. To a very good first approximation, the rule we employ is that factors should be controlled for that are suggested by those logically possible theories that are themselves projectable. Again our judgments of projectability turn out to be theory-dependent judgments relying on the accounts of unobservable causal factors represented by our best confirmed theories.

For each of the theory-dependent principles of scientific methodology we can ask what explains its contribution to the instrumental reliability of scientific practice. In each case, the only plausible explanation is given by the realist conception that in making such judgments we employ the approximate knowledge of observable and unobservable causal factors reflected in existing theories in order to establish methods for improving our knowledge of both observable and "theoretical" entities. Theoretical understanding of unobservable causal factors enjoys a dialectical relationship with the development and improvement of methods for improving theoretical understanding itself. In particular, judgments of projectability require knowledge of unobservable causal factors. Thus, the appeal to projectability in the Humean definition of causation deprives that definition of its Humean content and hence of its only philosophical justification.

It will be useful to consider one important rebuttal to the position I have just taken. In "Natural Kinds" (1969), Quine sometimes describes the natural kinds in mature sciences as issuing from theory "full-blown." When Quine writes in this way, his account of natural kinds (and thus of projectability) seems very close to the realist and anti-Humean conception just discussed. In other places he seems to prefer to treat the identification of projectable theories or predicates as involving "second-order induction about induction." He says: "We establish the projectability of some predicate, to our satisfaction, by trying to project it. . . . In induction, nothing succeeds like success" (p. 129). This formulation suggests that projectability judgments might be thought of as *a posteriori* judgments involving consideration only of observable phenomena. After all, the instrumental reliability thus far displayed by some particular inductive strategy (with its particular judgments of projectability) is an observable phenomenon, and Quine's suggestion appears to be that we can (at least tentatively) identify projectable theories or predicates by looking at which ones have figured in successful inductive inferences in the past. No consideration of unobservable phenomena appears to be involved.

It is important to see what the philosophical consequences would be if this conception of projectability judgments could be sustained. We have already seen that projectability judgments play an essential epistemic role in establishing the instrumental reliability of scientific methodology, and that therefore it is not adequate to treat projectability judgments as purely conventional or to offer a purely pragmatic account of their rationale. As Quine and Goodman both recognized, projectability judgments must have some sort of empirical basis in order for their epistemic role to be explicable. The proposal that Quine appears to be making would, if it were successful, provide an adequate account of that epistemic role without invoking knowledge of unobservable causal factors. The Humean conceptions of causation and explanation would therefore succeed in offering a reductive analysis of causal notions. Moreover, a nonrealist account of the epistemic role of projectability judgments would undermine the arguments rehearsed earlier in this section to the effect that experimental knowledge of unobservable causal factors is possible. The Humean definition of causation and the D-N account of explanation would indeed be philosophically justified, and the project of the present essay would be misconceived.

I have discussed the second-order induction about induction interpretation of projectability at length elsewhere (Boyd 1972, section 2.3; forthcoming, part III; and especially 1983, section 8). Roughly, the flaw in the proposal lies not in the claim that projectability judgments can be thought of as a species of second-order induction about induction but rather in the conception (which may not have been Quine's) that such inductions are independent of knowledge of unobservable causal factors. The problem is that such inductions—like all inductions—depend upon projectability judgments, and the projectability judgments upon which they depend involve just the appeals to knowledge of unobservable causal factors that (the version we are considering of) second-order induction about induction is supposed to eliminate.

The simplest among several ways to see this is to realize that the inductive inferences about induction we are considering are supposed to take as premises claims that previous inductive inferences guided by certain standards of projectability have been largely successful and to conclude that future inductive inferences guided by the same standards will also tend to be successful. The premises of such inferences admit of two interpretations. On one reading, the premises state that, generally speaking, the theories that have been accepted in accordance with the relevant conception of projectability are instrumentally reliable; they make approximately true predictions about the behavior of observable phenomena. On this reading, knowledge of the premises requires knowledge about various theories that their observational predictions—future as well as past, untested as well as tested—are approximately true. On the second and weaker reading, all the premises state is that it has been largely true that those observational predictions that follow from theories accepted in accordance with the relevant conception of projectability and have been subjected to experimental test have proved approximately true. Only on the weaker reading do the premises report observed facts.

Pretty plainly, the actual structure of the inference from observable data to the conclusion that the relevant conception of projectability is epistemically reliable must be thought of as proceeding in two steps. From the information provided by the premises on the weaker reading (from the predictive successes in actually tested cases of the theories in question) we first infer the instrumental reliability of a suitable number of the relevant theories. These conclusions about instrumental reliability, together with the presumably uncontroversial premise that the standards of projectability in question dictated the acceptance of those theories, then confirm the premises we are considering on their stronger reading. It is from those premises on their stronger reading that we reasonably conclude (in the second step) that the relevant conception of projectability is epistemically reliable.

There is nothing whatsoever wrong with this pattern of inductive inference about inductive procedures. Indeed, reliance on just such inferences is essential not only in science but also in the practice of epistemologists of science. It is true, however, that the first step in this inference consists simply in a number of cases of inferring the instrumental reliability of a theory on the basis of experimental evidence. It is just this sort of inference that as we have seen, depends upon projectability judgments grounded in knowledge of unobservable causal factors. Second-order induction about induction thus presupposes such knowledge and cannot form the basis for a Humean reconstruction of projectability judgments.

The Humean definition aims to rationally reconstruct causal notions in noncausal terms. The philosophical justification for this project rests upon the epistemological premise that experimental knowledge of unobservable causal factors is impossible. The epistemological premise is false, and the rational reconstruction is in any event unsuccessful. There is no reason to accept the Humean definition. The D-N account of explanation is—for those cases in which it is most plausible—simply an application of the Humean definition, and thus it is also without philosophical justification. There is no reason to reject the preanalytic conception that, for a wide and central class of cases, to explain an event or a recurring phenomenon is to say something about how it is caused. Nor is there any reason to think that the empiricist analyses of causation and explanation rest on, or provide, an even approximately accurate conception of the nature of causal knowledge in science or in any other area of inquiry.

The Semantics of "Cause" and "Explain"

It is a puzzling fact that many philosophers who reject the empiricist conclusion that knowledge of unobservables is impossible and who are sympathetic to scientific realism rather than to verificationism or instrumentalism nevertheless employ the Humean definition of causation or the D-N account of explanation, even in cases in which the phenomena caused or explained are unobservable. The principal explanation of this phenomenon, I believe, is that the Humean definition and the D-N account were so widely accepted during the time when empiricism dominated the philosophy of science that they now have the status of established philosophical maxims whose initial justification has been forgotten. The fact that the rejection of an empiricist conception of experimental knowledge in favor of a realist conception leaves these positions without any philosophical justification may have gone largely unnoticed. I am inclined to think, however, that there is an additional explanation for the durability of these two pieces of philosophical analysis in the face of widely accepted criticisms of their empiricist foundations. I think that philosophers may believe that we need to have some analysis of the meaning of terms such as "cause" and "explain" and that the definitions that arise from the Humean tradition may serve as good first approximations to such meaning analyses. Of course, that such definitions fail to be reductive and thus fail to meet empiricist standards would not, by itself, show that they are inappropriate as nonreductive philosophical analyses.

It nevertheless seems plain that the Humean definitions are strikingly inadequate. In the case of the Humean definition of cause, what seems to be the primary causal notion gets defined in terms of the highly derivative causal-epistemological notion of projectability. Instead, it would seem that the revealing definition would go more nearly in the opposite direction, projectability being defined in terms of knowledge of causal powers and mechanisms. The D-N account of explanation of course inherits these difficulties; moreover, there are notorious difficulties in assimilating clear-cut cases of explanation to the D-N model. It would appear that neither of these definitions is a very promising beginning for a philosophical analysis of the relevant concept. It might be thought that they have the advantage of reminding us that our concept of causation is related to something like a conception of determinism. (It had better not be exactly like one if our conception is to correspond to something real; that is why there are statistical versions of both definitions.) The D-N account of explanation might also be thought of as setting a standard for complete explanations appropriate to the conception that something like determinism is involved in causation.

Against these claims one may reply, first, that the technical criticism of the Humean definitions presented above show that in any event they embody the wrong analysis of determinism (see Boyd 1972). Moreover, in any event we want the question of the relationship between causation and determinism to be spelled out by research in the various sciences and social sciences rather than by so abstract a definitional specification. In a similar way, we should want the relevant methodological standards of completeness of explanations to be determined (in a theory-dependent way) by the aims and the findings of the various special sciences. (Indeed, it is difficult to imagine what scientific activity would, even with a suitable idealization, require us to seek explanations complete in the D-N sense even in a deterministic world.) Finally, as we shall see in the next section, in order to account for the evidential import of explanatory power we need not assimilate explanation to the sort of retrodiction provided by "complete" D-N explanations.

But, someone might ask, if the Humean definitions of causation and explanation are rejected in the name of scientific realism then what does the realist propose as an alternative account of the semantics of these and other causal notions? The answer dictated by the realist considerations offered here has several components. First, of course, it is not to be expected that any significant causal notions are adequately definable in noncausal terms. That is just what the critique of the Humean definition establishes. Second, it is quite doubtful that there are philosophically interesting analytic definitions of scientifically important causal notions, even in terms of each other. As the change in our conception of the relation between causation and determinism induced by the acceptance of quantum mechanics indicates, there is no reason to believe that proposed philosophical analyses or definitions of causal notions will be immune in principle from amendment in the light of new theoretical discoveries.

It is nevertheless clear that informative philosophical analyses of many causal or partly causal notions are possible. I take it that such analyses are in some sense empirical because they depend upon empirical facts about causal phenomena and about our practices regarding them and because they are revisable in the light of new discoveries in these areas. Nevertheless, they appear to lie squarely within the province of philosophy. Analyses of such causal notions as explanation, projectability, reference, and knowledge I take to be in this category. About the less derivative causal notions, such as "(total) cause," "causal power," "interaction," "mechanism," and "possibility," it seems less likely that informative analyses of the sort that philosophers typically seek are available; it might be that "cause" and "causal power" are somehow interdefinable, but it is doubtful that whatever definition might be available would prove very informative to someone who wanted to know, e.g., what causal powers are. Informative definitions or analyses in these latter cases, I suggest, are not primarily a matter of conceptual analysis, even on the understanding suggested above according to which conceptual analysis is a kind of empirical enterprise. Instead, the informative analyses or definitions of more basic causal notions are to be established by theoretical inquiry in the various sciences and social sciences. What causation is and what causal interaction amounts to are theoretical questions about natural phenomena (to reject the Humean project is just to admit that causal relations, powers, and interactions really are features of nature), so it is hardly surprising that answers to them should depend more upon the empirically confirmed theoretical findings of the various sciences than should answers to more abstract (and more typically philosophical) questions about the nature of knowledge, reference, or explanation.

The distinction between the two sorts of questions is one of degree. "Conceptual analysis," when done well, has an ineliminable empirical component, and the more foundational questions in the various sciences are typically philosophical questions as well, often requiring the special analytical techniques of philosophy for their resolution. But even though the distinction is one of degree, the fact that definitional questions about fundamental causal notions fall on the side nearer to the various empirical sciences dictates an important conclusion: that such notions as "(total) cause," "causal power," and "interaction" are like the notions of various natural kinds in that they possess no analytic definitions, no "nominal essences." They are defined instead by natural definitions or "real essences" whose features are dictated by logically contingent facts about the way the world is. From a realist perspective, this is hardly surprising. Natural kinds and categories lack stipulative *a priori* definitions precisely because, in order to play a reliable role in explanation and induction, natural kinds and categories must be defined in ways that reflect the particular causal structure the world happens to possess (Putnam 1975a, 1975b; Boyd 1979, 1982, 1983, forthcoming). For exactly the same reason, of course, the definitions of our causal notions must also reflect *a posteriori* facts about the nature of causation.

It follows that the reference of terms referring to fundamental causal notions is not fixed by analytic definitions; there are none. Instead, such terms are like natural-kind terms, theoretical terms in the particular sciences, and other terms with "natural" rather than analytic definitions, in having their reference fixed by epistemically relevant causal relations between occasions of their use and instantiations of the causal phenomena to which they refer. (For discussions of the epistemic character of reference see Boyd 1979, 1982.) It follows that in order to account for the semantics of causal terms we need no such analyses of their meaning as the Humean definitions provide. To hold that some largely *a priori* conceptual analysis must provide definitions for such terms is to fall victim to an outmoded empiricist conception of the semantics of scientific and everyday language.

Explanation and Evidence

At least for many central cases (and for the cases the D-N account is designed to fit), an explanation of an event is an account of how it was caused. In all but the most atypical cases the account will be partial: Not all the causally determining factors will be indicated, nor will the relevant mechanisms be fully specified. The D-N account is typically extended to cover the cases of explanations for laws or regularities in nature. On the D-N conception, to explain a law or a regularity is to deduce the law or a statement of the regularity from other laws, together with statements of appropriate boundary conditions. It is not entirely clear that this standard extension of the D-N account is really appropriate to the Humean task of reducing causal notions to noncausal ones. One should argue that, inasmuch as the possible knowledge reflected in the explained theory is supposed to be exhausted by its observational predictions, all the consistent Humean should require by way of an explanation is the deduction of those observational consequences from the explaining theory. In any event, the realist conception of explanation also generalizes (even more naturally) to cases of the explanation of laws or regularities: To explain a law or a regularity is to give an account (presumably partial) of the causal factors, mechanisms, processes, and the like that bring about the regularity or the phenomena described in the law.

It is a consequence of the Humean account that all explanations of particular events have a certain level of generality "built in" in virtue of reference to the relevant laws; of course, this is just what the Humean conception of causation requires. The realist conception of causation and explanation does not rule out the possibility of singular causal relations that are not instances of more general patterns; it leaves such issues to the findings of the various special sciences. Nevertheless, it does appear that, given what we know about causal relations and about the sorts of causal explanations that are actually discovered, the Humean conception is in this respect right or very nearly right. Scientific explanations of individual events do, almost always, extend to cover similar cases, actual and counterfactual. In consequence, our conception will be appropriate for the central cases of causal explanation if we think of explanations as being provided by small theories describing the causal factors that determine, or the causal mechanisms or processes that underlie, some class of phenomena. I will use the term *explanation* to refer to such theories. Explanations will of course differ considerably in the extent to which they are complete in their identification of causative factors, in the specificity with which they describe underlying mechanisms or processes, in the level of numerical precision with which they characterize the relations between such factors, and in other respects. Part of the task of a theory of causal explanations is to say how the epistemological significance of an explanation is influenced by factors such as these.

Suppose that a theory E is an explanation for some phenomenon p. It would be natural to understand the terms "explains" and "explanatory power" so that it is just E that is therefore said to explain p and so that it is just the explanatory power of E that is thereby demonstrated. Neither scientific usage nor scientific practice conforms to this picture. E might well be said to explain p, but scientific practice dictates our taking the explanatory success of E as grounds for saying of other more general theories that they explain p. Indeed, under the circumstances envisioned, we would not ordinarily speak of the explanatory power of E being manifested at all; instead, E's being an explanation of p would ordinarily be taken to indicate the explanatory power of those other, more general theories and to provide evidential support for them.

Consider for example the explanation of the "law" of fixed combining ratios, according to which in a certain class of reactions chemical elements combine in fixed ratios by weight. An explanation is provided by a theory that says that this phenomenon is produced by the underlying tendency for atoms to combine in fixed ratios by number, together with the claim that atoms of an element all have the same weight. (I ignore here the issue of isotopes.) What chemists and historians of chemistry correctly say is that this explanation indicated that the atomic theory of matter could explain the phenomenon in question; it demonstrated the explanatory power of the atomic theory and thus provided evidence for it. Similarly, consider the occurrence of subcutaneous degenerate hind limbs in the larger constrictors. An explanation for this phenomenon is provided by a theory according to which these limbs are the vestigial remnants of the ordinary hind limbs of reptiles ancestral to the snakes, which were gradually lost through the process of natural selection. Insofar as this explanation is accepted, it indicates the explanatory power of the Darwinian conception of the origin of species and provides evidence for it.

In these cases we can see a pattern that is utterly typical. An explanation for a particular phenomenon will typically draw upon the resources of some more general theory. It will appropriate the theoretical resources of the broader theory (entities, mechanisms, processes, causal powers, physical magnitudes, and so on), and it will employ, and often elaborate upon, these resources in describing how the phenomenon in question is caused. The dependence of minor theories and explanations

upon the theoretical resources of larger theories has been amply doc-
umented by Kuhn (1970), who describes the dependence of research
in "normal science" upon the ontological picture dictated by the most
general theories in the theoretical tradition or "paradigm." (Note that
in thus agreeing with Kuhn one need not accept the constructivist
conception of scientific knowledge he so ably defends; see Boyd 1979,
1983.) Adequate explanations of particular phenomena are taken to be
indicative of the explanatory power of the more general theories whose
resources they exploit and to provide additional evidence for those
theories. The evidential relevance of explanations does not depend
upon its being possible to retrodict or deductively subsume the explained
phenomena from the explanation itself or from the explanation together
with the relevant general theory(ies) (together with auxiliary hypotheses). The examples of explanations just mentioned illustrate the
last point. In the case of the explanation of degenerate limbs in the
large constrictors (as in the case of almost all similar evolutionary explanations) we lack altogether the resources for retrodiction, but the
fact that the Darwinian theory provides the resources for explanations
in such cases properly counts as evidence for it nevertheless.

The case of the "law" of fixed combining ratios is more complicated,
because the more precise formulations of the "law" were developed
simultaneously with its explanation. Nevertheless, it seems clear that
the capacity to predict previously unnoticed instances of the "law" or
to deduce an adequate formulation of it emerged quite slowly as the
explanation became more detailed, and probably not until Mendelev's
work on the periodic table was anything remotely approximating the
sort of explanation anticipated by the D-N model available. Despite
this fact, it is also clear that the cogency of earlier versions of the atomic
explanation, from Dalton's early-nineteenth-century work on, were
rightly taken to indicate the explanatory power of the atomic theory
of matter and to constitute some evidence in its favor.

There are indeed cases in which the explained phenomenon (or rather
a statement describing it) is better thought of as a premise than as a
conclusion in the testing of the theory that constitutes its explanation
and thereby indicates the (evidentially relevant) explanatory power of
the theory upon which it itself is based. The following sort of situation
is commonplace (though perhaps not typical): Some general theory T,
which is projectable (and thus already supported by some theory-
mediated empirical evidence), postulates mechanisms of a certain sort
as causally relevant in a broad class of related phenomena; however,

T is not sufficiently well developed to permit prediction of such phenomena. T itself does not specify in detail the mechanisms underlying particular cases of the phenomena in question, but it does specify theoretically important general descriptions under which (if T is right) such mechanisms will fall. For some phenomenon p of the relevant sort, an explanation E is proposed. E says that p is produced by certain more precisely specified mechanisms of the general sort prescribed by T. Despite its greater precision, even E is inadequate (given available auxiliary hypotheses) to reliably predict occurrences of p. Nevertheless, it is predictable from E that if an instance of p occurs then various experimentally distinguishable symptoms of the operation of the mechanisms specified by E will be present. Experimental confirmation of E consists in producing or finding occurrences of p and testing for the relevant symptoms. The experimental test of E consists, not in finding occurrences of p where and when E predicts their occurrence (since it makes no such predictions), but in finding the relevant symptoms when and where they are predicted to occur, given E and the occurrence of p (together, presumably, with other auxiliary hypotheses) as premises! The success of such predictions tends to confirm E and, less directly, T.

Cases of this sort are routine where T is a general chemical theory about complex and predictively intractable reaction mechanisms, E is a proposed application of T to the case of a particular sort of reaction, and the symptom in question indicates the presence of a reaction by-product that can, on the basis of well-established chemical theories, be taken to be distinctive of the particular reaction mechanisms postulated by E. What all the sorts of cases we have examined suggest is that when an explanation E of a phenomenon p provides evidence for a more general theory T by indicating that T has explanatory power, what is crucial is that E be testable largely independent of T and that the approximate truth of E constitute good reason for believing the approximate truth of T. It appears not to matter very much whether the occurrence of p is itself otherwise significantly confirmatory of E or T, much less whether it is predictable from E or T.

What we need to know is how this sort of confirmatory evidence for theories upon whose resources successful explanations are based is related to the sort of confirmatory evidence provided by the experimental confirmation of observational predictions made from the theories themselves. Understanding this will be easier if we understand better the confirmatory relationship between theories and those of their observational predictions whose experimental confirmation supports them.

In mature sciences all theory confirmation is theory-mediated. As we have seen, theories are not confirmable at all unless they are projectable, and projectability judgments are theory-dependent judgments of plausibility. The confirmation of an observational prediction of a projectable theory does not count significantly toward its confirmation unless theory-determined considerations indicate that it is a relevant test (that is, roughly, unless it tests the theory against a projectable rival). Particular experiments do not count as well designed (and thus are not potentially confirmatory or disconfirmatory) unless there are appropriate controls for the possible experimental artifacts that are indicated as relevant by previously established background theories. No piece of experimental evidence counts for or against a theory except in the light of theoretical considerations dictated by previously established theories. (For discussions of these and the following points see Boyd 1972, 1973, 1979, 1982, 1983, forthcoming.)

The theoretical considerations that thus bear on theory confirmation are themselves evidential considerations. The fact that a theory is plausible in the light of well-confirmed theories is evidence that it is approximately true. This is so because the evidence for the well-confirmed theories that form the basis for the plausibility judgment is evidence for their approximate truth (as scientific realists insist) rather than just for their empirical adequacy (as empiricists typically maintain) and because the inferential principles by which conclusions about the plausibility of proposed theories are drawn from the previously established theories are themselves determined by previously acquired theoretical knowledge. The dialectical development of theoretical knowledge and of methodological principles extends to the principles by which plausible inferences are made from wholly or partly theoretical premises to theoretical conclusions. Judgments of theoretical plausibility reflect inductive inferences at the theoretical level (that is, inferences from previously acquired theoretical knowledge to inductively justified theoretical conclusions). These inferences proceed according to theory-determined assessments of projectability, just as inferences from observational data to theoretical conclusions do. (See especially Boyd 1983, Boyd forthcoming.)

The evidence for a theory provided by its being plausible in the light of previously established background theories is every bit as much empirical evidence as is the evidence provided by experimental tests

of the theory's observational predictions; the empirical basis for this
evidence consists of the various observations involved in the confirmation of the relevant background theories. Call empirical evidence
for a theory *direct* if it is provided by experimental tests of observational
predictions drawn from the theory itself, and *indirect* if it is obtained
by inductive inferences at the theoretical level from other theories that
have themselves been confirmed by experimental tests. The important
consequence of a realist conception of scientific epistemology is that
the distinction between direct and indirect empirical evidence is of no
fundamental significance. The observations that provide direct experimental evidence for a theory provide significant evidence at all only
because of indirect evidential considerations in support of the theory
itself (viz. the bases of the judgment that it is projectable) and in support
of various other theories (those of its logically possible rivals also judged
projectable, the theoretically plausible accounts of possible experimental
artifacts, and so on). Thus, direct experimental evidence is only su-
perficially direct. Moreover, indirect empirical evidence can be very
strong evidence indeed. The fact that a theory provides theoretically
plausible accounts of a very large number of phenomena in a way in
which none of its plausible rivals do can, under appropriate circumstances, constitute genuinely confirmatory evidence for it even though
almost no evidence for it is provided by direct tests of its observational
predictions. This was the situation of the Darwinian theory of the origin
of species until very recently, and it is the current situation of many
astronomical theories. Observations can often provide striking confirmation of a theory not by confirming a prediction of the theory but
by ruling out its theoretically plausible alternatives. It may be true,
nevertheless, that in many important cases direct empirical evidence
for a theory plays an especially important confirmatory role, and there
may even be some general methodological reason why this should be
so. But the distinction between direct and indirect evidence cannot be
an epistemologically fundamental one. They are two closely related
and interpenetrating cases of the same epistemological phenomenon.

It will now be clear, I believe, how the evidence for a theory that
arises from a demonstration of its explanatory power is to be understood.
Let T be a theory, E an explanation that draws upon the resources of
T, and p the phenomenon that E explains. Evidence (direct or indirect)
for E will demonstrate the explanatory power of T just in case (given
the available background theoretical knowledge and the inductive standards it determines) the way in which E draws on the theoretical re-

sources of *T* is such that *E*'s being approximately true provides inductive reason to believe that *T* is also approximately true. It will not matter in any fundamental way whether or not the evidence for *E* includes successful prediction of *p*. In any event, the evidence thus provided for *T* will be a perfectly ordinary case of indirect empirical evidence of the sort we have just been examining. Assessments of explanatory power are just one species of assessment of indirect theory-mediated empirical evidence. There is nothing going on when we prefer explanatory theories over and above what goes on in all cases in which we prefer theories that are supported by (necessarily partly indirect) empirical evidence.

Let us return to the D-N account of explanation. Its three most attractive features were that it rested upon an appropriate account of causation, that it indicated that the explanatory power of a theory depends upon its integration into a larger body of well-confirmed theories, and that it portrays the preference for explanatory theories as a special case of the preference for theories supported by observational evidence. In all these respects the D-N account is right. What we have seen is that the weakness of the D-N account lies not in the unworkability of the above three features but in the mistaken Humean conceptions of causation and evidence upon which the D-N account rests. When we adopt a realist conception of causal relations and causal powers as real features of the world, an account of the integration of theories that countenances inductive integration of theoretical knowledge as well as conjoint deductive prediction, and an account of empirical evidence that recognizes the crucial methodological role of such inductive integration, we are able to preserve the best features of the D-N account while avoiding the insuperable difficulties to which empiricist accounts of scientific methodology invariably fall victim.

Other Epistemological Issues

Simplicity and Parsimony

Traditional logical-empiricist accounts assimilate the methodological preference for explanatory theories to a preference for empirically tested theories but typically treat the other nonexperimental standards for the acceptability of a theory as purely conventional or pragmatic. At least, that is the typical "official" empiricist position. In applied philosophy of science, logical empiricists often treated considerations of "simplicity" and "parsimony" as though they had evidential weight. In this latter case, I believe, empiricists were basically right.

One of the striking things about the methodological judgments philosophers of science assimilate to the categories of "simplicity" and "parsimony" is the extent to which they are more complex than those descriptions would suggest. Scientists do not, as a general rule, prefer the simplest from among the empirically unrefuted theories about some natural phenomenon. They quite often—and without any misgivings— reject theories as too simple (or perhaps as too simpleminded) even when they fit the data that have already been examined. There are whole disciplines in which "single-factor" theories are held up to methodological derision, and there are even more disciplines in which this would be true were single-factor theories seriously proposed. Similarly, the principle of parsimony, or Occam's razor, seems to be applied quite unevenly. In many fields, at particular moments in their histories, scientists quite cheerfully postulate new entities in order to account for new empirical discoveries rather than making other theoretical accommodations equally compatible with the data in question. What plainly happens in these cases is that theoretical reasons legitimate the unsimple or unparsimonious theoretical choices. Thus, judgments of simplicity and parsimony are—like judgments of explanatory power—theory-dependent.

We know, moreover, that if (as seems plausible) judgments identified as simplicity or parsimony judgments are important factors in judgments of projectability then such judgments cannot be merely conventional or pragmatic; they must play an epistemic role in scientific practice. What I suggest is that judgments of "simplicity" and of "parsimony" are simply special cases of judgments of theoretical plausibility. When a proposed theory assimilates new data into our existing theoretical framework via a modification that is (according to the evidential standards dictated by that framework) warranted by those data, we see the modification as a simple one (in the sense that it does not introduce epistemologically needless modifications into theories we already take to be well confirmed) and we somewhat misleadingly describe the theory itself as simple. Similarly, we reject proposed theories that accommodate new data by postulating theoretically implausible new entities, and we misleadingly characterize our preference as being for parsimonious theories in general.

If this suggestion is basically right (I invite the reader to consider various actual cases), then the methodological preferences we typically misdescribe in terms of simplicity and parsimony are simply special cases of the methodological preference for theories that are supported by inductive inferences at the theoretical level from the approximate theoretical knowledge we already have. But that principle, as we have seen, amounts to a preference for theories supported by indirect experimental evidence. In the case of the principles of "simplicity" and "parsimony" just as in the case of the principle that we should prefer explanatory theories, all that is really going on is a recognition of the role of indirect evidence in science. The nonexperimental criteria of theory acceptability, which initially appear puzzling, turn out to be nonexperimental only in the sense that they do not reflect the assessment of direct experimental evidence. The logical positivists were right in their applied philosophy of science when they took these principles to be evidentially relevant, but their anti-realist Humean conception of scientific knowledge prevented them from seeing why they were right.

Contexts of Discovery and Confirmation

I have suggested above that three characteristic features of the D-N conception of explanation that help to explain its philosophical plausibility actually represent important insights of the empiricist tradition in the philosophy of science—insights that can be extended to the cases of other nonexperimental criteria as well, but insights that an empiricist as opposed to a realist conception of scientific knowledge cannot successfully assimilate. The development of a consistent realist conception of the epistemic role of nonexperimental (or, better, indirectly experimental) criteria of theory acceptability permits us to examine the cogency of another distinctive feature of empiricist philosophy of science: the traditional logical-empiricist claim that the epistemology of science need concern itself with the logic of confirmation but not with the principles of reasoning by which scientific theories are invented or discovered. On the logical-empiricist conception, the latter issue belongs to psychology and to the social study of science but not to the philosophy of science.

Part of what empiricists meant when they held that issues about the context of discovery were irrelevant to the philosophy of science was that philosophers of science need not develop a formal inductive logic to account for the discovery of theories. No doubt they were right in this respect; there is no reason to believe that what is ordinarily meant by an inductive logic would provide an even remotely adequate account of theory discovery in science. They also meant that philosophers of science need not concern themselves with all the details of psychological theories about how theories are discovered. Here too they were no doubt right. What is striking, however, is that some quite important empirical issues about theory discovery are irremediably central to an adequate epistemology of science.

It is a central part of the business of the philosophy of science to answer the fundamental epistemological question of why the methods of science are epistemically reliable. We have just had the occasion to examine in some detail two important features of those methods. First, the problem of sampling in experimental design is solved by the requirement that a proposed theory be tested under experimental circumstances that pit it against those alternatives to it that are theoretically plausible (and thus evidentially supported by inductive inference) given the body of previously established theories. Second, it can sometimes count as overwhelmingly confirmatory evidence for a theory that it is the basis for theoretically plausible explanations of a wide variety of phenomena that none of its otherwise plausible rivals can explain equally well. In each of these sorts of cases, the epistemic reliability of the relevant methodological practice depends on its being true in the (not too) long run that, when a proposed theory is in fact seriously mistaken, among its theoretically plausible rivals there will be theories that are relevantly closer to the truth and that can serve to identify the errors in the first theory or to challenge its exclusive claim to explanatory power with respect to the relevant class of natural phenomena.

It is, of course, impossible to assess the theoretical plausibility of theoretical proposals unless someone thinks them up. Failures of theoretical imagination can thus render the methodological practices we are discussing epistemologically unreliable in particular cases. The scientists who test a proposed theory against all the available theoretically plausible alternative theories will be employing an epistemically reliable testing strategy only on the assumption that the imaginative capacity of the scientific community is sufficient, so that theories near the truth in relevant respects will appear among those theories. Similarly, the scientists who accept a theory because it displays an apparent explanatory capacity utterly unmatched by any of the available plausible rivals will be reasoning reliably only if the imaginative capacity of the scientific community is up to the task of inventing a rich enough class of theoretically plausible rivals. Rival theories that would be theoretically plausible if we were only able to invent them and to understand them well enough to assess their theoretical plausibility play no methodological role unless we actually possess and display the relevant imaginative and cognitive capacities.

It is true, even on an empiricist conception of the matter, that successful science depends upon facts about our intellectual and imaginative capacities. Even if theory confirmation did not depend upon those capacities, we would not succeed in science unless we were able to think up suitably accurate theories to test. What we have just seen is that the same dependence of success upon our imaginative and cognitive capacities infects our ability to reliably confirm or disconfirm the theories we have already invented. The epistemic reliability of our scientific practices depends not only upon our possession of a suitably approximately true body of background theories but also upon our having quite contingent psychological capacities for exploiting these theoretical resources. This fact has three quite different implications for the philosophy of science.

First, it seems plausible that something somewhat like an "inductive logic" of theory invention may be epistemically important in science. It is probably true that theory invention (and creativity in general) involves finding new combinations of previously understood ideas and concepts. It is also true that inductive inferences at the theoretical level favor theoretical proposals that are relevantly similar (where the relevant respects of similarity are themselves theory-determined) to proposals that have already been established. It would be quite surprising if the respects of similarity to previous theories involved in theory invention were not fairly closely related to the respects of similarity determined to be epistemically relevant by the previous theories themselves. Indeed, if there were no relevant relations between the two it would be hard to see how our methodological practices thus far would have been epistemically reliable. The "logic of confirmation" must be somehow related to psychologically real inductive procedures for theory invention if scientific practice is to be epistemically reliable at all. The question "Just what is the relationship?" is simultaneously a question in empirical psychology and a question in the epistemology of science.

Second, a recognition of the role of theoretical imagination in epistemically reliable scientific practice opens up important possibilities in applied philosophy of science. Consider the question of the role of social prejudice in the practice of scientists. It has been traditional, on discovering that some figure in the history of science reached conclusions on scientific matters that we can, in retrospect, see as having been determined by inaccurate racial or sexual stereotypes, to conclude that the figure in question must have failed to employ the scientific method conistently. No doubt this is right in many cases, but reflection upon the crucial epistemic role of theoretical imagination in the evidential assessment of theories suggests an alternative hypothesis that may prove more accurate in many actual historical cases. The scientist may well have adhered scrupulously to the dictates of sound scientific methodology; all the available theoretically plausible alternatives to the now objectionable conclusions may have been taken quite seriously in assessing the evidence for them. The epistemic unreliability of the scientist's procedures may have stemmed, not from a failure to be methodologically scrupulous, but rather from socially determined failures of imagination on the part of the scientific community as a whole. In cases where this explanation is the right one, there may be no culpable methodological failure at all. Avoidance of socially prejudiced conclusions in such cases will depend either on political and social changes affecting the imaginative capacity of researchers or (perhaps) on extraordinary leaps of imagination, which are not part of normal scientific practice. In important ways, then, good scientific methodology is not prejudice-proof even when practiced with the greatest possible care, which is not to say that good methodological practice does not in the very long run help to overcome social prejudice. It is an instructive exercise to see how well or badly this model fits the various cases of social prejudice in biology described by Gould (1981).

Third, the importance of scientists' imaginative capacity for the epistemic reliability of scientific methodology illustrates in a striking way what is perhaps the most surprising feature of the realist conception of scientific knowledge. The epistemic reliability of scientific methods is logically contingent. It depends upon the historically contingent emergence of relevantly approximately true theoretical traditions (Boyd 1982, 1983, forthcoming) and also upon logically contingent features of our individual and collective capacities for theoretical imagination. Thus, principles of scientific methodology are not defensible *a priori* but have empirical presuppositions. The philosophy of science is an empirical discipline, not an *a priori* one. Indeed, this is probably true of philosophical inquiry generally. Here again is a conclusion with which Hume might have agreed, although it is true for distinctly non-Humean reasons.

References

Boyd, R. 1972. "Determinism, Laws and Predictability in Principle." *Philosophy of Science* 39: 431–450.

Boyd, R. 1973. "Realism, Underdetermination and a Causal Theory of Evidence." *Nous* 7: 1–12.

Boyd, R. 1979. "Metaphor and Theory Change." In *Metaphor and Thought*, ed. A. Ortony. Cambridge University Press.

Boyd, R. 1980. "Materialism Without Reductionism: What Physicalism Does Not Entail." In *Readings in Philosophy of Psychology*, vol. 1, ed. N. Block. Cambridge, Mass.: Harvard University Press.

Boyd, R. 1982. "Scientific Realism and Naturalistic Epistemology." In *PSA 80*, vol. 2. East Lansing, Mich.: Philosophy of Science Association.

Boyd, R. 1983. "On the Current Status of the Issue of Scientific Realism." *Erkenntnis* 19: 45–90.

Boyd, R. Forthcoming. "Lex Orenci est Lex Credendi." In *Images of Science: Scientific Realism Versus Constructive Empiricism*, ed. Churchland and Hooker. University of Chicago Press.

Carnap, R. 1950. "Empiricism, Semantics and Ontology." *Revue Internationale de Philosophie* 3: 20–40.

Feigl, H. 1945. "Operationism and Scientific Method." *Psychological Review* 52: 250–259.

Goodman, N. 1973. *Fact, Fiction and Forecast*, third edition. New York: Bobbs-Merrill.

Gould, S. J. 1981. *The Mismeasure of Man*. New York: Norton.

Hanson, N. R. 1958. *Patterns of Discovery*. Cambridge University Press.

Hempel, C. G. 1965. "Aspects of Scientific Explanation." In Hempel, *Aspects of Scientific Explanation*. New York: Free Press.

Hempel, C. G., and P. Oppenheim. 1948. "Studies in the Logic of Explanation." *Philosophy of Science* 15: 98–115.

Kripke, S. 1972. "Naming and Necessity." In *The Semantics of Natural Language*, ed. G. Harman and D. Davidson. Dordrecht: Reidel.

Kuhn, T. 1970. *The Structure of Scientific Revolutions*, second edition. University of Chicago Press.

Popper, K. 1959. *The Logic of Scientific Discovery*. London: Hutchinson.

Putnam, H. 1975a. "The Meaning of 'Meaning.' " In Putnam, *Mind, Language and Reality*. Cambridge University Press.

Putnam, H. 1975b. "Explanation and Reference." In Putnam, *Mind, Language and Reality*. Cambridge University Press.

Boyd   94

Quine, W. V. O. 1969. "Natural Kinds." In Quine, *Ontological Relativity and Other Essays*. New York: Columbia University Press.

Salmon, W. 1971. *Statistical Explanation and Statistical Relevance*. University of Pittsburgh Press.

Smart, J. J. C. 1963. *Philosophy and Scientific Realism*. London: Routledge and Kegan Paul.

van Fraasen, B. 1980. *The Scientific Image*. Oxford: Clarendon.

4   Probability and the Art of Judgment

R. C. Jeffrey

Probabilism

In the middle of the seventeenth century, philosophers and mathe-
maticians floated a new paradigm of judgment[1] that was urged upon
the educated public with great success, notably in a prestigious "how
to think" book, *The Port-Royal Logic* (Arnauld 1662), which stayed in
print, in numerous editions, for over two centuries. We judge in order
to act, and gambling at odds is the paradigm of rational action. ("Anything you do is a gamble!") That was the new view, in which judgment
was thought to concern the desirabilities and probabilities of the possible
outcomes of action, and canons for consistency of such judgments were
seen as a logic of uncertain expectation. In time (ca. 1920), "The probability of rain is 20 percent" came to be possible as a weather forecast,
the point of which is to identify 4:1 as the "betting odds" (ratio of gain
to loss) at which the forecaster would think a gamble on rain barely
acceptable, e.g., the gamble a farmer makes when he decides to harvest
his alfalfa.[2]

Such is probabilism. *Radical* probabilism adds the "nonfoundational"
thought that there is no bedrock of certainty underlying our probabilistic
judgments. The 20:80 probability balance between rain and dry need
not be founded upon a certainty that the air has a certain describable
feel and smell, say, and that when it feels and smells that way it rains
just 20 percent of the time. Rather, probabilistic judgment may be
appropriate as a direct response to experience, underived from sure
judgment that the experience is of such and such a character. However,
"direct" does not mean spontaneous or unschooled. Probabilizing is a
skill that needs learning and polishing.

There will be more about this soon. For the moment, I only wish to
identify the position briefly and locate it in relation to the antithesis

between dogmatism and skepticism—an antithesis that probabilism is meant to resolve.

Dogmatism

Descartes's dogmatism, as in this defense of the second of his *Rules for the Direction of the Mind*, is familiar:

He is no more learned who has doubts on many matters than the man who has never thought of them; nay he appears to be less learned if he has formed wrong opinions on any particulars. Hence it were better not to study at all than to occupy one's self with objects of such difficulty, that, owing to our inability to distinguish true from false, we are forced to regard the doubtful as certain; for in those matters any hope of augmenting our knowledge is exceeded by the risk of diminishing it. Thus . . . we reject all such merely probable knowledge and make it a rule to trust only what is completely known and incapable of being doubted.

I call this dogmatism because it admits of no degree of belief short of utter certainty. The contemporary weather forecaster who simply gives the probability of fair weather tomorrow as 80 percent would be seen by Descartes as not having a belief about the matter until he is prepared to state baldly that tomorrow will be fair or that it will not. In order to have what Descartes would see as opinions in such cases, we are "forced to regard the doubtful as certain." This usage is implicit in ordinary talk, where, dogmatically, "belief" refers equally to the mental attitude and its propositional object. In that usage, my belief about tomorrow's weather (if I have one) may be the proposition that it will be fair or the proposition that it will not, but it cannot be the nonpropositional judgmental balance of 4:1 between those possibilities.

Realism

By projecting probability from the mind into the world, one might try to place probabilism within the dogmatic framework. Is your judgment about tomorrow's weather describable by probability odds of 1:4 between rain and fair weather? If so, you fully believe the proposition that the probability of rain is 20 percent. So the move goes. It continues: If someone else thinks the odds even, then you two disagree about an objective question (i.e., "What is the real probability of rain tomorrow: 20 percent or 50 percent?"). On this view, today's probability of rain tomorrow is a physical magnitude, like today's noontime barometric pressure.

This move fails, not because the notion of real probability is bogus but because your current 1:4 judgment about rain versus fair weather need not represent your view about what the real probability of rain tomorrow is, now. Perhaps you lack information—say, about the current barometric pressure—in the presence of which your judgmental probability for rain would go from 20 percent to either 10 percent or 70 percent, depending on what the barometer reveals. If that is how you think matters stand, you do not think the real probability of rain is 20 percent; what you might think is "It's surely 10 percent or 70 percent, with odds of 5:1 on 10 percent."[3]

Meteorologists have long used percentages of probability as a way of "expressing the degree of assumed reliability of a forecast numerically" (Hallenbeck 1920, p. 645). This is clear in the earliest report on proto-probabilistic weather forecasting (Cooke 1906):

> All those whose duty it is to issue regular daily forecasts know that there are times when they feel very confident and other times when they are doubtful as to the coming weather. It seems to me that the condition of confidence or otherwise forms a very important part of the prediction, and ought to find expression. It is not fair to the forecaster that equal weight should be assigned to all his predictions and the usual method tends to retard that public confidence which all practical meteorologists desire to foster. It is more scientific and honest to be allowed occasionally to say "I feel very doubtful about the weather for to-morrow, but to the best of my belief it will be so-and-so;" and it must be satisfactory to the official and useful to the public if one is allowed occasionally to say "It is practically certain that the weather will be so-and-so to-morrow."
> With a view to expressing various states of doubt or certainty, as simply as possible, I now assign weights to each item of the forecast.

Cooke's scheme falls short of probabilism, because he still sees himself as making a propositional forecast, for example "rain" or "fair," with the weight indicating his confidence in the side he has chosen. Thus, he describes the third of his five weights as follows: "Very doubtful. More likely right than wrong, but probably wrong about four times out of ten."

In speaking of probability as "the degree of assumed reliability of a forecast," Hallenbeck, too, sounds as if he has stopped short of probabilism; however, it is clear from the overall context that his proposal is fully probabilistic. Cooke's "Precipitation (3)" differs from Hallenbeck's "60 percent probability of precipitation" in that Hallenbeck, but not Cooke, would regard "40 percent probability of fair weather" as the same forecast in different words. Cooke's scheme is a halfway house between dogmatism and probabilism: He first gives a quasi-dogmatic forecast and then qualifies it probabilistically. (On the other hand, I do not see Cooke's reference to the probable frequency of error as realistic deviationism. According to the calibration theorem given below, a probabilist should estimate that 40 percent of the propositions to which he attributes probability 60 percent will be false.)

Skepticism

Descartes died four years before the start of the correspondence between Fermat and Pascal that initiated the development of probabilism in its modern, quantitative form. Like Montaigne before him, Descartes took the only alternative to dogmatism to be the utter suspension of belief that the Hellenistic epistemologists had associated with the philosopher Pyrrho. Here is Montaigne's statement of the case in his "Apology for Raymond Seybond" (1580):

I can see why the Pyrrhonian philosophers cannot express their general conception in any manner of speaking; for they would need a new language. Ours is wholly formed of affirmative propositions, which to them are utterly repugnant; so that when they say "I doubt," immediately you have them by the throat to make them admit that at least they know and are sure of this fact, that they doubt. Thus they have been constrained to take refuge in this comparison from medicine, without which their attitude would be inexplicable: when they declare "I do not know" or "I doubt," they say that this proposition carries itself away with the rest, no more or no less than rhubarb, which expels evil humors and carries itself away with them.
   This idea is more firmly grasped in the form of interrogation; "What do I know?"—the words I bear as a motto, inscribed over a pair of scales. [from *The Complete Essays of Montaigne*, tr. Donald M. Frame (Stanford University Press, 1957), pp. 392–393]

The motto is Montaigne's famous verbalization ("Que sçay-je?") of the Pyrrhonian shrug. The classical verbalization, known to him through the writings of Cicero and Sextus Empiricus, had been the paradoxical affirmation "I affirm nothing."
   Probabilism is sometimes classified as a kind of skepticism, since the probabilist seeks to get along without dogmatic belief. And here comes the grab at the throat: "At least you know and are sure of this fact, that your probability odds between rain and fair are 1:4." On this telling, probabilists have dogmatic beliefs about their own attitudes.

Maybe so; for example, maybe you are sure that you have judgmental odds between rain and fair, and that they are 1:4; but there is nothing in probabilism to require you to be quite clear about your own judgmental states, or even to have definite judgmental states in all cases to be clear about.

Probabilism does not say that the extreme probabilities, 0 and 1, are never to be used. Rather, it seeks to provide alternatives to such extreme judgmental states for use on those many occasions where they are inappropriate. Full dogmatic belief is represented in the probabilistic framework by odds of 1:0 (or 2:0, or 100:0—it makes no difference), and more moderate precise judgments are represented by odds such as 4:1. But probabilism does not insist that you have a precise judgment in every case. Thus, a perfectly intelligible judgmental state is one in which you take rain to be more probable than snow and less probable than fair weather but cannot put numbers to any of the three because there is no fact of the matter. (It's not that there are numbers in your mind but it's too dark in there for you to read them.)

"The Guide of Life"

Cicero and Sextus had been aware of an alternative to both dogmatism and Pyrrhonism: the "probabilism" that informed the Academy during the second century B.C. under the leadership of Carneades and his successor Cleitomachus. (Some scholars hold it to be only linguistic confusion that connects this sense of probabilism with what emerged in the seventeenth century.[4]) Cicero himself was such a probabilist, but Sextus counted it as a species of dogmatism: ". . . as regards the End (or aim of life) we differ from the New Academy; for whereas the men who profess to conform to its doctrine use probability as the guide of life, we live in an undogmatic way. . . ." (*Outlines of Pyrrhonism* I, p. 231) This translation echoes the familiar slogan that Bishop Butler was to introduce some fifteen centuries later at the end of this passage from the introduction to his *Analogy of Religion* (1736): ". . . nothing which is the possible object of knowledge, whether past, present, or future, can be probable to an infinite intelligence; since it cannot but be discerned absolutely as it is in itself, certainly true, or certainly false. But to Us, probability is the very guide of life."

What does this slogan mean? Butler himself took it to mean that, in the absence of conclusive reasons for belief in any of the alternatives, we should believe the most probable of them (e.g., he thought, the truth of the Christian religion). This is quite antithetical to probabilism, which would have us withhold full belief in such cases. Thus, after formulating his slogan, Butler goes on to say

From these things it follows, that in questions of difficulty, or such as are thought so, where more satisfactory evidence cannot be had, or is not seen; if the result of examination be, that there appears upon the whole, any the lowest presumption on one side, though in the lowest degree greater; this determines the question, even in matters of speculation; and in matters of practice, will lay us under an absolute and formal obligation, in point of prudence and interest, to act upon that presumption or low probability, though it be so low as to leave the mind in very great doubt which is the truth. For surely *a man is really bound in prudence, to do what on the whole appears, according to the best of his judgment, to be for his happiness, as what he certainly knows to be so.* [emphasis added]

This last statement, though true, is far from justifying the adoption of dogmatic belief in the most probable alternative. On the contrary, a bet can be judged favorable even where the probability of winning is less than that of losing, but such a judgment requires joint consideration of probabilities and desirabilities; it is not enough to consider the probabilities alone (as Butler would have us do) or the desirabilities alone (as do those who buy lottery tickets simply because it would be so lovely to win). That was the point of Pascal's famous "wager," his prudential argument in favor of belief in God even if the probability of the truth of that belief is very slight. Whereas Butler urged belief in the Christian religion as the most probable of the alternatives, Pascal argued that, although sober judgment may find the odds unfavorable in that sense, that same judgment must think it worthwhile to try to believe[5] in view of the infinite desirability that belief would have in the unlikely case that God does exist.

The Geometry of Choice

Perhaps the clearest general statement of the above point came at the end of the *Port-Royal Logic*: ". . . to judge what one ought to do to obtain a good or avoid an evil, one must not only consider the good and the evil in itself, but also the probability that it will or will not happen; and view geometrically the proportion that all these things have together. . . ." This geometrical view can be understood in familiar terms as a matter of balancing a seesaw. If we refer to the numerical measures of the good and the evil as their *desirabilities*, and if the

Figure 1
Positions on the seesaw represent desirabilities, weights represent probabilities, and the overall desirability of the course of action is represented by the position (f) of the fulcrum about which the opposing turning effects of the weights exactly cancel each other. As the weight (probability) at g is 4 times that at e, the position f of the fulcrum must divide the interval from e to g in the ratio 4:1 if the seesaw is to balance. Thus, if the desirabilities are $e = 0$ and $g = 100$, the point of balance is $f = 80$.

course of action under consideration is thought to bestow a probability on the good that is 4 times the probability it would bestow upon the evil, then we can balance these considerations as in the seesaw of figure 1.

The point of balance identifies the desirability of the chancy prospect of getting the good or the evil, with probabilities p and $1 - p$. The rule is that the ratio of distances from f to the good and to the evil must be the inverse of the ratio of the weights of the good and the evil. If we think of f as the desirability the gambler attributes to his situation before the outcome is determined, then the distances from f to g and to e will be the gain and the loss that the gambler faces, measured in units of desirability (as he sees it): gain $= g - f$, loss $= f - e$. If we call the ratio $(g - f):(f - e)$ of gain to loss the "betting odds" and call the ratio $p:(1 - p)$ of probabilities of winning and losing the "probability odds," then the rule is this:

Probability odds are inverse to betting odds.

That is, one would as willingly take one side of the bet as the other. One regards the wager as advantageous (or disadvantageous) if the probability odds are larger (or smaller) than the inverse betting odds.[6] An equivalent statement is that the desirability f of the wager is a weighted average

$$f = pg + (1 - p)e$$

of the desirabilities of winning (g) and of losing (e), where the weights are the probabilities of winning (p) and of losing ($1 - p$).

The Wager

The founders of modern probability theory floated this balancing of probabilities against desirabilities as the right way to think about all worldly decisions. This is clear in *The Port-Royal Logic* (Dickoff and James translation, pp. 356–357), where the new way of thinking, given the place of honor at the end, is introduced by a discussion of the wisdom of buying lottery tickets but is quickly generalized:

These reflections may appear trifling, and indeed they are if we stop here. We may, however, turn them to very important account: Their principal use is to make us more reasonable in our hopes and fears. . . . We must enlighten those persons who take extreme and vexatious precautions for the preservation of life and health by showing that these precautions are a much greater evil than is the remote danger of the mishaps feared. We must reorient many people who conduct their lives according to maxims like these:

There is some danger in that affair.
Therefore, it is bad.
There is some advantage in this affair.
Therefore, it is good.

We ought to fear or hope for an event not solely in proportion to the advantage or disadvantage held for us but also with some consideration of the likelihood of occurrence.

The new paradigm, however, was good only for worldly deliberation:

Infinite things alone—for example, eternity and salvation—cannot be equaled by any temporal advantage. *We ought never to place them in the balance with any things of the world.* Consequently, even the slightest chance of salvation is worth more than all the goods of the world heaped together; and the slightest peril of being lost is more serious than all temporal evils, considered simply as evils. [ibid., emphasis added]

Technically, this is a better version of Pascal's wager than the one in the script his friends found after his death and published in the *Pensées* (1670). Here, but not there, we are explicitly forbidden to use the new paradigm to evaluate gambles between finitely and infinitely desirable goods. That is an essential restriction, for the new paradigm would see nothing to choose between a long shot and a sure thing when salvation is at stake. The reason is that any positive fraction of infinity, no matter how slight, is the same size as any other positive fraction and as the whole of infinity. Thus, where the fulcrum in figure 1 is a finite distance to the right of e but infinitely far to the left of g, a single grain of probability at g will tip the scales to the right no less decisively if there are a million grains at e than if there are none there at all.

Acceptance

Dogmatism and probabilism are best seen as alternative forms of language, supporting different sorts of practice and different sorts of interpretation and criticism of common practices. As Montaigne suggests
in the "Que sçay-je?" passage, natural languages are originally dogmatic,
although there have been probabilistic accretions over the millennia
(especially, in the past three centuries). From a dogmatic point of view,
a decision problem is most naturally seen as as problem of what to
believe; for example, a decision to leave one's umbrella at home is
seen as expressing a belief that it will not rain, since that is the decision
one would make if one were sure it would not rain. But, of course, it
would not be an utter disaster to be caught in the rain without an
umbrella, and so even dogmatists admit that the practical sort of acceptance of hypotheses (the acceptance they see as underlying a decision
to leave one's umbrella at home) need not indicate certainty.

From a probabilistic point of view it is evident that the dogmatist's
question of how close to unity the probability must be in order to make
it reasonable to "accept" a hypothesis depends on what is at stake—
on how important it is to be right. (To "accept"a hypothesis in a particular practical situation is to act as one would if one were quite sure
the hypothesis were true.) Although an 80 percent probability of fair
weather might warrant leaving one's umbrella at home on a particular
occasion, it might not be enough to warrant sparing the expense of a
canopy at a particular outdoor wedding reception. Thus, no one figure
can be named that will serve as a probability just high enough for
acceptance irrespective of what is at stake in the particular decision
problem in question, and therefore probabilists will see practical acceptance as a bogus notion, a screen to hide the inadequacy of dogmatical
decision theory. (For more about this, see Jeffrey 1956.)

This is also the point that probabilists see in Kyburg's (1961, p. 197)
"lottery paradox," where we see that no number $(n - 1)/n$ can be
close enough to 1 to warrant acceptance irrespective of what is at stake
because, if it were, then in a fair n-ticket lottery one would accept,
concerning each ticket, the hypothesis that *that* ticket will not win,
since that hypothesis would have probability $(n - 1)/n$; but one would
also accept the hypothesis that *some* ticket will win, since that hypothesis
would have probability 1. Thus, one would accept a logically inconsistent
collection of hypotheses: (0) One of the n tickets will win, but (1) it
will not be ticket 1, and (2) it will not be ticket 2, and . . . (n) it will

not be ticket *n*. (By the way, Kyburg, who finds no fault with the notion
of acceptance, takes the paradox to have a very different point, viz.,
that it need not be unreasonable to accept a collection of hypotheses
that one knows to be logically inconsistent.)

Acceptance of that sort is thought to be a matter of belief, where
"belief" ambiguously signifies both the proposition believed and the
believer's attitude toward it. Conflating the two, we easily think of
beliefs as presences of particular propositions in particular minds. In
deciding that the butler did it, Poirot accepts a certain proposition as
a guest in his mind, where it will reside alongside other such guests
and numerous gatecrashers that he believes willy-nilly. Of course, one
might speak in terms of representatives of propositions, e.g., sentences
(French? English? Mentalese?) or, better, eletrochemical configurations
of the little gray cells that encode both propositional content and status
as belief (rather than as hope, fantasy, or hypothesis). In any case,
Poirot's present state of belief can be identified by identifying a certain
bag of propositions or sentences, and Poirot's inferences can be identified
via changes in the contents of the bag. One could simulate that aspect
of Poirot's life by keeping a list of his beliefs in computer storage and
updating it from time to time as he changes his mind. However, I find
that an unattractive approach to methodology. Even if the electro-
chemical state of Poirot's little gray cells were eventually decoded in
terms of sentences or propositions, the gesture toward neurology and
computer science would be empty, I think. It is not into our heads (our
"hardware" or "wetware") that I would look, but into our history and
our lives, to see where the ideas of acceptance and probability come
from and what they are good for. Methodologies are shared cultural
artifacts—"software" in the public domain.

Approvability

Before the middle of the seventeenth century, the term "probable"
(Latin *probabile*) meant *approvable*, and was applied in that sense, univocally, to opinion and to action. A probable action or opinion was
one such as sensible people would undertake or hold, in the circumstances. The writings of Descartes (d. 1650) are rich in examples of
this archaic use of "probable" as *approvable*. Here is one: ". . . when
it is beyond our power to discern the opinions which carry the most
truth, we should follow the most probable. . . ." That is from Descartes's
comments on his second maxim in the *Discourse on the Method*. He

continues: "and even though we notice no greater probability in the one opinion than in the other, we at least should make up our minds to follow a particular one and afterwards consider it as no longer doubtful in its relationship to practice, but as very true and very certain. . . ."

Probability in this sense is a rough and ready notion. Thus, we are not to quibble about how we know which opinion is approvable. If there is a real question as to which side of an issue sensible people in our circumstances would take, then (in Descartes's words) "we notice no greater probability in the one opinion than in the other."

Recall what Descartes was commenting on: "My second maxim was that of being as firm and resolute in my actions as I could be, and not to follow less faithfully opinions the most dubious, when my mind was once made up regarding them, than if these had been beyond doubt." It seems that an action is to count as a case of following a particular opinion if it is such as one would perform who had no doubt of that opinion (= belief = proposition). In that sense, carrying an umbrella counts as following the opinion that it will rain, and leaving it at home counts as following the opinion that it will not rain. I suppose that to consider a doubtful opinion as "very true and very certain" in relation to practice is simply to act on that opinion no less decisively than one would have done in the absence of doubt. In such ways one can make sense of what would be dark sayings if one were to forget that Descartes was a dogmatist. He failed to see why it might be wise to "follow" the less likely of two opinions, e.g., by gambling on rain when you take the probability odds to be 1:4, provided you see the betting odds as better than 4:1. But that is just to say that he failed to anticipate the new paradigm of judgment that emerged in the Fermat-Pascal correspondence four years after his death.

Probabilistic Methodology

Betting odds were an old instrument on which the founders of modern probability theory took a hopeful hold, keen to use it on everything in sight—not least the new science. New methodologies were abroad for which there were large claims: Bacon's in England, Descartes's on the continent. Perhaps the instrument that had only been put to trivial uses, might prove to be a new organon for rightly conducting the reason and seeking truth in the sciences. In the foreword to his textbook *On Calculating in the Games of Luck* (1657), Huygens expressed some such hope: ". . . I would like to believe that in considering these things more attentively, the reader will soon see that the matter here is not a simple game of chance, but that we are laying the foundations of a very interesting and deep speculation." Thirty-three years later he made the claim quite explicitly and confidently, in introducing his *Treatise on Light*:

There will be seen . . . demonstrations of those kinds which do not produce as great a certitude as those of geometry, and which even differ much therefrom, since, whereas the geometers prove their propositions by fixed and incontestable principles, here the principles are verified by the conclusions to be drawn from them; the nature of these things not allowing of this being done otherwise. It is always possible thereby to attain to a degree of probability which very often is scarcely less than complete proof. To wit, when things which have been demonstrated by the principles that have been assumed correspond perfectly to the phenomena which experiment has brought under observation; especially when there are a great number of them, and further, principally, when one can imagine and foresee new phenomena which ought to follow from the hypotheses which one employs, and when one finds that therein the fact corresponds to our prevision. But if all these proofs of probability are met with in that which I propose to discuss, as it seems to me they are, this ought to be a very strong confirmation of the success of my inquiry; and it must be ill if the facts are not pretty much as I represent them.

This thought, that something like moral certainty is attainable "when things which have been demonstrated by the principles that have been assumed correspond perfectly to the phenomena which experiment has brought under observation," is no novelty of probabilism. Indeed, it is urged by Descartes in the same sort of context at the conclusion of his *Principles of Philosophy* (part IV, principle CCV): ". . . they who observe how many things regarding the magnet, fire, and the fabric of the whole world, are here deduced from a very small number of principles, although they considered that I had taken up these principles at random and without good grounds, they will yet acknowledge that it could hardly happen that so much would be coherent if they were false."

What probabilism does here is explain a methodological precept that dogmatists accept, unexplained, in a not-quite-valid form, i.e., what I shall call the converse entailment condition.

Converse entailment condition   If H entails E, then truth of E confirms H.

[The terminology is designed to fit with that of Hempel (1965, p. 31).] In dogmatic terms of acceptance and rejection, the project of repairing and then justifying that precept looks unpromisingly like the project

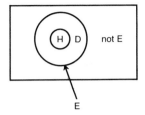

Figure 2
Proof of Huygens's rule. Areas represent probabilities. Prior odds on H are the ratio
of $p(H)$ to $p(D) + p(\text{not } E)$, and posterior odds are the ratio of $p(H)$ to $p(D)$. If $p(H)$
and $p(\text{not } E)$ are both positive, the second ratio is the greater.

of tinkering with the following invalid rule of inference so as to get a
rule that is both valid and useful:

Fallacy of affirming the consequent   If H entails E, and E is true, then
H is true.

However, the project is straightforward in probabilistic terms, where
we understand confirmation as improvement of the odds on truth:

Definition of confirmation E confirms H if and only if the posterior
odds on H are better than the prior odds, i.e., $p(E$ and $H):p(E$ but not
$H) > p(H):p(\text{not } H)$.

Now, if H entails E (see figure 2), the proposition E *and* H is H itself,
while E *but not* H is the donut, D. Then, under the hypothesis of the
converse entailment condition, the posterior odds on H will be $p(H):p(D)$,
which will be greater than the prior odds $p(H):p(\text{not } H)$ as long as the
common numerator, $p(H)$, is not 0 and the first denominator, $p(D)$, is
less than the second, $p(\text{not } H)$—as it must be, if $p(E)$ is not 1. Thus we
have proved the validity of the following (cleaned-up) probabilistic
version of the converse entailment condition:

Huygens's rule   If H implies E, then E confirms H, unless the prior
probability of H was 0 (H was written off as false, in advance) or that
of E was 1 (E is no news).

The term "Huygens's rule" (new here) seems apt—not because that
one rule exhausts Huygens's methodological remarks, quoted above,
but because it represents a salient, central feature of them.

The foregoing was a case of probabilistic domestication of what was
first seen as an unrationalized bit of dogmatic methodology. Before

leaving the topic, let me note a similar domestication of the (rationalized) core of Karl Popper's skeptical methodology. Taking confirmation to be an idle conceit (he sees no alternative to skepticism but dogmatism), Popper holds that the aim of science can only be the refutation of false hypotheses. Therefore his central methodological rule is to test less probable hypotheses first, since lower probability means a better chance for a definitive result of the test, i.e., a refutation of the hypothesis. (Popper thinks that the converse consequence condition is a howler, viz., the fallacy of affirming the consequent.) Now it seems apt to attach Popper's name to the following probabilistic rule, which identifies the analog of Popper's basic idea (albeit within a framework that he rejects).

Popper's rule   The least probable consequences of a hypothesis are (1) the most confirmatory for it if they prove true, and (2) the most likely to refute it, by proving false. [For (1), the hypothesis must have positive prior probability.]

The comparative notion of confirmation used in (1) is to be understood in the obvious way: that the most strongly confirmatory evidence is the evidence that gives H the greatest posterior odds.

Proof of (1)   In figure 2, if the H circle remains fixed, H's posterior probability increases as E (and, so, D) shrinks.

These two valid rules illustrate in detail what correct scientific hypothesis testing looks like from the probabilistic point of view. (For some further detail, see Jeffrey 1975 and 1983.) But equally important is the other side: the probabilistic critique of methodological fallacies. Here are four such fallacious rules, with counterexamples referring to a simple eight-ticket lottery.

Special consequence condition   If E confirms G, and G implies H, then E confirms H.

Counterexample E: the winner is ticket 2 or 3. G: It is 3 or 4. H: It is neither 1 nor 2. Then $p(G|E) = 1/2 > p(G) = 1/4$, but $p(H|E) = 1/2 < p(H) = 3/4$.

Conjuction condition I   If E confirms G, and E confirms H, then E confirms G-and-H.

Counterexample E: It is 1 or 3. G: 1 or 2. H: 3 or 4.

Conjuction condition II   If E confirms H, and F confirms H, then E-and-F confirms H.

Counterexample E: 1 or 2. F: 2 or 3. H: 1 or 3.

Conjuction condition III If E confirms G-and-H, then either E confirms G or E confirms H.

Counterexample G: 1, 2, or 5. H: 3, 4, or 5. E: 5 or higher.

Probability and Frequency

Information about frequencies and averages can have a special sort of influence on probabilistic judgment: It can make judgmental probabilities and expectations agree numerically with actual counts and their averages. That is what led von Mises (1928) and others to hold that probabilities are objective magnitudes, i.e. frequencies. But that position has proved surprisingly difficult to defend in detail; indeed, I think that frequentism has been shown to be an unworkable program (Jeffrey 1977). Even so, the problem remains of showing how the judgmental view of probability takes frequency data adequately into account.

It strikes me that this has been done by Bruno de Finetti (1937); I would call special attention to his understated treatment of frequencies in finite sequences in chapter 2. (This is not to deny the importance of the better-known treatment of the question in subsequent chapters, where the law of large numbers and the notion of exchangeability are used; however, chapter 2 has striking simplicity and unexpected power.) The basic fact is what I shall call the finite frequency theorem.

Finite frequency theorem In any finite sequence of propositions, the expected number of truths is the sum of the probabilities.

This is very easily proved—so easily, indeed, that de Finetti is at pains to avoid applying to it any such grand term as "law" or "theorem."[7]

In proving the finite frequency theorem it is useful to adopt de Finetti's practice of identifying propositions with functions that assign truth values, 0 (falsity) and 1 (truth), to "possible worlds."[8] Then the function that assigns to each world the number of truths in that world among the propositions $A_1, \ldots, A_n$ will simply be the sum of separate functions[9] (propositions): $A_1 + \cdots + A_n$. By the *expectation* (or "expected value") of a function, f, of possible worlds is meant the probability-weighted average of the different values the function can assume. If there are only finitely many possible values, say the $n + 1$ values x_k ($k = 0, \ldots, n$), this will be a sum

$$E(f) = x_0 p_0 + x_1 p_1 + \cdots + x_n p_n,$$

where p_k is the judgmental probability that f assumes the value x_k.[10]Note that if f is a proposition, then n is 1, and x_0 and x_1 are 0 and 1, respectively. It follows that $E(f)$ is p_1, i.e., that

the expectation of a proposition is its probability.

Now the the finite frequency theorem takes the form

$$E(A_1 + \cdots + A_n) = p(A_1) + \cdots + p(A_n),$$

where A's are used instead of f's to emphasize that the functions are propositions, and where the p's on the right can be rewritten as E's. The proof is immediate, from the additivity of the expectation operator.[11]

If we divide both sides of this last equation by n and, on the left, apply the fact that $[E(f)]/n = E(f/n)$, we have the following theorem.

Finite relative frequency theorem The expected relative frequency of truths among the propositions in any finite sequence is the (unweighted) average of their probabilities.

As de Finetti points out, this goes some way toward explaining why we often use observed frequencies as probabilities.

Example 1: The Next Trial
Knowing the number of successes on the past trials of an experiment, you need to form a judgment about the outcome of the next trial. If you judge all trials, past and next, to have the same probability of success, then the probability of success next will equal the relative frequency of past successes. (*Proof*: As you know the relative frequency of successes on past trials, your estimate of that magnitude will be its actual value, and so the finite relative frequency theorem identifies that value as your judgmental probability for success next.)

In no way does this result depend on any assumption of independence of trials from one another. All that is required is that probability of success be judged constant from trial to trial. The lack of any assumption of independence is illustrated by the following example, where all propositions in the sequence are the same.

Example 2: Only Constancy Is Assumed, Not Independence
Suppose the sequence is simply $\langle S,S,S \rangle$, with $S =$ success on the first trial. The theorem still applies, for in this case the expected relative frequency of success is, by arithmetic,

$E[(S + S + S)/3] = E(S)$,

which equals $p(S)$ since expectations of propositions are their probabilities. (That was a proof of this quirky special case of the finite relative frequency theorem.) Here the trials are as far from independent as they could be.

If the theorem is to be applied precisely as in example 1, you must not know too much about the past trials. You need to know the overall statistics—say, s successes in n trials—without knowing which particular ones were successes and which were failures (unless it happens that s is 0 or n). The reason is that if you know the truth value (0 or 1) of the kth proposition in the sequence, that number will be your judgmental probability for the proposition that the kth trial was a success, and the theorem will be applicable only if you are sure that all past trials had the same outcome, for only then can you be attributing the same probability to all trials.

Still, the theorem may be brought to bear indirectly when you know the detailed history of the trials but regard the details as unimportant.

Example 3: Irrelevant Detail
You must name odds at which you will bet on the ace turning up next after a sequence of probative trials. You toss the die 100 times, writing 1 or 0 after each trial to record the outcome as ace or not. Suppose that if all you had known was the number s of past successes then your judgmental probability of success would have been the same on all 101 trials, and so the theorem would have given probability s% for ace as outcome of the 101st toss. But in fact you are perfectly aware that (say) the first number you recorded was 0 and the fifth was 1, so that your judgmental probabilities are not the same for all 101 trials. The theorem is not directly applicable as in example 1. Still, it applies indirectly, for to regard the details as irrelevant is just to regard as correct the judgment about the next toss that you would have made if you had known only the overall statistics.

Note that the determination of which statistics are relevant is a probabilistic judgment (example 4), and that there will be cases where that judgment is simply negative, i.e., available statistics are seen as irrelevant (example 5).

Example 4: The Odd-Numbered Trials Have Succeeded
In 100 probative trials, the successes have occurred precisely on the odd-numbered ones. Then, although the relative frequency of success

has been 50%, the judgmental probability of success on the (odd) 101st trial may well be 100%, and will be if the finite frequency theorem is judged applicable in the manner of example 1 to the subsequence consisting of the odd-numbered trials.

Example 5: Too Small a Sample
If the number of past trials is very small (1, let us say), then there is little tendency to have the same judgmental probability for success on the next trial as on the past trial (i.e., 0 or 1). Even with a larger number of probative trials, other considerations may discredit the past statistics as a good measure of future probability, as when a questionable die is tossed six times and no aces appear. Here we are unlikely to use 0 as the probability of ace next, and there is no subsequence of past trials that will serve as in example 4.

Finally, note an application of these ideas to the evaluation of expertise.

Calibration theorem $x\%$ is your estimate of the percentage of truths among propositions to which you assign $x\%$ probability.

The proof is immediate, from the linearity of the operator E: If you assign probability $x\%$ to each of (say) 100 propositions, then your estimate of the number of truths among them must be x, and so you must estimate that $x\%$ of them are true.

You are said to be perfectly calibrated if, for every x, precisely $x\%$ of the propositions to which you assign probability $x\%$ are true. Calibration is obviously a good thing. So is *refinement*, that is, sensitivity to differences between cases. Example: Suppose that it rains every day in September and never rains any other time. Then I would be perfectly calibrated if I attributed probability $30/365$ to rain on each of the 365 days of the year. So would you be if you attributed 100% probability to rain on each day in September and 0% probability to rain on each other day. The difference between us would be one of sensitivity: maximum for you, minimum for me. (See DeGroot and Fienberg 1982 for more about this.)

The Emergence of Odds

Ian Hacking's *Emergence of Probability* (1975) is an attempt "to understand a quite specific event that occurred around 1660: the emergence of *our* concept of probability" (p. 9, Hacking's emphasis). Our concept:

". . . the probability emerging in the time of Pascal is essentially dual. It has to do both with stable frequencies and with degrees of belief."[12] (p. 10) Hacking holds (p. 1) that "neither of these aspects was self-consciously and deliberately apprehended by any substantial body of thinkers before the time of Pascal."[13]

However, the duality is only superficial if it is true (as I argued in the preceding section) that when probability is clearly understood as a mode of judgment the frequency aspect is thereby understood as well. If so, the specific event that occurred around 1660 was just the emergence of betting at odds as a paradigm of all action, within which probabilizing replaces believing.[14]

It is as a paradigm of judgment that probability odds were surely new in seventeenth-century Europe. I do not suggest that the very practice of betting at odds was new there in the seventeenth century; I do not imagine that at Byzantine chariot races betting was always a matter of supporters' tying their fortunes to their teams' fortunes by backing them at equal terms.[15] See what I mean: There is a great difference between the supportive sort of bet, through which you demonstrate solidarity with a cause or opinion by betting on it so as to embrace its fate, and the cagey Damon Runyon sort, where you pit your judgment against others' by fine tuning of odds. In Runyon's world, only a chump would back a team out of party loyalty. Whether he is Harry the Horse or the Chevalier de Méré, the shrewd bettor follows his judgment, not his heart. There were plenty of shrewd bettors in Paris, in the circles in which Pascal moved.

New or not, judgmental betting struck Pascal, the gentlemen of Port-Royal, Huygens, the Bernoullis, and many others as a likely paradigm for action of all kinds. But that sort of betting was probably not new to Europe, and was probably not peculiarly Indo-European either. Geertz (1973, chapter 15) gives a fascinating account of an entrenched, seemingly indigenous institution of cagey betting in Bali, where the two kinds of bets have names and have definite places around the cockpit:

... there are two sorts of bets, or *toh*. There is the single axial bet in the center between the principals (*toh ketengah*), and there is the cloud of peripheral ones around the ring between members of the audience (*toh kesasi*). The first is typically large; the second typically small. The first is collective, involving coalitions of bettors clustering around the owner; the second is individual, man to man. The first is a matter of deliberate, very quiet, almost furtive arrangement by the coalition members and the umpire huddled like conspirators in the center of the ring; the second is a matter of impulsive shouting, public offers, and public acceptances by the excited throng around its edges. And most curiously, . . . *where the first is always, without exception, even money, the second, equally without exception, is never such.* [Geertz's emphasis]

In Balinese terms, the native ground of the new paradigm is the periphery. Probabilism's paradigmatic act is the judgmental bet, *toh kesasi.*

Radical Probabilism

Part of our knowledge we obtain direct; and part by argument. The Theory of Probability is concerned with that part which we obtain by argument, and it treats of the different degrees to which the results so obtained are conclusive or inconclusive. (Keynes 1921, p. 3)

I see the seventeenth-century emergence of probability as a fission of the concept of approval into judgmental probability and judgmental desirability (utility), with the second element coming into full view only lately, in work of Ramsey (1931).[16] In that same work Ramsey enunciated the further idea that I mark by calling his probabilism radical.

Ramsey's 1931 paper is in part a rejection of Keynes's (1921) view of knowledge and probability, according to which our probable knowledge is founded on certainties (i.e., truths known by direct experience). If the proposition E reports everything I know directly, and H is some doubtful hypothesis, then the probability I ought to attribute to H is determined by the logical relationship between E and H. If E logically implies that H is true, then my odds between H's truth and H's falsity ought to be 1:0. If E logically implies that H is false, then the odds ought to be 0:1. If E fails to determine whether H is true or false, then the odds ought to be x:y, where x is the a priori, "logical" probability that H and E are both true and y is the logical probability that H is false although E is true. Ramsey finds these Keynesian probabilities elusive:

. . . there really do not seem to be any such things as the probability relations he describes. He supposes that, at any rate in certain cases, they can be perceived; but speaking for myself I feel confident that this is not true. I do not perceive them . . . moreover I shrewdly suspect that others do not perceive them either, because they are able to come to so little agreement as to which of them relates any two given propositions. (Ramsey 1931, p. 161; 1978, p. 63)

But furthermore—and this is his radical probabilism—Ramsey denies that our probable knowledge need be based on certainties. This second point is independent of the first. Thus, one might modify Keynes's scheme by taking E to be merely the latest proposition that one has come to believe fully through direct experience, not the totality of all such direct knowledge over one's whole life to date. One might then take the *a priori* probabilities in Keynes's scheme to be simply one's judgmental probabilities as they were before one came to believe E— not mysterious "logical" probabilities. Keynes's foundationalism could still be formulated in these more modest terms, as the thesis that the only rational changes in judgmental probabilities are those prompted by fresh certainties. Here is Ramsey's rejection of that more modest foundationalism:

> A third difficulty which is removed by our theory is the one which is presented to Mr Keynes' theory by the following case. I think I perceive or remember something but am not sure; this would seem to give me some ground for believing it, contrary to Mr Keynes' theory, by which the degree of belief in it which it would be rational for me to have is that given by the probability relation between the proposition in question and the things I know for certain. He cannot justify a probable belief founded not on argument but on direct inspection. In our view . . . there is no objection to such a possibility, with which Mr Keynes' method of justifying probable belief solely by relation to certain knowledge is quite unable to cope. (Ramsey 1931, p. 190; 1978, p. 92)

What Ramsey calls "a probable belief founded not on argument but on direct inspection" is what I would call direct probabilistic judgment: the probability odds as you take them to be, all things considered, but without breaking that consideration down into (a) a proposition E that you have just come to believe fully and (b) your judgmental probabilities as they were before you became certain of E. Sometimes such a breakdown is feasible, and in such cases it may be well to perform it, to obtain Keynes's "knowledge based on argument." But even then, the judgment that such a breakdown is possible will involve a direct probabilistic component.

Where considerations are broken down into (a) and (b) as above, the odds after becoming certain of E are obtained from the odds prior to that certainty by conditioning on E, i.e., by applying the following rule:

Your odds between H and *not* H after becoming certain of E are the same as your prior odds between E *and* H and E *but not* H.

Is that rule correct? Not necessarily; e.g., not if E is the information that this is an eight, H is the hypothesis that it is a heart, and the thing you saw that convinced you of E's truth was that the eight of spades was drawn. In this case your odds between E and H (eight of hearts) and E but not H (eight of some other suit) were 1:3 before the experience, but your odds between H and not H after the experience are 0:1. Evidently, the rule is correct only if E conveys all the information that prompted your change of judgment. But, as Ramsey points out, the prompt may have been an experience that you do not have the words to convey. His example is the particular auditory quality of whatever it was that you heard. ("Led?" "Red?") Another example might be the visual experience that prompts odds of 3:2 between Persi and Chico as originals of the image just flashed on the screen: Facial recognition goes on in the right hemisphere, verbalization in the left, and the lines between are sparse.

Radical probabilism is a "nonfoundational" methodology. The intended contrast is with dogmatic empiricism of Keynes's sort, in which the foundations underpinning all our knowledge are truths known by direct experience. (In this contrast the other part of Keynes's view— the view of "logical" probabilities as the framework through which the upper stories of our knowledge are made to rest on the foundation— is unimportant.) This foundationalism is not peculiar to Keynes's empiricism; e.g., see Lewis 1946, p. 186: "If anything is to be probable, then something must be certain. The data which eventually support a genuine probability, must themselves be certainties. We do have such absolute certainties, in the sense data initiating belief and in those passages of experience which later may confirm it." Here Lewis accepts the foundationalist thesis that the only rational changes in judgmental probabilities are those prompted by fresh certainties. (His basis for that acceptance seems to be a prior commitment to conditioning as the only rational way of changing probability judgments in response to experience.) Where probability is seen as a basic mode of judgment—i.e., where probabilism is radical—that thesis loses its plausibility. It was the felt need for a certainty to condition upon that dogmatized Lewis's empiricism, and Keynes's.

Psychology and Probability Logic

. . . a precise account of the nature of partial belief reveals that the laws of probability are laws of consistency, an extension to partial beliefs of formal logic, the logic of consistency. (Ramsey 1931; 1978, p. 84)

Dissatisfaction with the view of Poirot's judgmental state as representable by an assignment of truth values to the propositions or sentences he believes and disbelieves should be displaced, but not cured, by the substitution of probabilities for truth values. The new difficulty is that the ability to remember the probability of a single proposition requires a capacity to store no end of digits if the probability might be just any real number in the unit interval. What makes truth values manageable is that there are only two, and so each represents just one bit of information. If there are infinitely many values that a probability might take, each represents an infinity of bits.[17]

The most obvious way out is to make the probabilities only approximate. This amounts to increasing the possible numbers of values assignable to single propositions from Poirot's two to larger finite numbers so that the stored information stays finite. Now of course our probability judgments are often only approximate; however, if probabilizing were only or primarily a matter of assigning approximate probabilities to propositions, it would be a far cruder and less useful technique that I take it to be.

I think that we seldom have judgmental probabilities in mind for the propositions that interest us. I take it that what we do have are attitudes that can be characterized by conditions on probabilities, or (what comes to the same thing) by the sets of probability assignments that satisfy those condiions, where the members of those sets are precise and complete: Each assigns exact values to all propositions expressible in some language.[18]

Approximate probability assignments themselves are conditions on precise assignments. To say that $p(E) = 0.25$ to two decimal places of accuracy is to impose on any precise, complete assignment p the condition of assigning to E a value between $2/10$ and $3/10$. (If p violates that condition it cannot be a completion of the judgmental state in question.) But these are by no means the only cases, or the most useful, in which probabilistic judgment is a matter of adopting conditions that are satisfied by an infinity of precise, complete probability assignments. Here are some further examples, two plausible and two not.[19]

Example 1: Constant Probability
As in example 1 above, you would view as unacceptable any assignment p that failed to attribute the same probability to all the propositions in the set $S = \{$success on trial 1, success on trial 2, ... $\}$.

Example 2: Exchangeability
In addition to the condition in example 1, you require that any acceptable assignment p attribute equal probabilities to conjunctions of equal numbers of members of S.

Example 3: Bernoulli Trials
In addition to the condition of example 1, you require that p assign to any conjunction of distinct members of S the product of the values it assigns to them separately. (Perhaps the trials are drawings from an urn of unknown composition, and success is a matter of drawing a black ball.)

Example 4: Independence
You take the truth or falsity of proposition E to have no evidentiary import for that of proposition H.

Example 3 is implausible because in it the set of precise, complete probability assignments is naturally parametrized by the unknown proportion r of black balls in the urn, concerning which you are likely to have views (either in the shape of a definite judgmental probability distribution for r or in the shape of a set of such distributions). In the presence of any one such distribution, your judgments about trials are characterized not by the parametrized set but by a single exchangeable probability assignment: the weighted average of the parametrized assignments, with weights determined by your judgmental probability distribution for the parameter.[20] In the presence of a set of such judgmental distributions of r, your judgmental state will be represented, as in example 2, by a set of exchangeable assignments to the outcomes of trials.[21]

Probabilism does not suppose that we have particular probability assignments in mind, or that (whether we know it or not) our current states of mind are characterizable by single probability assignments, exact or approximate. Rather, probabilism characterizes our (judg)mental states by conditions on probability assignments—conditions that are typically satisfied by infinite sets of exact assignments. In these terms, revising judgment is a matter of revising each of those exact assignments, e.g. by conditioning on some certainty.

Your judgmental state will be represented by a region in the space of all exact probability assignments. Moving every point in that region has the effect of moving your judgmental state to a new region in probability space: Revision of judgment maps regions into regions. Does that sound like too much work, because it is described in terms of the individual points that make up the regions? It should not. When I wave my hand, I move all the points in it from certain positions in ordinary space to certain others, but moving all those points in this way is the same easy gesture as waving the hand.

Formal probability logic is the familiar elementary calculus of probabilities. Probability logic uses complete probability assignments just as deductive logic uses complete truth-value assignments.[22] For example, an assignment of truth (probability) values to some propositions is deductively (probabilistically) consistent if and only if it is extendable to all propositions that are expressible in the language.[23]

Just as complete truth-value assignments would be suitable dogmatic judgmental states for gods, not humans, complete probability assignments could not be human probabilistic judgmental states. No more than human dogmatic judgment accepts complete truth-value assignments does human probabilistic judgment adopt complete probability assignments. Rather, our dogmatic judgment accepts propositions (i.e., sets of complete truth-value assignments) and, similarly, our probabilistic judgment adopts what one might call "probasitions" (i.e., sets of complete probability assignments). To accept a proposition (probasition) is to reject as incompatible with one's current judgment all the superhumanly complete truth-value (probability) assignments that fall outside it.

Relative to a complete probability assignment p, the degree of confirmation of H by D would be defined as[24]

$$pc(H|D) = p(H|D) - p(H).$$

This is the basic concept of probabilistic methodology, not only for superhumans whose judgmental states are complete probability assignments p but also for humans whose judgmental states are probasitions (i.e., sets of such complete assignments). Relative to any such set, D confirms H when $pc(H|D)$ is positive for each p in it, and D confirms H more than C confirms G when $pc(H|D)$ exceeds $pc(G|C)$ for each p in it.[25]

Probasitions corresponding to human judgmental states form sets of points that we find manageable *en bloc*. (See the remarks on hand waving four paragraphs back.) Here the parallel between the roles of propositions and probasitions in dogmatic and probabilistic methodology remains exact. Thus, if there are n "atomic" propositions whose truth values are of interest, it is unlikely that a dogmatist will be in the happy position of having a particular truth-value assignment in mind for them. (In that position, a capacity of n bits suffices to record whichever of the $N = 2^n$ assignments the dogmatist might accept.) Rather, dogmatic opinion must be expected to be represented by one of the 2^N molecular propositions that can be compounded out of the n atomic ones. However, a storage capacity of N bits would be needed to provide for all those possibilities, and, as Harman (1980) points out, a capacity of N bits is physically unattainable for fairly modest values of n (e.g., emphatically, for $n = 300$).[26] Of course, dogmatists do not provide for all those possibilities when $n = 300$; since in any one notation nearly 100% of the 2^N molecular propositions will be represented by sentences each too long to record, dogmatic acceptance can address only the remaining tiny fraction. Similarly for probasitions: We can deal with those that are easily parametrized or otherwise accessible, but those are only a tiny fraction of the mathematical possibilities. Where we can go is where there are roads, and though we can build any one of millions of new roads as needed we cannot build millions.

Acknowledgments

For suggestions and criticisms that have prompted what I take to be improvements over earlier drafts, thanks are due to John Burgess, Loraine Daston, Bas van Fraassen, Alan Gibbard, Gilbert Harman, and, I fear, some I'm forgetting.

Notes

1. This began with the famous Fermat-Pascal correspondence (1654) that Huygens reported in his widely read textbook (1657). Four high points in the subject's first century: James Bernoulli 1713, de Moivre 1718, Daniel Bernoulli 1738, Bayes 1763. Laplace supplied the "classical" philosophical (1795) and mathematical (1812) statements. The view that what emerged in 1654 was a new paradigm of judgment stems from Ramsey (1931) and de Finetti (1931).

2. Hallenbeck 1920, p. 647: "There probably are no more than 10 days during the year when fair weather can be forecast for the Pecos Valley with certainty. . . . Knowing this, the farmers of this district naturally wish to choose occasions for certain operations when the rain hazard is least. In the cutting and curing of alfalfa, most of them will accept a risk of 20 per cent—a good deal, however, depends on the state of the crop, the press of other work, etc."

3. This is because 20 splits the interval from 10 to 70 in the ratio 1:5. Algebraically: As your judgmental probability for rain is 20%, your probability odds (p:1 − p) between the two hypotheses (10%, 70%) about the real probability must be such that $10p + 70(1 − p) = 20$. Thus, p is 5/6, and p:1 − p is 5:1.

4. Cicero used the Latin *probabile* as a translation of the Greek *pithanon* (persuasive). Sixteenth-century Latin translators of Sextus followed him in this, and so English translators have used *probable* where it would be bare anachronism to read the original as referring to probability in the sense that emerged in the mid seventeenth century. Yet the seventeenth century was so taken with these very texts of Cicero and Sextus that it is plausible to see the putative confusion as a congenital part of the new probabilism—i.e., to see the seventeenth century as having read the new probability concept into (and, in part, out of) the "probabilism" of the second century B.C. But see Burnyeat's (forthcoming) argument that "Carneades was no probabilist."

5. He suggests acting as if you believed: "taking the holy water, having masses said, etc. Even this will naturally make you believe, and deaden your acuteness." [*Pensées*, tr. F. W. Trotter (Dutton, 1958), p. 68] The wager is item 233 here, item 418 in the Krailsheimer translation.

6. For example, advantageous: $p/(1 − p) > (f − e)/(g − f)$, or, after some algebra, $f < pg + (1 − p)e$. Since the fulcrum is to the left of the center of gravity, the seesaw inclines to the right.

7. On the other hand, the proof is no more trivial than that of Bayes's theorem, which is an immediate consequence of the definition of conditional probability.

8. See de Finetti 1972, pp. xviii–xxiv. Possible worlds (not de Finetti's term) are simply ways things might go (and might have gone), in all relevant detail. (Mathematical probabilists call them elementary events. They call propositions events.) More commonly, propositions are identified with sets of possible worlds: $A = \{w : A$ is true in $w\}$. De Finetti's way comes to the same thing if sets are identified with their indicators (i.e., functions that assign 1 to members and 0 to nonmembers).

9. This is defined as the function that assigns to each world w the sum of the values that the terms of the sum assign it.

10. Thus, $p_k = p\,(\{w : f(w) = x_k\})$. If f is a proposition, then $x_k = k\,(= 0,1)$, and p_1 will be the set of worlds where f is true (i.e., the proposition itself, in the more usual representation).

11. Additivity is provable from the definition of expectation in terms of probabilities, above.

12. Degrees of belief are judgmental probabilities.

13. See Garber and Zabell 1979 for a contrary argument.

14. Yet it was not until our century that judgmental probabilizing was clearly and persistently distinguished from the ambient dogmatism (Ramsey 1931; de Finetti 1931).

15. For example, with three teams in the race a supporter's betting odds will be 2:1 even if his probability odds are less than 1:2.

16. See the essay "Truth and Probability," presented to the Moral Sciences Club of Cambridge University in 1926 and published after Ramsey died (Ramsey 1931, 1978). The fragment of Ramsey's construction that has to do with utility was reinvented by von Neumann and Morgenstern, whose 1943 book marks the beginning of any broad awareness of the very general concept of utility that we have today. It was Savage (1954) who gave Ramsey's 1931 work its due.

17. For Carnap this is no problem: What your judgmental probability for H at any stage ought to be is a conditional logical probability, a number $c(H|E)$ that is determined by finitely many bits of information in the shape of the conjunction E of all the *Protokollsätze* that you have been remembering, Poirot-style. But I find that doubly implausible; I can swallow neither protocol statements nor logical probabilities.

18. This idea of characterizing our probabilistic attitudes by sets is no novelty here. See, e.g., Good 1952, 1962 and Levi 1974, 1980. Note that the conditions determining such a set can involve other magnitudes beside probabilities— e.g., the condition that truth of A be preferred to truth of B, which I read as an inequality between expected utility conditionally on A and on B (Jeffrey 1965).

19. The implausibility of example 3 illustrates Levi's (1980) contention that judgmental conditions must determine convex sets of probability assignments. However, as conditions on conditional probabilities and expectations can determine nonconvex sets, I doubt that convexity will do as a general restriction on sets that represent judgments. Example 4 is a case in point. There the set $\{p : p(H|E) = P(H)\}$ is not convex, for while it contains both q and r where

$$q(H) = q(E) = 0.5, \quad q(H \text{ and } E) = 0.25$$

and

$$r(H) = r(E) = 0.3, \quad r(H \text{ and } E) = 0.09,$$

it does not contain $s = (q + r)/2$, since $s(H|E) = 0.425 \neq s(H) = 0.4$.

20. For example, with the uniform distribution for r, the weighted average turns out to be the assignment Carnap called m^*.

21. Thus, in example 3, suppose you are convinced that all balls have the same color. Then the set of exchangeable assignments will be $\{qx + r(1 - x) : 0 \leq x \leq 1\}$, where $q(r)$ is the probability assignment corresponding to the hypothesis that the color is (is not) the one associated with success. [Thus, for each x in the unit interval, $p = qx + r(1 - x)$ is exchangeable, with $p(\text{success on trial } n) = x$ and $p(\text{success on } n \mid \text{success on } m) = 1$.]

22. To be more precise: The role of models in deductive logic is played by the probability models of Gaifman (1964).

23. For deductive logic, this is Lindenbaum's lemma. For probability logic, see section 5.9 of de Finetti 1972 and section 3.10 of de Finetti 1974.

24. pc is what Carnap (1962, p. xvi) called D (i.e., degree of increase in firmness). Here I go along with Popper: Degree of confirmation (pc) is not a probability measure.

25. Finer-grained representations of probabilistic judgment would support subtler notions of confirmation. In note 21 above, nuance of judgment might be represented by what is formally a probability distribution for x. In general, such finer-grained representations make distributions over whole spaces of complete judgmental assignments do the work of probasitions. Degrees of confirmation of H by D would be distribution-weighted averages of the values $pc(H|E)$ for the various possible p's. Savage (1954, p. 58), endorsing an argument of Max Woodbury's, rejects these distributional representations on grounds that a weighted average of all complete judgmental assignments would be just one of the complete assignments in the average, but I see no reason to carry out that averaging operation here. [In Jeffrey 1983 (pp. 142–143) I mistakenly attributed the Woodbury argument to I. J. Good. As Issac Levi has pointed out to me, Good finds no fault with the sort of averaging that Savage and Woodbury reject.]

26. Harman 1980, p. 155. Note that the argument in Harman's section III ("Why We Don't Operate Purely Probabilistically") does not address the probasitional representation of judgment.

References

Arnauld, Antoine. 1662. *Logic, or, The Art of Thinking ("The Port-Royal Logic")*. tr. J. Dickoff and P. James. Indianapolis: Bobbs-Merrill, 1964.

Bayes, Thomas, 1763. "An Essay Towards Solving a Problem in the Doctrine of Chances." *Philosophical Transactions of the Royal Society of London* 53.

Bernoulli, Daniel. 1738. "Specimen Theoriae Novae de Mensura Sortis" (Exposition of a New Theory of the Measurement of Risk). St. Petersburg Academy of Sciences. Translated in *Econometrica* 22 (1954): 23–36; reprinted in *Utility Theory: A Book of Readings*, ed. Alfred N. Page (New York: Wiley, 1968).

Bernoulli, James. 1713. *Ars Conjectandi*. Basel.

Burnyeat, Miles. Forthcoming. "Carneades Was No Probabilist." *Riverside Studies in Ancient Skepticism*.

Butler, Joseph. 1736. *The Analogy of Religion, Natural and Unrevealed, to the Constitution and Course of Nature*. London.

Carnap, Rudolf. 1950, 1962. *Logical Foundations of Probability*. University of Chicago Press.

Cicero. *De Natura Deorum* and *Academica*. Loeb Classical Library. There is a good Penguin edition of the first and an excellent out-of-print translation of the second [*The Academics of Cicero*, tr. James S. Reid (London: Macmillan, 1880)].

Cooke, W. Ernest. 1906. "Forecasts and Verifications in Western Australia." *Monthly Weather Review* 34: 23–24.

de Finetti, Bruno. 1931. "Sul significato soggetivo della probabilita" (On the Subjective Significance of Probability). *Fundamenta Mathematica* 17: 298–329.

de Finetti, Bruno. 1937. "La prévision: ses lois logiques, ses sources subjectives." *Annales de l'Institut Henri Poincaré* 7: 1–68. Translated in *Studies in Subjective Probability*, ed. H. E. Kyburg, Jr., and H. E. Smokler (Huntington, N.Y.: Krieger, 1980).

de Finetti, Bruno. 1972. *Probability, Induction, and Statistics.* New York: Wiley.

de Finetti, Bruno. 1970. *Teoria delle Probabilita.* Torino: Giuli Einaudi. Translated as *Theory of Probability* (New York: Wiley, 1974, 1975).

DeGroot, Morris H., and Stephen E. Fienberg. 1982. "Assessing Probability Assessors: Calibration and Refinement." In *Statistical Decision Theory and Related Topics*, vol. 3. New York: Academic.

De Moivre, Abraham. 1718, 1738, 1756. *The Doctrine of Chances.* Reprints. New York: Chelsea, 1967.

Descartes, René. ca. 1628. *Rules for the Direction of the Mind.*

Descartes, René. 1637. *Discourse on the Method of Rightly Conducting the Reason and Seeking Truth in the Sciences.* In *Philosophical Works of Descartes*, ed. E. S. Haldane and R. R. T. Ross (New York: Dover, 1955).

Fermat, Pierre de. Correspondence with Pascal. See F. N. David, *Games, Gods, and Gambling* (London: Griffin, 1962).

Gaifman, Haim. 1964. "Concerning Measures in First-Order Calculi." *Israel Journal of Mathematics* 2: 1–18.

Garber, Daniel, and Sandy Zabell. 1979. "On the Emergence of Probability." *Archive for History of Exact Sciences* 21: 33–53.

Geertz, Clifford. 1973. *The Interpretation of Cultures.* New York: Basic.

Good, I. J. 1952. "Rational Decisions." *Journal of the Royal Statistical Society* B 14: 107–114.

Good, I. J. 1962. "Subjective Probability as the Measure of a Non-Measurable Set." In *Logic, Methodology, and Philosophy of Science*, ed. E. Nagel et al. Stanford University Press.

Hacking, Ian. 1975. *The Emergence of Probability.* Cambridge University Press.

Hallenbeck, Cleve. 1920. "Forecasting Precipitation in Percentages of Probability." *Monthly Weather Review* 48: 645–647.

Harman, Gilbert. 1980. "Reasoning and Explanatory Coherence." *American Philosophical Quarterly* 17: 151–157.

Hempel, Carl G. 1965. *Aspects of Scientific Explanation*. New York: Free Press.

Huygens, Christiaan. 1957. *De Ratiociniis in Aleae Ludo* (On Calculating in Games of Luck). Reprinted in Huygens, *Oeuvres Completes* (The Hague: Martinus Nijhoff, 1920).

Huygens, Christiaan. 1690. *Treatise on Light*. Translation: New York, Dover, 1962.

Jeffrey, Richard C. 1956. "Valuation and Acceptance of Scientific Hypotheses." *Philosophy of Science* 23: 237–246.

Jeffrey, Richard C. 1965. *The Logic of Decision*. Second edition, revised: University of Chicago Press, 1983.

Jeffrey, Richard C. 1975. "Probability and Falsification: Critique of the Popper Program." *Synthese* 30: 95–117, 149–157.

Jeffrey, Richard C. 1977. "Mises Redux." In *Basic Problems in Methodology and Linguistics*, ed. R. E. Botts and J. Hintikka (Dordrecht: Reidel).

Jeffrey, Richard C. 1983. "Bayesianism with a Human Face." In *Testing Scientific Theories*, ed. John Earman. University of Minnesota Press.

Keynes, John Maynard. 1921. *A Treatise on Probability*. London: Macmillan.

Kyburg, Henry E., Jr. 1961. *Probability and the Logic of Rational Belief*. Middletown, Conn.: Wesleyan University Press.

Laplace, Pierre Simon, Marquis de. 1795. *Essaie philosophique sur les probabilités*. Translated as *A Philosophical Essay on Probabilities* (New York: Dover, 1951).

Laplace, Pierre Simon, Marquis de. 1812, 1814, 1820. *Theorie analytique des probabilités*.

Levi, Isaac. 1974. "On Indeterminate Probabilities." *Journal of Philosophy* 71: 391–418.

Levi, Isaac. 1980. *The Enterprise of Knowledge*. Cambridge, Mass.: MIT Press.

Lewis, Clarence Irving. 1946. *An Analysis of Knowledge and Valuation*. La Salle, Ill.: Open Court.

Pascal, Blaise. 1670. *Pensées*. Tr. A. J. Krailsheimer (New York: Penguin, 1966); F. W. Trotter (New York: Dutton, 1958). See also Fermat reference, above.

Pyrrho et al. For background on Hellenistic epistemology, see David Sedley, "The Protagonists," in *Doubt and Dogmatism*, ed. M. Schofield, M. Burnyeat, and J. Barnes (Oxford: Clarendon, 1980).

Ramsey, Frank Plumpton. 1931. "Truth and Probability." In *The Foundations of Mathematics* (London: Kegan Paul); also in *Foundations*, ed. D. H. Mellor (London: Routledge and Kegan Paul, 1978).

Savage, L. J. 1954. *The Foundations of Statistics*. New York: Wiley.

Sextus Empiricus. See Loeb Classical Library, no. 273 (*Outlines of Pyrrhonism*) and no. 291 (*Against the Logicians*).

von Mises, Richard. 1928. *Probability, Statistics and Truth*. Published in German. Second revised English edition: London: Allen & Unwin, 1957.

The Method of Hypothesis: What Is It Supposed to Do, and Can It Do It?

Peter Achinstein

The method of hypothesis, which in contemporary circles is usually called the hypothetico-deductive method, seems to be widely used in the sciences. It is, of course, controversial, although even its critics exhibit mixed feelings about it. Inductivists such as Newton and Mill professed to reject it, yet Newton seems to be using a version of it in the *Optics*, and Mill (1959, p. 325) allows that although the method of hypothesis is not sufficient to prove a hypothesis it can be useful in "suggesting a line of investigation which may possibly terminate in obtaining real proof." Descartes advocated a form of rationalism based on deduction from self-evident principles, yet in part III of *The Principles of Philosophy* he employs a method of hypothesis that is radically different from the former.

Versions of the Method of Hypothesis

What exactly is the method of hypothesis? As might be expected with any broad doctrine, there are different versions. Here is a particularly simple formulation, by the physicist Richard Feynman:

In general we look for a new law by the following process. First we guess it. Then we compute the consequences of the guess to see what would be implied if this law that we guessed is right. Then we compare the result of the computation to nature, with experiment or experience, compare it directly with observation, to see if it works. If it disagrees with experiment it is wrong. In that simple statement is the key to science. (1965, p. 156)

Despite the simplicity of Feynman's formulation, one can distinguish in it three topics that are addressed by the method of hypothesis. First, this method has something to say about the origin of a hypothesis, a law, or a theory—about how the scientist arrives at the idea in the first place (in the "context of discovery," to use Reichenbach's expression).

Second, it speaks of determining some relationship between the hypothesis and other propositions—something is "computed." Third, it speaks about testing the hypothesis or law by experiment or observation. Let me comment briefly on each of these.

As far as origin is concerned, hypothesists (as I shall call them) are in agreement. Scientific hypotheses are not arrived at in the first place by deductive or inductive reasoning. Rather, they are guesses. In Popper's (1965, p. 192) words, they are *"free* creations of our own minds, the result of an almost poetic intuition." A century earlier, William Whewell wrote:

The conceptions by which Facts are bound together are suggested by the sagacity of discoverers. This sagacity cannot be taught. It commonly succeeds by guessing; and this success seems to consist in framing several tentative hypotheses, and selecting the right one. But a supply of appropriate hypotheses cannot be constructed by rule, nor without inventive talent. (Butts 1968, pp. 129–130)

How remarkably similar this is to what Hempel wrote more than 100 years later:

Scientific hypotheses are not derived from observed facts, but invented in order to account for them. They constitute guesses at the connections that might obtain between the phenomena under study. . . . (1966, p. 15)

When we turn to the computational part of the method, some differences emerge. Feynman's statement that "we compute the consequences" strongly suggests that we derive consequences deductively, using logic and mathematics. This emphasis on deduction, which is common to many writers, is no doubt responsible for the label "hypothetico-deductivism." But there are other relationships stressed by some who defend a method of hypothesis, viz. explanation and prediction. For example, Whewell writes

The hypotheses which we accept ought to explain phenomena which we have observed. But they ought to do more than this: our hypotheses ought to *fortel* phenomena which have not yet been observed. (Butts 1968, p. 151)

To Hempel and Popper, this distinction between deductive derivation, on the one hand, and explanation and prediction, on the other, is no real distinction at all. But I do not believe that hypothesists must be committed to this identity. So far as I can determine, Whewell does not have a deductive model of explanation. And, of course, predictions from a theory can be inductive as well as deductive. In brief, then, the "computational" part may consist in deductively deriving consequences, in explaining phenomena, or in predicting them. Depending on one's viewpoint, these may or may not be the same things.

It is in the case of the third part of the method—the testing of hypotheses—that we have the most controversy. Here it is useful to make two distinctions among hypothesists.

First, there are those who think that when a hypothesis is tested according to the method, and passes the test, the hypothesis is proved (or at least is shown to have reasonably high probability, or to be worthy of acceptance). In this category I would put Whewell, Braithwaite, and Hempel. By contrast, Mill, Popper, and indeed Feynman claim that the method cannot be used to prove hypotheses or to show that they have high probability. For Popper and Feynman the method can only disprove, never prove. Here is Feynman's pithy version:

. . . with this method we can attempt to disprove any definite theory. . . . but notice that we can never prove it right. Suppose that you invent a good guess, calculate the consequences, and discover every time that the consequences you have calculated agree with the experiment. The theory is then right? No, it is simply not proved wrong. (1965, p. 157)

Mill would agree that the method of hypothesis is not a legitimate one for inferring the truth or high probability of a hypothesis, yet he does hold that the method can test whether the hypothesis is worth pursuing.

The second distinction is between those who think that the testing of a hypothesis involves some form of nondeductive reasoning and those who believe that it involves only deductive reasoning. In the latter category I would place Popper and Feynman, in the former Whewell, Mill, Braithwaite, and Hempel. Among those who allow some form of nondeductive reasoning there is a division. Consider the following simple type of nondeductive argument:

1
O
Therefore (probably)
h, in the light of b,

where O is a set of observation reports, h is the hypothesis in question, and b is a set of background theories and beliefs. On one version, the method of hypothesis provides the following rule for arguments of form 1:

A

A nondeductive argument of form 1 is valid if O is deductively entailed by h together with b, but not by b itself.

Rule A, or some modification of it, is found in Braithwaite and in Hempel. On a second version, the method of hypothesis provides the following type of rule:

B

A nondeductive argument of form 1 is valid if h together with b *explains* O, but b by itself does not.

This sort of rule seems to be present in Whewell (provided that the observations that h explains are varied).

In a nondeductive argument of form 1 we infer h with (high) probability. However, as already mentioned, there are hypothesists who allow nondeductive arguments but will not permit an inference to h with high probability. For them we might formulate the following type of nondeductive argument:

2

O

Therefore,

h is worth considering, in the light of b.

The method of hypothesis might then be construed as providing the following rule for arguments of form 2:

A'

A nondeductive argument of form 2 is valid if O is deductively entailed by h together with b, but not by b itself.

Mill, perhaps, would have supported A' rather than A.

To conclude this string of distinctions, we have the following:

B'

A nondeductive argument of form 2 is valid if h together with b explains O, but b by itself does not.

Retroductivists such as Peirce and Hanson support B' or some variant of it. Perhaps these philosophers can be classified as hypothesists. (Peirce did use the terms "hypothesis" and "retroduction" interchangeably.) I hesitate a little in so classifying them because their views about the origin of hypotheses differ sharply from those expressed earlier. However, their positions on "computation" and "testing" (parts 2 and 3) would place them with the hypothesists.

I have formulated assorted versions of the method, which are fairly simple ones. More complex varieties can be found, especially in recent philosophical literature. However, what I want to stress in concluding this section on versions of the method of hypothesis is the universality of the method—in any of its versions. By this I have in mind four related features that are common to all the versions with which I am acquainted:

• The method incorporates no empirical assumptions in what it says about the origin of hypotheses, about computation, or about testing. There are no strictures that if one invents a physical hypothesis it must satisfy the principles of conservation of energy and momentum, or that if one invents an astronomical hypothesis it must be tested by using the telescope. Such strictures would obviously imply or presuppose empirical claims.

• The method does not vary from one hypothesis or theory to another, from one science to another, or from one period to another. It does not give different directions to the physicist, the biologist, or the psychologist. It is a method all of them should follow, in this century and any other.

• It makes no "pragmatic" or contextual references. To apply the method, one need not take into account the particular knowledge or interests of the scientist or the scientific community. It does not say, e.g., that if the scientist or the community knows or believes X and is interested in achieving understanding of sort Y, then hypotheses of type Z should be proposed. What it says is supposed to hold for any scientist or scientific community. There is no need for relativization to particular ones or types.

• The method, if it needs justification at all, is not to be justified on empirical grounds. Perhaps the idea is that following the method is what it *means* to proceed scientifically. The method is definitive of the criteria to be used in constructing and evaluating scientific theories. If so, the method has an *a priori* justification.

Now, turning to the questions raised in the title of the chapter, let me consider various specific tasks that a scientific method might be thought capable of performing, and ask whether the method of hypothesis—in any of its versions—can perform them.

Can the Method of Hypothesis Generate New Hypotheses?

The answer of any hypothesist I am familiar with is No! Scientists invent new hypotheses by guessing them. Hypotheses are *"free* creations of our own minds" (Popper). Whewell says explicitly that hypothesis generation is a matter of "sagacity" and "guesswork," and that "a supply of appropriate hypotheses cannot be constructed by rule." It is clear that the method of hypothesis offers no rules that will suffice for thinking up, say, the hypothesis that evolution occurs in sudden spurts, or the hypothesis that protons contain three quarks. Hypothesists are right about this characterization of their method, but one of the main reasons they offer for this is not convincing.

They claim that hypothesis generation involves guesswork, and so, they conclude, hypotheses cannot be generated by any set of rules. But that is a non sequitur. Here are some rules for generating hypotheses, yet the hypotheses generated might well be classified as guesses:

If you hold fewer than half the tickets in a lottery, then consider the hypothesis that you will lose.

If thin, wavy tracks form in a bubble chamber, then consider the hypothesis that electrons are traveling through the chamber.

These rules generate specific hypotheses for you to consider under certain circumstances, yet the hypotheses, if inferred in those circumstances, are guesses or conjectures, albeit "educated" ones.

The point is not that hypothesis generation involves guesswork, but that the method of hypothesis provides no rules for guessing. It remains silent on what hypotheses to consider in what circumstances. It does so for two reasons. First, it wants to be extremely liberal and impose no constraints at all on hypothesis invention (by contrast with hypothesis acceptance). Scientists should be free to consider any hypothesis they wish. Second, for the most part the kinds of rules that would suffice to generate specific hypotheses would be empirical ones, or would contain empirical assumptions (e.g., that electrons produce thin, wavy tracks). And, we recall, hypothesists do not want their method to be subject to empirical justification.

Can the Method Evaluate Hypotheses? If So, How?

The answer hypothesists give to the first question is an emphatic Yes. Although the method will not churn out new hypotheses for the scientist to consider, it will enable the scientist to evaluate ones that have been proposed. The question that needs pressing, however, is what sort of evaluation this is. Do hypothesists seek a "purely epistemic" evaluation that asks whether, or to what extent, it is reasonable to believe the hypothesis, or at least take it seriously, on the basis of the observations (or how much the evidence supports it)? Or do they seek a more general type of evaluation that asks whether, given the observations, the hypothesis is a good one for scientific purposes (whatever those are)? A weak hypothesis (e.g., that at least some bodies exert forces on some other bodies) might be very reasonable to believe on the basis of the observations. (It may be supported strongly by the evidence.) However, it might not be judged particularly good for scientific purposes (since, e.g., it is not sufficiently universal, not quantitative, and not specific about the kind of force exerted). With this latter type of evaluation, the epistemic aspect may be a component but it is not the only one.

Some hypothesists are concerned primarily with this broader type of evaluation. Indeed, Popper would reject the purely epistemic evaluation—at least as I have characterized it—since he repudiates the idea that hypotheses are reasonable to believe. So long as the hypothesis is not falsified by the observations, Popper's primary concern is with the "scientific value" of the hypothesis—its generality, simplicity, and falsifiability (which Popper takes to be the same things).

If this broader type of evaluation is to be made, then the method of hypothesis will need additional constraints. For a hypothesis to be a good one, it should do more than simply entail or explain observations. As just noted, Popper insists that it satisfy some criterion of generality. Whewell, using the expression "consilience of inductions," suggests that the hypothesis be capable of explaining phenomena "of a *kind different* from those which were contemplated in the formation of our hypothesis" (Butts 1968, p. 153). This is an insistence on a kind of unification. Now generality, unification, and related criteria such as simplicity, high information content, and precision are universal criteria in the sense indicated earlier. They incorporate no empirical or contextual assumptions; their applicability is not supposed to vary from one hypothesis or science to another; and such criteria would be justified if at all on *a priori* grounds.

No doubt such universal criteria are utilized in evaluating hypotheses. But I want to make two claims about them. The first I will simply assert without argument: These criteria serve only as a general guide to the kinds of hypotheses and explanations scientists should try to achieve at some point. They are not necessary conditions for classifying every scientific hypothesis as a good one. (For arguments in defense of this, see chapter 4 of Achinstein 1983.) The claim I do propose to argue for here is that, contrary to the usual hypothesist line, the set of such universal criteria is not sufficient for determining the worth of a hypothesis. It is not sufficient precisely because it is, or purports to be, universal.

Let me defend this claim by invoking an example I have used on some other occasions: hypotheses invented to account for the hydrogen spectral lines. In the latter part of the nineteenth century it was known that when hydrogen is excited by heat or electricity it emits light. When this light is analyzed spectroscopically it is seen to consist of a series of discrete lines at various wavelengths. Now, if you are a hypothesist, what you want is a hypothesis or a set of them on the basis of which the discreteness and the particular values of the observed wavelengths can be deduced and/or explained. Here is one set, containing two hypotheses, which I put in the form of a hypothetico-deductive argument.

H_1
Whenever hydrogen is excited thermally or electrically, it emits radiation whose spectrum contains lines satisfying the Balmer formula $1/\lambda = R(1/2^2 - 1/n^2)$, where $n = 3,4,5,6$ (for each line).

H_2
$R = 109,677.581$ cm^{-1}.

Therefore, the hydrogen spectral lines are discrete and have the wavelengths $\lambda_1 = 6562.08 \times 10^{-7}$ cm, $\lambda_2 = 4860.8 \times 10^{-7}$ cm, $\lambda_3 = 4340 \times 10^{-7}$ cm, and $\lambda_4 = 4101.3 \times 10^{-7}$ cm.

Will this satisfy the hypothesist? Well, we do have a set of hypotheses from which the discrete, observed values can be derived. Indeed, if we let $n = 7,8,\ldots$, then these hypotheses can be used to predict new lines (which were later observed). What about the additional universal criteria some hypothesists impose?

H_1 and H_2 are general hypotheses—at least in the sense of generality usually required of laws. True, they are restricted to hydrogen, but I doubt that hypothesists will exclude them on these grounds. What about unification? They do satisfy Whewell's demand that they generate observations different from those on which they are based. (They yield predictions of new lines.) Whewell does say that the hypothesis should generate phenomena of a "different kind" from the original ones, although unfortunately he has no clarification to offer for the idea of "different kind." Let me suggest a revision of this hypothetico-deductive argument that may go at least some way toward satisfying Whewell and will also make it more general. Replace H_1 with the following.

H_3
Whenever hydrogen is excited thermally or electrically, it emits radiation whose spectrum contains lines satisfying the Rydberg formula $1/\lambda = R(1/n_2{}^2 - 1/n_1{}^2)$, where $n_2 = 2$ and $n_1 = 3, 4, 5, \ldots$ for the Balmer series; $n_2 = 1$ and $n_1 = 2, 3, 4, \ldots$ for the Lyman series (in the ultraviolet region); and $n_2 = 3$ and $n_1 = 4, 5, 6, \ldots$ for the Paschen series (in the infrared region).

From H_3 and H_2 the discrete wavelengths of the lines in the visible spectrum are again derivable. But now we can also generate two new series of lines, one in the ultraviolet region and the other in the infrared region. We have a significant increase in generality and unification. (We have subsumed three distinct series under one formula.) Can we then evaluate these hypotheses as good ones for scientific purposes? Should we give high marks to the derivation and/or explanation they offer for the discreteness of the hydrogen spectral lines? My response is that in the abstract—without focusing on a particular context in physics—no answer can be given. Without introducing some contextual and empirical assumptions, general criteria of the sort hypothesists invoke will not suffice to determine whether a set of hypotheses is a good one for generating and explaining the observations. This, I think, is a limitation on the usefulness of the method of hypothesis (and, indeed, on the usefulness of other universal methods as well) in providing a general evaluation of hypotheses.

To defend this claim, let me mention that in 1913 Niels Bohr offered a set of hypotheses—very different from those above—for generating and explaining the spectral lines. Bohr postulated that the hydrogen atom contains a nucleus around which a single electron revolves in various possible energy states. When the hydrogen atom absorbs radiation, the electron jumps from one stable orbit to a higher one and then back again, emitting energy whose wavelengths are those of the lines in the hydrogen spectrum. Bohr introduced quantum hypotheses regarding the angular momentum of the electron and the radiation absorbed or emitted by the atom. From these general laws together with various classical principles from mechanics and electricity he derived the wavelengths of the observed lines in the hydrogen spectrum.

By the time Bohr's paper appeared, the atomic theory of matter was widely accepted in physics, as was the idea that the atom itself is not atomic but has an internal structure. It was also thought reasonable to suppose that the spectral lines were produced at the subatomic level by the motions of excited electrons in the atom, rather than by the vibration of the atom as a whole. The question was how to work this out quantitatively using some consistent picture of the atom. Based on scattering experiments, Rutherford had proposed a solar model of the atom according to which an atom contains a positive charge concentrated in a small nucleus around which orbit negatively charged electrons. In explaining the spectral lines of hydrogen, Bohr took this model as a basis. His task—the set of constraints he was operating under—was to derive the hydrogen spectral lines from Rutherford-type assumptions about the hydrogen atom.

Now obviously such constraints on the sorts of hypotheses to be introduced are not universal, in the sense I have given earlier. They contain specific empirical assumptions (e.g., that there is a hydrogen atom and that it has an internal structure); they are not invariant for theories and problems; and their use cannot be justified on *a priori* grounds. Yet one reason that Bohr's hypotheses are regarded as such good ones for him to have offered in accounting for the spectral lines is that they satisfied such constraints. They derived the lines from assumptions about the internal structure of the hydrogen atom, which was what the community of physicists was interested in doing. Furthermore, the Balmer-formula hypotheses H_1 and H_2 or the Rydberg-formula hypotheses H_3 and H_2 would not have been regarded as good ones for Bohr to have given, despite the fact that they too generate the wavelengths of the lines. The reason is that such hypotheses fail to satisfy constraints calling for a subatomic explanation. (Indeed, it was the fact that Bohr could explain the lines at the subatomic level that made physicists take his model of the atom seriously.)

More generally, I suggest, to decide whether hypotheses are good ones for a scientist to use for purposes of derivation and/or explanation requires appeal to constraints that are not universal. Such constraints impose empirical conditions on the kinds of hypotheses it is appropriate to offer. To know what empirical conditions are appropriate requires appeal to contextual considerations—the beliefs and interests of the scientific community. If we abstract completely from the latter, as hypothesists do, we will not really see why Bohr's explanation was so much better than the Balmer-formula and Rydberg-formula hypotheses.

Perhaps Bohr's explanation is more general and unifying than those offered by the alternatives I mention. (I am not convinced that it is, since Bohr's explanation was pretty much restricted to the hydrogen atom and could not be extended very well to more complex lines.) But even if it is more general and unifying, my point remains. Bohr's explanation was good not simply because it derives the spectral lines from general, unifying (simple, quantitative, etc.) assumptions, but because it does so at the subatomic level of the hydrogen atom—a level at which physicists at the time of Bohr were interested in understanding the spectral lines. But this consideration is too specific—too "theory dependent" and too pragmatic or contextual—for the kind of universal criteria hypothesists have in mind.

I conclude that the method of hypothesis, or indeed any method advocating only universal criteria, cannot provide a set of sufficient conditions for determining the worth of a hypothesis. This is not to say that the universal conditions it supplies are irrelevant (a point I will return to in the last section). My claim is simply that, contrary to what hypothesists advocate, no purely universal criteria will suffice in the "context of justification" to determine whether a hypothesis is a good one.

Can the Method of Hypothesis Provide "Purely Epistemic" Evaluations?

So far I have considered a general type of evaluation that asks whether, given the observations, the hypothesis is a good one for general scientific purposes. Let me now focus more narrowly on what I have called purely epistemic evaluations—those that address only the question of whether, or to what extent, it is reasonable to believe a hypothesis, or take it seriously, on the basis of the observations. Here it may be possible to propose universal criteria and to abstract from contextual matters. At least the kinds of problems I will raise have nothing to do with contextual matters. I will assume, with the hypothesists (except Popper and Feynman), that there is an objective, noncontextual, nondeductive relationship between observation statements and hypotheses in virtue of which the former provide some reason for believing, or at least considering, the latter. The question is whether hypothesists have adequately captured this relationship in their theory. I think they have not, although I believe it may be possible to do so.

Earlier I distinguished the following types of nondeductive arguments:

1

O

Therefore (probably)

h, in the light of *b*.

2

O

Therefore,

h is worth considering, in the light of *b*.

Hypothesists who believe that scientists make nondeductive inferences of form 1 may provide rules of one of these types:

A

A nondeductive argument of form 1 is valid if *O* is deductively entailed by *h* together with *b*, but not by *b* itself.

B

A nondeductive argument of form 1 is valid if *h* together with *b* explains *O*, but *b* by itself does not.

Analogous rules are possible for arguments of form 2 if "1" is changed to "2" in A and B. Now, I believe that rules A and B are invalid in the cases of both 1 and 2. The problem with A is one that Mill pointed out a long time ago. Numerous conflicting hypotheses—many of them absolutely crazy—can be constructed that entail *O*, yet we are certainly unwilling to infer that all such hypotheses are probable. Indeed, I would extend this to the weaker argument of form 2, which Mill himself may have accepted as being governed by the analog of rule A. Take a completely implausible hypothesis: that last night a group of monkeys escaped from the zoo, reached my car, siphoned off the remaining gas in the tank, and substituted crushed bananas. This *h*, together with the background information that a car will not start without gas in the tank, entails the proposition *O* that my car will not start. Yet from the fact that *O* is true I cannot conclude that the monkey hypothesis is probable, or even that it is worth considering.

Turning to rule B, we should clear up an ambiguity: When we say that *h* (together with *b*) explains *O*, do we mean that it explains *O* correctly, or only that *h*, if true, would explain it correctly? (Hanson and Peirce opt for the latter.) In either case, rule B will not be valid for arguments of form 1 or 2. Suppose we mean "correctly explains." Consider once again my monkeys, which, *mirabile dictu*, did escape from the zoo last night and did siphon off the remaining gas in my tank and substitute crushed bananas. Then this h does correctly explain why O is the case, where O is "my car will not start." However, if I were given just O and "normal" background information, it would be absurd for me to conclude that h is probable or worth considering.

Now take "explains" in rule B to mean "would explain correctly if true." Once more my monkey hypothesis, if it were true, would explain correctly why my car will not start. However, from the fact that my car will not start and "normal" background information, again it would be absurd to conclude that the monkey hypothesis is probable or worth considering.

To be sure, A and B are not particularly sophisticated rules. Following Whewell's suggestion one might be tempted to require that O consist of not one observed event or type of phenomenon, but many. If hypothesis h can entail and/or explain a variety of observation reports, then an inference to h may become reasonable in light of such reports.

This ploy will not work. Let O contain as many and varied observation reports as you like. For example, let it contain reports such as "the sky is blue," "grass is green," and "the sea is salty." Now construct a hypothesis h that is a conjunction of two propositions. Let the first proposition postulate the existence of X, where X is anything you like, however implausible, be it God, the Devil, or the Loch Ness monster. Let the second proposition in hypothesis h be of the form "If X exists, then X causes it to be the case that O_1 and $O_2, \ldots$, and O_n," where these are the many and varied observation reports that make up O. (For example, the first proposition in h might be "God exists," and the second "If God exists then God causes it to be the case that the sky is blue, that grass is green, and that the sea is salty.") With these constructions, we have a hypothesis h that entails observations that are as many and varied as we like. Yet from these observation reports it is not reasonable to infer h with probability, or indeed to infer that h is worth considering. We can even concoct an h that will yield predictions that have not yet been tested. Take some very successful scientific theory T that makes a set of observational predictions O' that have not yet been tested. To the second proposition in h add "If X exists, then X causes it to be the case that O'." Now we have an h that will yield not just varied observational claims that we already know to be true but others we do not yet know to be true. Strictly analogous arguments are possible if we focus on explanation instead of implication and require that h explain a variety of observations.

I am sure that I have not seen all the rules of inference (or theories of evidence) proposed by hypothesists. The ones I have seen are subject to these or analogous difficulties. However, the matter is not hopeless, I believe. Elsewhere I have suggested a definition of the concept of evidence that can be pressed into use (Achinstein 1983, chapter 10). This is not the occasion to argue in depth for this proposal, but let me note what it is.

The rule governing arguments of form 1 that I would suggest is this:

C

A nondeductive argument of form 1 is valid if and only if (i) $p(h, O\&b) > k$, (ii) p(there is an explanatory connection between h and O, $h\&O\&b) > k$. (Let $k \geq 1/2$.) (For arguments of form 2 to the conclusion that h is worth considering, we might change i and ii by requiring simply non-negligible probabilities.)

Rule C requires the (high) probability of h, given O and b, and the (high) probability that there is an explanatory connection between h and O, given h and O and b. (Roughly, there is an explanatory connection between h and O if h correctly explains why O is true, or O correctly explains why h is true, or some common explanation correctly explains why both h and O are true.) Thus, I would favor those hypothesists who focus on explanation rather than mere deduction, although I incorporate the explanatory connection into probability condition ii. However, unlike the typical hypothesist, I do not regard the (probability of an) explanatory connection between h and O as sufficient for an inference from O to h. In addition, the hypothesis h must be "plausible"—an idea reflected in the probability condition i. Let me make two comments about this.

First, the counterexamples I presented against the standard rules A and B involve the introduction of extremely implausible hypotheses. They are hypotheses that, given the observations and the background information, have very low probabilities, even though they entail and explain the observations. These hypotheses would be excluded by my probability condition i.

Second, when a method of hypothesis is used by a scientist, and when an inference is made from the fact that the observations entailed or explained are true to the probability of the hypothesis, then (often, at least) it is independently assumed or argued that h is plausible. Let me offer an illustration.

Herapath, Waterston, and Maxwell were three nineteenth-century contributors to the kinetic theory of gases, each of whom used a method of hypothesis (in some version). Herapath begins with five "postulata," including "Let it be granted that matter is composed of inert, massy perfectly hard, indestructible atoms. . . ." and "Let it be granted that gaseous or uniform bodies consist of atoms, or particles moving about, and among one another, with perfect freedom." (Boorse and Motz 1966, vol. 1, p. 198) Herapath writes that he "purposely put these hypotheses . . . into the form of postulata, to avoid being obliged to establish them by direct demonstration," and he proceeds to show that these postulates can be used to explain observed phenomena, such as the relation between pressure and volume of a gas, and the diffusion of gases. This seems to be a simple employment of a method of hypothesis: the proposing of hypotheses that explain (if not predict) observable phenomena. Yet there is reason to question whether the method of hypothesis, as reflected in standard rules such as A and B, is being employed. Immediately after introducing his five hypotheses, Herapath writes in a parenthetical remark ". . . if indeed we can call those things hypotheses which can be deduced from the analysis of phenomena." In the paragraphs that follow, although he does not "deduce" them from phenomena, he does argue for their plausibility by appeal to considerations of simplicity.

Similarly, Waterston believes it is necessary to argue for the plausibility of one of the central postulates in his theory, viz. that heat is molecular kinetic energy. He does so on empirical grounds, citing experiments of Forbes and Melloni showing that the mode of heat radiation is similar to that of light, which Waterston (following many others of his time) takes to be "undulatory" in nature. Given this and other considerations, Waterston writes: "It seems to be almost impossible now to escape from the inference that heat is essentially molecular vis viva." (Boorse and Motz 1966, vol. 1, p. 223)

Maxwell gives the most explicit statement of the idea that hypotheses must have some independent warrant and cannot be introduced or defended simply on the grounds that observed phenomena can be explained or deduced. Maxwell begins by describing the method of hypothesis:

The method which has been for the most part employed in conducting such inquiries is that of forming an hypothesis, and calculating what would happen if the hypothesis were true. If these results agree with actual phenomena the hypothesis is said to be verified, so long, at least, as some one else does not invent another hypothesis which agrees still better with the phenomena. (1965, vol. 1, p. 419)

But Maxwell complains that those who use this method tend either to introduce vague hypotheses or to fill in details by "the illegitimate use of the imagination." Instead Maxwell proposes a more legitimate "method of physical speculation," which "proposes to deduce from the observed phenomena just as much information about the conditions and connections of the material system as these phenomena can legitimately furnish." (1965, vol. 1, p. 420) Maxwell requires that the hypotheses introduced have independent experiential warrant, and not be just any from which the observed phenomena can be derived.

I would not claim that scientists always independently argue for or assume the plausibility of the hypotheses they use to derive or explain the observations, but I believe that they often do. More strongly, I believe that this must be the case if they are to infer the probability of their hypothesis. Otherwise, to use Maxwell's words, we have an "illegitimate use of the imagination." Otherwise, we can generate hypotheses of the sort illustrated earlier. This is why I think that the probability condition i in C, or something like it, is necessary in addition to an explanatory connection condition.

So far as I am aware, hypothesists have not provided an adequate basis for a purely epistemic evaluation of a hypothesis. I believe they cannot do so without adding some plausibility requirement such as i in C. However, some hypothesists may balk at this, because it introduces something in addition to deducibility and explanation without informing us how the plausibility or the probability of h, given O and b, is to be determined. I have no general theory to offer about this, only the claims that it is often assumed or argued for and that unless it is the method of hypothesis will not do the job it is supposed to do.

Can the Method of Hypothesis Serve as a Guide to Scientists in Suggesting Types of Hypotheses to Consider?

In the second section above, the question was "Can the method of hypothesis generate new hypotheses?" The answer given by hypothesists is No. They claim that hypotheses are guesses or conjectures not generated by rules. The method can be used for evaluating hypotheses once they have been discovered, but not for discovering them in the first place. Is this the only legitimate way to view the method?

A chess master may write a book on chess that does not prescribe particular moves but only general strategies (e.g., "At the beginning of the game develop pieces in the center of the board rather than the sides"). Such a book can serve as a useful guide in discovering what moves to make in appropriate circumstances even though it does not provide a list of specific moves. It serves as a guide by suggesting broad types of moves that, in the given circumstances of play, may help a player discover some particular move to make. Similarly, it may be reasonable to view the method of hypothesis as providing a general strategy for discovering hypotheses by telling the scientist in a general way what kind of hypothesis to look for, without indicating any specific ones.

Is this a reasonable way to view the method? If it is, then the method will be of more use to scientists than even its proponents allow, for it will be of use not only in the context of justification but in the context of discovery as well.

How helpful could the method be in suggesting types of hypotheses to consider? Is it really as helpful as chess strategies are in suggesting types of moves to consider? I can imagine a skeptic answering No. It will not really help the scientist very much, he will say, to be told "Introduce a hypothesis that entails and/or explains your data." It would be much more helpful to say things like "Derive the observed data from Rutherford-type assumptions about the hydrogen atom" or "Make sure your hypotheses satisfy conservation of energy and momentum." These constraints could be helpful in thinking up new hypotheses. But they are not constraints of the sort hypothesists can propose as part of their method, since they are empirical.

My inclination is to be more generous than the skeptic. To be sure, empirical constraints would be very useful for the scientist in thinking up new hypotheses. And admittedly the method of hypothesis offers none. But it does offer some significant *a priori* constraints, or at least it can be developed to do so. Thus, following Whewell, one might seek a hypothesis that yields a "consilience of inductions," i.e., that explains not just the data at hand but other data of different kinds as well. The more the hypothesist can say about what counts as explanation, unification, generality, simplicity, and so forth, the more useful such general criteria become for the scientist in suggesting types of hypotheses to consider.

I am not saying that these criteria provide either necessary or sufficient conditions for specific scientific hypotheses. General chess strategies do not provide such conditions for specific chess moves. In applying such strategies one must consider contextual matters, such as the skills and styles of the players and the positions of the pieces on the board. Similarly, as I argued earlier with the Bohr example, in science one must consider contextual matters pertaining to the interests and knowledge of the scientific community. In some contexts it will not be appropriate to impose one or more of the general constraints on hypotheses to be considered. Indeed, the main point of the third section above was to argue that such constraints fail to provide sufficient conditions for the evaluation of scientific hypotheses. Nor, I think, do they supply necessary ones. But it does not follow that they are irrelevant—any more than the strategic principles in a good chess book are irrelevant for evaluating chess moves, even if they do not constitute a set of necessary and sufficient conditions for being good chess moves.

Paul Feyerabend (1970, p. 18) has claimed that a scientific method furnishes only "rules of thumb, useful hints, heuristic suggestions rather than general laws." I can accept this characterization (provided we are thinking of general evaluations rather than purely epistemic ones), but I do not accept Feyerabend's more radical "anarchism," according to which the criteria provided by a scientific method are mere "verbal ornaments" and "anything goes" in science. The criteria provided by the method of hypothesis are "rules of thumb," or "heuristic suggestions," or, as I prefer, general strategies—rather than unalterable commandments—for the simple reason that these criteria do not take into account contextual matters pertaining to the specific interests and knowledge of the scientist and the community. The latter can vary from one situation to another, so that while, e.g., unification and precision may be particularly appropriate for one situation they may be less so for another. However, the general criteria provided by a scientific method are not mere verbal ornaments; it is not true that "anything goes." They do constitute genuinely normative conditions. To provide a general evaluation of a scientific hypothesis it is appropriate to consider such criteria as well as specific beliefs and interests of the scientific community. We evaluate Bohr's 1913 hypotheses highly because we combine the general methodological principle that understanding phenomena at a unifying, law-governed, micro level is scientifically valuable with the specific belief held by the scientific community that the spectral lines of hydrogen are produced by events within the hydrogen atom and the specific interest in explaining those lines in this manner.

Finally, then, suppose there are general principles concerning explanation, unification, simplicity, and so forth that are relevant in the context of justification in nonepistemically evaluating hypotheses. Then these principles can also be used in the context of discovery to provide general criteria for the types of hypotheses to be thought up. This is a possible and indeed quite useful role for the method of hypothesis. Hypothesists, I think, are too modest in their own goals.

Acknowledgment

I am indebted to the National Science Foundation for support of research.

References

Achinstein, Peter. 1983. *The Nature of Explanation.* New York: Oxford University Press.

Boorse, Henry A., and Lloyd Motz, eds. 1966. *The World of the Atom.* New York: Basic.

Butts, Robert E., ed. 1968. *William Whewell's Theory of Scientific Method.* University of Pittsburgh Press.

Feyerabend, Paul. 1970. *Against Method: Outline of an Anarchistic Theory of Knowledge.* Minneapolis: University of Minnesota Press.

Feynman, Richard. 1965. *The Character of Physical Law.* Cambridge, Mass.: MIT Press.

Hempel, Carl G. 1966. *Philosophy of Natural Science.* Englewood Cliffs, N.J.: Prentice-Hall.

Maxwell, James Clerk. *Scientific Papers.* New York: Dover, 1965.

Mill, John Stuart. 1959. *A System of Logic.* London: Longmans.

Popper, Karl. 1965. *Conjectures and Refutations.* London: Routledge and Kegan Paul.

6 Idealizations and the Testing of Theories by Experimentation

Ronald Laymon

Idealizations and the Hypothetico-Deductive Method

In order to be testable a scientific theory must make some claim about the world where the truth or falsity of this claim is detectable by available experimental methods. This is truism. But the generation by human scientists of such verifiable claims from theory requires practical computability—that is, that the computation or derivation of a claim or a prediction be completable in real time and with real resources. By real resources I mean our available analytic skills as well as our computational power. It is in order to achieve practical computability that scientists use idealizations and approximations in their derivations of verifiable claims. This creates problems for the confirming and the disconfirming of theories. The use of idealizations and approximations is also required when there is an absence of necessary data and required auxiliary theories.

In this chapter I shall be concerned primarily with deficits of analytic and computational power. Classic examples of what I have in mind are Newton's demonstration of Kepler's second law and the use of the Schwarzschild solution of the relativistic field equations to calculate the bending of starlight by the sun. Newton assumed that there exists only a single planet and that its mass is negligible compared with that of the sun. The situation is similar in the Schwarzschild case, since here we assume the existence of only a single body, of simple constitution (e.g., it is static and symmetrical), and light particles of negligible mass.

An elementary hypothetico-deductive model of these derivations will be sufficient to indicate the confirmational problem caused by the use of idealizations and approximations. Let T represent the underlying

theory; in our examples this will consist of Newton's laws or the relativistic field equations. Let I represent the idealizing assumptions
made. Include in I the required parameter or initial condition values.
Let P be the practically derivable prediction: Kepler's second law will
hold true for planetary orbits; light rays will deflect according to a
hyperbolic law with an ordinate intercept value of 1.75″ at the solar
radius. That P is a consequence of T and I can be represented in a
standard way: $T \ \& \ I \Rightarrow P$. There are now two possibilities: P is true or
it is false, i.e., P is or is not correct to within calculated or estimated
experimental error. Philosophical and scientific common sense has it
that in the first case there is confirmation (or at least the satisfaction
of a necessary condition for confirmation) and in the second case there
is disconfirmation. Consider first the case of disconfirmation, which
may be formally represented in our model as a *modus tollens*:

$$\frac{T \ \& \ I \ \Rightarrow \ P \\ \text{not } P}{\text{not } T \text{ and/or not } I} \ .$$

Simple inspection reveals the problem immediately. Even if we assume as unproblematic the truth of the premises (i.e., that the theory
and the idealizations have P as a logical consequence, and that the
experimental result is correctly described as inconsistent with the truth
of P), nothing logically follows about the truth or the falsity of the
theory. What follows is only that either or both the theory and the
idealizations are false. But we already know that the idealizations are
false, so nothing is gained. In other words, the falsity of I protects the
theory against refutation.[1] This is not a totally unwelcomed result, since
in both the Newton and the Schwarzschild case the predictions were
not correct to within experimental error. The situation is particularly
dramatic in the Newton case, since it is a consequence of universal
gravitation that planetary perturbations will cause deviation from Kepler's laws.[2] None of this should be surprising if we consider that the
idealizations (because they are false) introduce bias or distortion into
our computations.[3] Hence, given a true theory, the prediction cannot
be true. Laboratory students who attempt to fudge data by distributing
bogus experimental values "normally" about a predicted value de-
servedly fail because they have not assimilated this basic truth. Given
this perspective, we can also see why the standard hypothetico-
deductive account also fails to yield confirmation or confirmatory value:

Only a false theory, when conjoined with a biased idealization, could lead to a correct prediction. The use of idealizations in science therefore seems to put conventional hypothetico-deductive wisdom on its head. An incorrect prediction is now seen to be a necessary condition for confirmation; a correct prediction is a sure sign of the falsity of a theory. Unfortunately, this revised theory of confirmation will not work because of two sorts of complication. First, it is usually the case that several idealizations are used. In such cases there is always the possibility of a fortuitous cancellation of biases such that a correct prediction results when a true theory is used. But it is because such cancellations have been rare in the history of science that correct predictions usually arouse our suspicions and not our admiration. Second, theories differ in their sensitivity to idealizations used; there may be amplification or damping of distortion. This means that a biased idealization could distort a prediction in an amount smaller than observable error. Hence, a false theory and an idealization can yield a prediction correct to within experimental error.

The conclusion to be drawn from the discussion so far is that the use of idealizations shields theories from both praise and blame. Hence, if we are to understand the confirmation and the disconfirmation of theories, we must give up either the hypothetico-deductive format or traditional two-valued logic. It is the latter of these two options that I shall explore next.

Approximate Truth as a Truth Surrogate

A popular suggestion, especially among scientists, is that we use *approximately true* as a truth surrogate in hypothetico-deductive contexts.[4] The basic idea is that if our idealizations are approximately true this obliges our theory to produce approximately true predictions. What we would have then is a sort of three-valued logic that operates according to the following schemata:

CONFIRMATION		DISCONFIRMATION	
$I \& T \Rightarrow P$	true	$I \& T \Rightarrow P$	true
I	approximately true	I	approximately true
P	approximately true	P	not approximately true
T	true (or *may be* true)	T	false

Thus, in the case of disconfirmation we have that if it is true that a
theory and idealizations have P as logical consequence, and if the
idealizations are approximately true, then if the prediction is not approximately true the theory is false. Let us see how far this concept of
approximate truth can be taken.

The motivation for using approximate truth as a truth surrogate comes,
I believe, from a consideration of the calculus of errors. Consider, for
example, a simple "theory" of the form

$I = k \tan\theta.$

Think of I, if you like, as current intensity, θ as the deflection of say
a magnetic needle, and k as some constant that is a function of the
experimental particulars. Assume also that there is some independent
and confirmed means of determining I, so that our interest is to test
our little theory. Now,

$$\frac{dI}{d\theta} = \frac{k}{\cos^2\theta}$$

and from elementary calculus we know that (ignoring higher-order
terms)

$$\Delta I = (k/\cos^2\theta)(\Delta\theta),$$

where $\Delta\theta$ is to be understood as the error in the measurement of θ. If
we conceive of our make-believe theory as a model of scientific theorizing, we can interpret $\Delta\theta$ as the bias introduced by our idealizations.
Therefore, ΔI will be the error induced in our prediction by the error
or bias $\Delta\theta$. In the case of our little intensity theory the error ratio is
given by

$$\frac{\Delta I}{I} = \frac{k\Delta\theta}{k\cos^2\theta\tan\theta} = \frac{2\Delta\theta}{\sin2\theta}.$$

Maximum accuracy is achievable, then, when $\theta = \pi/4$.

There is an immediate and obvious lesson to be drawn here about
our proposed concept of approximate truth as a truth surrogate. Even
if we assume that there exists some measure of the degree to which
some input value or condition is approximately true, the theory is in
no way obliged to produce this particular degree of approximate truth
in its prediction. In fact, we must use the theory itself to compute its
obligations to the truth. This has a ring of paradox to it, since it suggests
a lurking circularity in our standards of empirical adequacy. In order

to display this "circularity" more prominently, consider in addition to our little theory some alternative or competitor theory of the form.

$$I = k' \tan(\theta + 0.2\pi),$$

and let $k' = k/\tan(0.45\pi)$ so that both theories give the same prediction for $\theta = \pi/4$. Assume now that I can be determined only in the vicinity of this value for θ, and with an accuracy that is only just sufficient to test the original theory. In such a situation our competitor theory is immune to falsification, because from its point of view[5]

$$\Delta I = \frac{k'}{\cos^2(\pi/4 + 0.2\pi)} \Delta\theta$$

and

$$\frac{\Delta I}{I} = \frac{2\Delta\theta}{\sin(0.9\pi)} \gg \frac{2\Delta\theta}{\sin[2(\pi/4)]}.$$

That there should be this difference is not, as it stands, a telling objection to the approach considered here. It is simply a corollary of the fact that, even given the degree to which our idealizations are approximately true, there is no independent standard for the degree of experimental fit that predictions must adhere to.[6] Modifying our schemata to reflect this observation, we have the following:

CONFIRMATION		DISCONFIRMATION	
I & T ⇒ P	true	I & T ⇒ P	true
I	approximately true to degree d	I	approximately true to degree d
P	approximately true to degree T(d)	P	not approximately true to degree T(d)
T	true	T	false

A theory receives confirmation, then, if it is true that it and the idealization I have P as a logical consequence, I is approximately true to degree d, and P is approximately true to some degree $T(d)$ computed on the basis of theory T.

As a model for scientific theorizing and testing, our little intensity "theory" is hopelessly simplified. Let us consider a slightly more complicated case: the standard calculation for the period of a pendulum. Assuming for the moment that the mass is concentrated at a point, that the supporting cord is massless, and that there is no air or pivot friction, it is a standard exercise in elementary physics to work out the equations of motion. In particular, where l is the length of the suspension cord, g is the gravitational field strength, and θ is the angular displacement, we find that

$$m\ell \frac{d^2\theta}{dt^2} = -mg\sin\theta.$$

Computing a solution to this equation is considerably easier if we assume that for "small" angles $\theta = \sin\theta$. With this approximation we obtain for the period P

$$P = 2\pi\sqrt{\ell/g}.$$

The claim that $\theta = \sin\theta$ (and note that strict equality is formally the relation that is used) can be conceived as an idealization that, although strictly false, is nevertheless approximately true. And, as was the case in our simple intensity theory, we can compute the degree or measure of approximation. Obviously this is not going to be the case generally, and we shall have to face up to this eventually. For the moment, though, let us continue with the example. The expansion of $\sin\theta$ in radian measure is

$$\sin\theta = \theta - \theta^3/3! + \theta^5/5! - \cdots.$$

Thus, for example, for $\theta = 3°$, $\sin 3° = 0.05234$ and $3° = 0.05236$ radian. Therefore, our idealized input, though false, is approximately true to within around one part in 2,000. However, we cannot compute the obligations to accuracy of our theory (as we could with our earlier intensity theory), since we have transformed it, via the equality $\theta = \sin\theta$, into something else. If we were to do an error analysis of $m\ell\ddot{\theta} = -mg\theta$ and its solution $P = 2\pi\sqrt{\ell/g}$, the expected accuracy for P, so calculated, would not transfer to our original differential equation because of the bias or distortion introduced by the assumption that $\theta = \sin\theta$. Of course, in this case (but not in general) we can calculate the exact solution to our original differential equation and hence compute the bias introduced by our angle assumption. If we stick with our original differential equation, the solution, though considerably harder to obtain, nevertheless takes a pleasing form:[7]

$$P = (2\pi\sqrt{\ell/g})(1 + \tfrac{1}{4}\sin^2\theta/2 + \tfrac{9}{64}\sin^4\theta/2 + \cdots).$$

Hence, the approximation $\theta = \sin\theta$ will serve to generate periods that are too small in a way that depends on the size of θ.

Now let us try to relate this pendulum example (still admittedly simple) to our original problem. We were able to calculate the effect of the bias only because we had the assumed correct theory.[8] In the absence of that theory we could not compute the biasing effect of the approximation $\theta = \sin\theta$. Here we have the crux of our problem. Idealizations are introduced precisely because we lack the analytic and computational skills to calculate "correct" theories, but in the absence of these correct theories we cannot compute the predictive bias introduced by our idealization. If this error calculation cannot be made (and remember that it is to be made on the basis of the theory to be tested), then we cannot satisfy the conditions of applicability for our confirmation schemata.

One case in which we are not able to rigorously compute anticipated error is the derivation using the Schwarzschild solution of the bending of light near the sun. The adequacy of the idealizations used in this derivation is usually justified with remarks such as "at infinity the metric becomes Minkowskian" and "[the gravitational field] on the earth's surface is more than 10^3 times weaker than that of the sun on the sun's surface" (Weinberg 1972, p. 191), and here adequacy means that errors introduced by the idealizations are relatively small and that those errors will translate into relatively small errors in the prediction. However, explicit arguments to demonstrate the cogency of remarks such as those above are rarely if ever given. The adequacy of these sorts of approximations and idealizations forms an important part of the traditional folklore of a scientific education. My suspicion is that, if pressed, the best that one could do in the light-bending case would be to make an appeal to the reduction of the general theory of relativity to Newtonian mechanics in special cases and to point out analogies of procedure with past successes. Birkhoff (1960, p. 4), in an interesting review of hydrodynamics (where similar problems arise), asserts that "progress would have been slower if rigorous mathematics had not been supplemented by various *plausible intuitive hypotheses*," such as the claim that "small causes produce small effects, and infinitesimal causes produce infinitesimal effects." Synge, the relativity theorist, describes the use of approximations as follows: "Approximations based on the neglect of small terms are very frequent in mathematical physics, and there is seldom any reason to object to them. One feels that if there is anything wrong, it will show up in some anomaly, and then one can revise the theory." (1960, p. 57)

If explicit error calculations cannot be made, however, there is no standard (even one that is theory-based) that will determine when a prediction is so far off the mark as to represent something gone "wrong." We can model this situation with our simple pendulum example if we consider the case of a beginning physics student who lacks the analytic skills needed to solve the original unsimplified differential equation. For such a student, the role of experiment is not to test the theory but to experimentally determine the bias introduced by the idealizing assumptions (and there are several in addition to the angle approximation) made in the elementary calculation. If we were to generalize here (i.e., identify this student's problem with that of the scientist who cannot practically compute an error analysis), we would have to say that the role of experiment in such cases is not to test theories but to determine the overall distortion caused by our idealized assumptions. Once again the use of idealizations and approximations serves to shield theories from both confirmation and disconfirmation. Our generalization had better be incorrect, then, if we are to avoid a radical form of conventionalism.

Before sketching a positive theory of confirmation, I want to remove a remaining simplification common to both the pendulum example and the intensity example: the assumption that we can meaningfully assign some degree or measure of the distance of our idealizations from the the truth. Clearly, this is an assumption that is not generally true. For example, how far from the truth, in quantative terms, is the assumption of Newton and Schwarzschild that the universe contains only one massy body? Or consider the case of the classical theory of the electron, where it is assumed that the electron is a point mass. There is no sensible or useful way of giving a measure of the distance from the truth of this assumption. Obviously, examples can be easily multiplied here. Synge notes that in the absence of practically computable alternatives there is no option but to work with our idealizations, even though we cannot determine either their distance from the truth or their biasing effect on our predictions:

The agonist needs no encouragement to work out, as a mathematical problem, the geodesics of space-time with the [Schwarzschild] metric. The realist, on the other hand, may have some doubts. Though convinced of the validity of the geodesic hypothesis for very small bodies [e.g., that photons follow geodesics], he may wonder just what 'very small' means—are the Earth and Jupiter very small? This question cannot be answered until a rational theory of the 2-body problem has been developed, and the only thing to do is to go ahead with the planetary motion and light rays. (1960, p. 290)

To summarize the discussion of this section: there are two problems associated with the use of idealizations. First, in contrast with the case of simple errors of measurement, there is usually no clear sense in which we can assign a metrical distance from the truth. Second, if we could assign a degree of approximation to our idealizations, mathematical complexities would often make it impossible to compute the effect of the use of such idealizations. These difficulties suggest that it is a mistake to begin with the concept of approximate truth. It looks as though approximate truth is something we should back into and not something we should start with. I shall now suggest a strategy for doing just that.

Confirmation as Improvability

Consider once again our simple intensity theory: $I = k \tan\theta$. If the theory is true, more accurate values for θ will lead to more accurate values for I. Alternatively, if more accurate values for θ do not lead to more accurate values for I, the theory is false. These are just simple restatements of our previous error analysis restricted though to comparative judgments of accuracy. This restriction will be an important difference. The idealization analog for our simple model will be that if more realistic idealizations do not lead to more accurate predictions then the theory is false. The advantage here is that no metric for degree of approximate truth is required; only a relative ranking is needed. (This can be further weakened to a requirement of partial ordering.) Now, although claims about the numerical degree of approximate truth are rarely if ever made about idealizations, comparative judgments about the relative realism of idealizations with respect to some background standard historically have been made. All this suggests the following theses about confirmation and disconfirmation:

A scientific theory is confirmed (or receives confirmation) if it can be shown that using more realistic idealizations will lead to more accurate predictions.

A scientific theory is disconfirmed if it can be shown that using more realistic idealizations will not lead to more accurate predictions.

We can conceive of this approach to confirmation in the following graphical terms. Consider the set of all logically possible descriptions of some particular phenomenon that are allowable according to some set of background standards. Assume that it makes sense to think of this set as ordered (perhaps only partially) with respect to some comparative background standard of being more realistic. This will be our space of idealized descriptions. Assume again that theoretical predictions about the case in question can be similarly ordered with respect to being closer to observed value. A theory then may be considered as a function that maps increasingly realistic but still idealized descriptions (i.e., models of the same phenomena) into members of the prediction space. Applying a theory to increasingly more realistic idealizations therefore will determine a trajectory through our idealization space, and the theory as function will generate a corresponding trajectory through the prediction space. Confirmation, then, depends on the character of the induced trajectory. Roughly, if in response to motion toward the truth the induced prediction trajectory also moves toward the truth, there is confirmation. If the induced trajectory moves away, there is disconfirmation. If there is failure to generate comparable predictions, there is neither confirmation nor disconfirmation.[9]

To see how this approach is supposed to work, consider calculating the pressure of a gas using kinetic theory. The most elementary calculation of pressure proceeds on the assumption that gas molecules are infinitesimal in size and that they exert forces on one another only in collision. On the basis of these simplifying assumptions, the ideal-gas law can be derived. However, as Jeans (1940, p. 63) notes, "neither of these assumptions is true for an actual gas, and so [we] must proceed to calculate the pressure for a real gas in which the molecules are of finite size, and exert forces of cohesion on one another even when they are not in contact." What is impressive about the approach of kinetic theory is that a more accurate equation of state (i.e., a more accurate prediction) can be generated when the simplifying assumptions are made more realistic. The standard example of this sort of improvability is the Van der Waals calculation of $pv = NRT(1 + bv)$ as the equation of state, where $b = 2/3N\pi r^3$ and where N is the number of molecules and r is their radius (Jeans 1940, pp. 64–66). The kinetic theory or program thus receives confirmation because the use of more realistic descriptive or initial conditions leads to more accurate predictions. This approach to confirmation does not require a particular measure of the degree to which one set of idealizations is more realistic than another; all that is required is a relative ranking.

Another suggestive example is Newton's *experimentum crucis*. Newton originally asserted that there is no color separation or dispersion after the second prism. But, as his critics correctly noted, this is not exper-

imentally correct; color separation and dispersion were within the capabilities of then-ordinary methods of observation. Newton responded
by arguing for the truth of this conditional: If the finite sizes of the
apertures are taken into account (and they were not in his original
treatment), the result is an improved prediction that allows for some
color separation and image dispersion.[10] With this possibility for improved prediction, the experiment was confirmatory of Newton's optical
theory. There is one suggestive feature of Newton's response that needs
to be emphasized here. Newton did not actually construct a more
realistic prediction computation for the experiment; he merely argued
that it could be done and that a correspondingly more accurate prediction
would result. This illustrates what I believe to be a general feature of
science: that because of an absence of analytical methods or a shortage
of computational power (or because of the lack of necessary auxiliary
theories and information) scientists cannot actually construct the better
calculations. In such cases, they resort to indirect arguments, the aim
of which is to show that if these shortcomings were to be overcome
then improvements of a specifiable sort in the idealizations used would
lead to a corresponding improvement in the prediction. Science proceeds
as much by argumentation about possible constructions as it does by
actually constructing more realistic computations. One caveat about
the example is required. It must be admitted that Newton's response
is not exactly of the type discussed, because with his analytical skills
it was possible for Newton to have given in fact the superior calculation.
In his case the cost in time and effort did not justify an explicit calculation, since anyone with experience in Euclidean ray optics would
immediately believe (perhaps on the basis of analogy with other problems) that such a computation was possible.

An example of disconfirmation based on nonimprovability is that of
specific heats and kinetic theory. On the basis of spectral lines and
other phenomena, it was known that gas molecules have internal structure. Adding structure to the billiard-ball molecules of kinetic theory
should then have afforded some improvement in the predictions about
specific heats. However, it was formally provable that adding degrees
of freedom of motion (a consequence of internal structure) would cause
a divergence from experimental values. Treating the molecules more
realistically would make the predictions about specific heats worse, not
better. Hence, there was a disconfirmation of kinetic theory; it was not
possible to improve the experimental fit.[11]

There is an important objection to be made against the view that the failure of a theory to produce a better prediction on the basis of a better idealization automatically disconfirms the theory. To generate a prediction from a theory usually requires a set of idealizations. It seems possible, then, that improving some of the members of this set could lead to a worsening of prediction because some additional feature necessary for the new combination to work was left out. Therefore, the appropriate response to a worsening of prediction would seem to be that it establishes *prima facie* disconfirmation; that is, it places the onus on proponents of the theory to explain why more realistic idealizations should lead to a worsening of prediction. And the sort of explanation required would be in terms of a conspiracy: that, in the absence of certain necessary additional improvements, partial improvements conspire (because of, e.g., fortuitously canceling forces or errors) to move predictions away from their target vectors. It is the perceived rarity of successful explanations of this conspiratorial sort that is responsible for the onus of defense being placed on proponents of theories that fail in the way described here. Certain aspects of the development of the classical theory of the electron may be analyzable in the terms suggested here. For example, the electromagnetic self-energy diverges only when structure terms are eliminated. However, because of quantum restrictions, classical improvements in structural descriptions did not lead to an adequate theory (quantum considerations being in a sense the missing necessary condition). The classical response, therefore, was not to seek relief from problems such as the "runaway solutions" to the Lorentz-Dirac equation by means of more realistic (though non-quantum) descriptions but to persevere with a point description of the electron.[12]

Being Approximately True

In our theory of confirmation we have used as a primitive the concept of one idealization being *more realistic* than another with respect to some background standard. That an idealization is *approximately true* is not a notion that we have had to use. Being approximately true was supposed to serve as a truth surrogate in the context of hypothetico-deductive reasoning. For example, an idealization was to be approximately true if it was good enough for the purposes of confirmation or disconfirmation. We can salvage this idea with the following analysis. The basic idea is that an idealization is approximately true if using more realistic idealizations leads to predictions that are or are not more accurate. That is, an idealization is approximately true if it is a part of a confirmational history. According to the graphical concepts of the preceding section, approximately true idealizations are those that are on trajectories that are mapped by theory toward or away from the truth. 'Approximately true' then is a term of confirmational approbation. It is not a concept that is prior to or more basic than confirmation and relative realism. This is why starting with approximate truth and then trying to develop a theory of confirmation is a mistake. We can also say that one idealization is a *better approximation* than another if it is more realistic than the other and both are approximately true. Initial conditions will fail to be approximately true if they are not part of a confirmational history—that is, if more realistic conditions do not lead to computable predictions, or if they lead to computable predictions that are not comparable with respect to accuracy. This latter possibility may occur if there are several dimensions of appraisal.

In the above discussion of Newton's *experimentum crucis* it was noted that confirmation and disconfirmation were possible even in the absence of predictions that were *practically* computable from more realistic idealizations. This would occur if it could be shown that if the impediments to practical computability were to be removed then more accurate predictions either would or would not result. Thus, the above analysis must be modified to include these sorts of *possible* confirmational histories. An idealization then will be approximately true if it can be shown that a more realistic idealization would lead to a more accurate prediction if the impediments to practical computability were to be removed. The same idea is to be applied *mutatis mutandis* to not being approximately true. I shall say more below about the sorts of arguments that are used to support claims of conditional computability.

Our analysis of being approximately true could, I believe, be used as part of an analysis of ordinary scientific usage. But here the situation is complicated by the fact that in circumstances when a theory is being used and not being tested, 'approximately true' means good enough for some practical purpose.[13] Some of the judgments of practical goodness, then, are taken over directly or to analogous cases where theories are being tested. Thus, whether a particular idealization is in fact called approximately true depends very much on historical particulars. Furthermore, any analysis of ordinary usage would also require a theory of reference that would distinguish the ontology of a theory from what was being claimed in an approximate or idealized way about the items of that ontology.

Confirmation in the Absence of Real Computability

As noted above, it is very often the case that scientists do not in fact construct more realistic models or sets of idealizations that have practically computable consequences. There are many reasons for this, including lack of necessary auxiliary theories, absence of known analytic methods, shortage of computing power, and lack of suitably precise instrumentation. What happens in such cases is that proponents of a theory argue that if the impediments were (in some appropriate sense) removed then more realistic models would have correspondingly more accurate predictive consequences. In terms of the concepts discussed above, the purpose of such arguments is to locate these "possible" but not actually computable models in the domain spaces of idealizations and predictions. Confirming trajectories are the targets of such arguments. Elsewhere (Laymon 1980b) I have dubbed such arguments *modal auxiliaries*.

Some initial types of modal auxiliary can be discussed and illustrated here. First, we can distinguish those that proceed by giving a recipe for the production of improved initial conditions and predictions from materials and procedures already in existence. The obvious analog is the metaproof in constructivist mathematics: If we have enough patience and symbols, the proof is constructible by means of the recipe. Newton's comparison of the acceleration of the moon with that of bodies falling at the surface of the earth utilizes such a constructive recipe. In proposition IV in Book III of the *Principia*, Newton demonstrates this counterfactual: *If* (a) the earth is stationary, (b) the earth-to-moon distance is 60 earth radii, and (c) the moon's period is $27^d7^h43^m$, *then*, assuming inverse square gravitational attraction, bodies should fall at the surface of the earth 15 1/2 feet in the first second. Bodies do not fall at this rate, but neither is the earth stationary, and the earth-to-moon distance is not 60 earth radii. These values were known to be outside the range of reasonable experimental error. Historically, Newton went on to show in various ways that if the counterfactuality of the antecedent were to be relieved then the consequent would approach a value that was correspondingly more accurate. In particular, Newton gave a constructive procedure for converting the stationary earth-moon system separated by 60 earth radii into a moving system separated by the more accurate value of 601/2 earth radii.[14]

Naturally, there are arguments to show the impossibility of such constructions. For example, Lorentz gave a proof (based on Huygens's principle and the calculus of variations) to show that more realistic and complete descriptions of the Michelson-Morley experiment would not lead (given aether theory) to other than a positive prediction. Some initial accounts suggested (mistakenly) that a null prediction would result if experimental complications were more accurately taken into account (Lorentz et al. 1929).

Another type of case is where there are perceived to be open or promising possibilities for improvement. An example is provided by the Einstein-Smoluchowski explanation of Brownian motion. The idealized models used by Einstein and Smoluchowski made predictions that considerably exceeded the range of experimental error. Nevertheless, the underlying statistical mechanics was seen as having been confirmed by the available experiments. My contention is that it was the existence of open possibilities for improvements that led to the judgment of confirmation. For example, Einstein assumed that each increment of the Brownian particle path is independent of previous increments. However, the theoretical framework of statistical mechanics gives reason to think that such an assumption is false. There was also the fact that Einstein, in the absence of a theory of statistical fluid mechanics, used Stokes's law for resistance in a continuous fluid.[15]

Finally, there are what Wimsatt (1981) calls robustness arguments. The basic idea is that if some property or feature (e.g., relative efficacy of group versus individual selection) is robust or insensitive to variations in the parameters of some model (or class of models) then that property or feature is likely to be true and should be incorporated in more realistic models.

A Role for Experimentation

All the modal auxiliaries discussed above are purely analytical. One wonders if experimentation can play any role here. For example, can experimentation be used somehow to determine or estimate the bias caused by the use of idealizations? If this were possible then we could correct the theoretical prediction on the basis of the estimated bias. Resurrect for one last time our simple theory about the variation of intensity. Not being able to compute the bias caused by an error in the determination of the angular quantity θ (the analog in that model for an idealization) would mean not being able to calculate the derivative $dI/d\theta$.[16] This derivative, however, could be determined or estimated experimentally by varying θ and measuring I. Once this determination was made, one could work "backward" and estimate the effect of an error in θ on the prediction for I. This speculation about method is not entirely fanciful, since it corresponds exactly to Newton's procedure in his use of the spectrum experiment to refute the then received laws of refraction.[17]

For Newton, "the received laws" were to include Euclidean ray optics, the thesis that light is a single homogeneous substance that when disturbed or modified produces color, and "the Hypotheses of the proportionality of the *sines* of Incidence, and Refraction" (i.e., Snell's law). In other words, there is only a single refractive index for light for each media pair. What Newton discovered was that sunlight produces an elongated colored image (the spectrum) when projected (by an aperture) through a prism. However, it follows from the received laws that a circular image should result if one assumes that the rays from the sun are parallel. This last assumption is, of course, an approximation or idealization, and so Newton undertook to recalculate image size on the more realistic assumption that the incoming rays are not parallel. In the special case of a prism oriented symmetrically with respect to the mean incoming and outgoing rays, he was able to prove that the image would continue to be circular. At other orientations, circularity would be lost and the image would become oblong. Newton did not derive an equation that would predict the proportions of the oblong image for nonsymmetrical orientations.

In his famous optics paper of 1671, Newton did not repeat his analytical proof *per se* but instead first reported the relevant experimental parameter values and then made an iterated application of Snell's law. Since the prism was symmetrically oriented in the experiment, the result of this iteration was (not surprisingly) the prediction of a circular image. On the other hand, the projected spectrum was measured to be about five times longer than it was wide. Now, every historical account of which I am aware stops the story here and gives this as an example of theory refutation by hypothetico-deductive methodology. The received theory made a prediction that was found to be false; hence, we are told, it follows that the theory is false. Unfortunately for such accounts, this is not Newton's mode of argumentation. He says in his paper:

... because this computation was founded in the Hypothesis of the proportionality of the *sines* of Incidence, and Refraction, which though by my own and others Experience I could not imagine to be so erroneous, as to make that Angle but 31', which in reality was 2 deg. 49'; yet my curiosity caused me again to take my Prisme. And having placed it at my window, as before, I observed that by turning it a little about its *axis* to and fro, so as to vary its obliquity to the light, more then by an angle of 4 or 5 degrees, the Colours were not thereby sensibly translated from their place on the wall, and consequently by the variation of Incidence, the quantity of Refraction was not sensibly varied. (Newton 1959, pp. 93–94)

Only after this variation on the experiment does Newton claim to have refuted the received laws. Explaining this delay creates a puzzle for most philosophical theories of confirmation, but in view of the analysis presented here it is clear that what Newton has done is anticipate the objection that his experimental determination of symmetrical orientation was in error. That is, it could be claimed that the image was oblong because the prism was not in the requisite orientation. (In fact, Pardies, missing the subtlety of Newton's experimental procedure, made exactly this objection.) Newton could have computed the anticipated image size for a range of incoming rays and hence shown that the amount of observed noncircularity was far greater than that predicted by the received view. But rather than engage in such a laborious calculation, Newton opted to, in effect, use his experimental apparatus as an analog computer. Assume that the spectrum's size and shape is some function f of incoming ray angles, prism size, aperture size and location (and whatever else is required) and θ, where θ is the angular deviation of the prism from symmetrical orientation. That is, assume that f is the correct but unknown theory of the spectrum. Therefore, $df/d\theta$ will also be unknown. However, this last function can be experimentally displayed by rotating the prism and noting the change in image size. By rotating his prism, Newton showed that large errors in the determination of symmetrical orientation would result only in imperceptible errors in image size and shape. The derivative $df/d\theta$ becomes large only when the prism is moved considerably from the symmetrical orientation. What all this shows is that the received laws cannot be saved by appeal to an error made in the determination of prism orientation. Improved experimental values will not result in improved experimental fit for the received laws.

We have found the calculus of errors to be a useful heuristic device for generating theses about the role idealizations play in scientific theorizing. The question posed by Newton's rotation of the prism, therefore, is whether this experimental variation can be generalized so as to yield a method of dealing with the biases introduced by idealizations. As a first step toward effecting this generation, I shall summarize Newton's methods so that we have a clear point of departure. In accordance with the view presented above, let us conceive of the received laws as a function f that takes values for the variables $x_1, \ldots, x_n$ to a predicted image size and shape, to be denoted I. Now, f is practically computable only for values $c_1, \ldots, c_n$, or some narrow range around these values. (This reflects Newton's proof that a circular image results only for a symmetrical prism orientation.) In addition, we have that it is not practically possible to compute $\partial f / \partial x_i$ for some or all of the variables. Finally, I as measured does not correspond to I as predicted. The hypothesis suggested by proponents of the received view is that this difference is due to an error Δc_m in the determination of the value for some variable x_m. Now, let f' be the true theory of the phenomenon. We can experimentally determine $\partial f' / \partial x_m$ (assuming that all the other relevant parameters are indeed kept constant) by varying the value for x_m and noting the change in I. What was discovered by Newton was a dependency curve that is very flat in the vicinity of the experimentally determined value c_m (the basis for the original calculation of I) and does not change greatly until we deviate far from this value for the variable x_m. Therefore, if f were the correct theory (i.e., if f were equal to f') then Newton would have had to have made a considerably greater error in the determination of x_m than is reasonably conceivable. (Newton may have been off by as much as 1°, but certainly not by 5°.)

From the viewpoint of the experimenter, the basic idea then is this: Vary the experimental parameters so as to bring them into agreement with what the measured values may be if in error, and then determine whether the to-be-predicted quantity moves into closer agreement with predicted values. With respect to idealizations, the method generalizes to varying experimental parameters or circumstances so that the idealizations become more realistic. If the theory is true, the prediction then should come into closer agreement with experimental determinations. A corollary to this methodology is to vary experimental circumstances away from the idealizations, that is, to make them less realistic. In this case a true theory should yield results further from experimentally determined values.

An elegant instance of our generalization of Newton's methodology is due to Robert Brown. From the viewpoint of phenomenological thermodynamics, a well-insulated and well-protected fluid can be described as being in thermodynamic equilibrium. Since, however, no insulation is perfect (and since internal variations require time to dampen), this description will be in varying degrees an idealization.

Furthermore, it is an idealization that is typically necessary in order to ensure practical computability. Brownian motion, as is well known, poses a major problem for phenomenological thermodynamics because, given the idealization of thermodynamic equilibrium, the particle should be stationary and not engaged in motion willy-nilly. If this idealization were correct, then the Brownian particle would be a perpetual-motion machine of the second type. The thermodynamical defense was to explain this motion as being due to varying external disturbances and, in the early stages of the debate, to "attractions and repulsions among the Brownian particles."[18] Predictive failure was due, then, to the bias introduced by the idealization of there being thermodynamic equilibrium. Brown's response was to bring the experimental particulars into closer agreement with idealization:

This experiment consists in reducing the drop of water containing the particles to microscopic minuteness, and prolonging its existence by immersing it in a transparent fluid of inferior specific gravity, with which it is not miscible, and in which evaporation is extremely slow. If to almond oil, which is a fluid having these properties, a considerably smaller proportion of water, duly impregnated with particles, be added, and the two fluids shaken or titurated together, drops of water of various sizes, from 1-50th to 1-2000dth of an inch in diameter, will be immediately produced. Of these, the most minute necessarily contain but few particles, and some may be occasionally observed with one particle only. In this manner drops, which if exposed to the air would be dissipated in less than a minute, may be retained for more than an hour. . . . But in all the drops thus formed and protected, the motion of the particles takes place with undiminished activity, while the principal causes assigned for that motion, namely, evaporation, and their mutual attraction and repulsion, are either materially reduced or absolutely null. (Brown 1829, p. 316)

Not only were many experiments of this sort conducted, but experiments instantiating the corollary methodology described above were conducted frequently as well. Interfering disturbances were magnified. Smoluchowski, in his review of these experiments (1906, p. 759) writes: "Very characteristic of this phenomenon is its independence from external circumstances. The various causal agents have been shown to be completely without influence."

As Smoluchowski's comment illustrates, the experimental methodology employed here is usually described in causal terms.[19] This should not be surprising, since the ignoring of interfering causes is usually the first sort of explicitly identified idealization to which we are introduced in our physics courses. The law of inertia predicts the possibility of ice hockey as a sport, but only in a variant form where pucks fly off the

face of the earth on tangents and with undiminished speed. To explain
the convergence of my idealization-based experimental methodology
with causality-based methodologies would require an analysis of causation and a theory of reference. That must await another time. My
intention here has been to generate some new perspectives on experimental method from a consideration of the necessary use of
idealizations.

A Bootstrapping Methodology

I close with the suggestion of what might be called a bootstrapping
methodology.[20] Consider a situation where the relative realism of idealizations I_1 and I_2 is unknown or indeterminate with respect to some
existing background standard. On the assumption that our theory T is
true, we would expect the more realistic idealization to produce a better
prediction.[21] Say that, with respect to phenomenon P, idealization I_2
produces the better prediction. Therefore, assuming the truth of T, our
judgment is that I_2 is the more realistic idealization. Now let it be the
case that T, I_1, and I_2 can be brought to bear on some other phenomenon,
P'. Then the judgment of the relative superiority of I_2 means that T, if
true, will produce a better prediction about P' with I_2 than with I_1. If
such a better prediction is not produced, we have reason to believe
that T is false. The method is appropriately called bootstrapping because
we first use the theory to generate an appraisal of relative realism; then
we test the theory using that appraisal. What is required is a coherence
of judgments about the relative realism of our idealizations. In this
way we could eliminate the need for a background standard of realism
or simply identify it with past applications of this bootstrapping
methodology.

Another feature of this proposed bootstrapping methodology is that,
in the absence of contravening reasons, we should expect theory T and
idealization I_2 to produce a prediction about a new or different phenomenon, P', that has the same or a similar degree of accuracy as was
produced by T and I_2 with respect to P. Thus, unless there are reasons
to think otherwise, we expect accuracy of fit to be transferable when
the same idealizations are used. If similar accuracy is not produced,
there is an onus on proponents of T to explain away this failure.[22]

The bootstrapping methodology proposed here casts new light on
the issues of variety of evidence and independent testability. The conventional wisdom is that one should maximize the differences among

the auxiliary hypotheses and theories that are used in conjunction with the idealizations and the theory to be tested to produce a prediction.[23] In this way one guards against the fortuitous cancellation of errors. The desire for such variety, however, should not blind us to the virtues of making only small variations in our experimental tests. An essential requirement of the bootstrapping methodology is that the same idealizations be applicable in "different" realms. But clearly this will require some similarity among auxiliary hypotheses and theories, for otherwise the idealizations would not attach logically to these auxiliary devices. Without multiple uses of the same idealizations we cannot test for coherence among our judgments of relative realism. In essence, then, what we require is a variety of testing contexts (i.e., sets of auxiliary hypotheses and theories), each of which allows for multiple uses of some idealizations.

For a historical illustration of the sort of bootstrapping discussed here, consider Stokes's (1851, pp. 1–141) development of the "theory" of viscous fluids. When Stokes gave his initial formulation of this theory, it was not known whether its predictions would be superior to those obtainable from the "ordinary hydrodynamics equations" (that is, the theory of nonviscous fluids). Think of these theories as consisting of a set of idealized descriptions conjoined with Newtonian mechanics, the underlying theory. Stokes obtained solutions to the proposed equations for viscous fluids for cylindrical and spherical pendulums. In the former case, the situation was reduced to a set of connected or constrained disks, each vibrating linearly. That is, using the proposed theory, Stokes was able to demonstrate more or less rigorously that certain aspects of a cylindrical pendulum could be ignored without observable consequence. For historical reasons, the quantity to be predicted was a correction factor n that when multiplied by the correction for buoyancy would give the true effect of the air (essentially an increase of period). On the basis of the theory of nonviscous fluids, Stokes had calculated that $n = 2$ for cylindrical pendulums. Experiments, however, showed that n is a quantity that varies, depending on experimental particulars, from about 2.3 to 8.0. With viscosity assumed, Stokes's solution for cylindrical pendulums gave predictions for n that were off by no more than one part in 30. (In one case, predicted $n = 2.364$; observed $n = 2.290$.) This was obviously an improvement over the results with nonviscosity assumed, and so the description of fluids as having viscosity (as defined by Stokes's equations) was (tentatively) a better idealization.[24] These improved predictions exceeded the anticipated experimental error. Stokes's response, in accordance with our methodology, was to give several arguments (modal auxiliaries) to show that various ignored complications would account for this discrepancy if they could be rendered computable (Stokes 1851, pp. 82–83). Of course, since it was the mid nineteenth century, Stokes did not feel the need to interpret these successes as confirming Newton's laws.

For the case of a spherical pendulum suspended by a cord, the "ordinary hydrodynamics equations" predicted $n = 1.5$. Experimental values in this case hovered around $n = 1.8$. Stokes's equations for viscosity therefore were obliged by our bootstrapping methodology to produce a value for n superior to 1.5. Furthermore, the predicted accuracy of these superior predictions was to be equal to that obtained in the case of cylindrical pendulums. Since both conditions were satisfied, we should agree with Stokes's historical judgment (1851, p. 124) that his theory had been "put to a pretty severe test."[25]

Summary

Practical computability is a central constraint on real science. Some numbers must be actually produced before the confirmational process can be started. In order to provide such computability, idealized or simplified initial condition descriptions must be used to input to scientific theories. Because of the bias introduced by these idealizations and simplifications, scientific theories in general will not produce either true predictions or predictions correct to within experimental error. If we assume the hypothetico-deductive account of confirmation, this means that theories are isolated from both praise and blame. Furthermore, the obligation of a theory to produce predictions of a certain accuracy cannot in general be calculated because of computational difficulties and the lack of a sensible metric for the distance of the idealized initial condition descriptions from the truth. Any philosophical theory of confirmation must take these features of real science into account. If we do not attend to the role played by idealization and simplification, much of the history of science will appear irrational and chaotic.

My thesis is that theories receive confirmation when they can be shown to produce better predictions on the basis of more accurate or more realistic initial condition descriptions. This is, of course, only a *prima facie* requirement, given the possibility of an unfavorable interaction among the various components of an idealized initial condition. In addition, because of computational difficulties, scientists often do not actually produce better predictions but instead argue that such predictions would be forthcoming if certain impediments to computation were removed. Theories therefore can be confirmed (or disconfirmed) in this indirect way.

Experiment does more than just provide a standard for the predictions produced given increasingly more accurate initial-condition descriptions. If, because of computational difficulties, a theory cannot be made to digest more accurate initial-condition descriptions, experimentation can be used to change actual initial conditions and thus make existing initial-condition descriptions more accurate. If the theory is correct, then the experimental result should be closer to the predicted value as well. As noted above, there is a formal similarity here with the method of concomitant variation. Finally, a bootstrapping or coherence test was proposed for determining the relative accuracy or realism of initial-condition descriptions.

Acknowledgments

Many of the ideas in this chapter were developed while I was a fellow at the Center for the Philosophy of Science at the University of Pittsburgh. I would like to take this opportunity to thank the past and current directors, Larry Laudan and Nicholas Rescher, for arranging my visit and for providing extremely stimulating accommodations. I could not have had a better or more productive visit. Research and development time was also made available by a faculty development grant awarded by The Ohio State University, my home institution, which is also to be greatly thanked for supporting faculty research.

Notes

1. This argument is similar to that used to show how theories are protected against refutation by the vulnerability of so-called auxiliary hypotheses. But, while the logic is similar, the point is different since my interest is in the use of known simplifications and idealizations and not in unnoticed assumptions that become prominent only when there is predictive failure. The classic case of such an auxiliary hypothesis is the Lorentz-Fitzgerald contraction hypothesis.

2. For more details on the Newton case see Glymour 1980, pp. 203–226, and Laymon 1983. For the light-bending experiment see von Kluber 1960, Earman and Glymour 1980, and Laymon 1982. Formally, Newton's assumption is that the reciprocity of gravitational attraction does not hold. There is a similar denial of reciprocity in the light-bending experiment. What this means is that the

idealizations also transform the theory to be tested. A simpler example of this sort of transformation is given below where we consider the effect on the simple theory of the pendulum of the assumption that $\theta = \sin\theta$.

3. The use of 'bias' in this context is due to Wimsatt (1981), who develops notions similar to those developed here. Another very important inspiration for this paper is Kuhn 1961.

4. For a good example, see Thomson and Tait 1912, pp. 1–5.

5. What I have done in the alternative theory is simply move the variable part of the theory to a steeper section of the tangent curve. If we were equipped with the appropriate statistical assumptions about error distribution, perhaps we could make the alternative theory testable. However, since this complication is not germane to my interests here, I shall not pursue it.

6. For a real example of the sort of competition envisioned here, see Laymon 1980a, pp. 22–24 and 33–34.

7. For the derivation, see Stephenson 1960, pp. 119–121.

8. There do exist formal estimation techniques that sometimes can be used in lieu of exact solutions. I have ignored mentioning these to keep the basic argument clear. Everything said, therefore, should be modified to include the existence or the nonexistence of such possibilities.

9. If we think there are several dimensions of appraisal of the realism of idealizations or the accuracy of predictions, it is natural to conceive of idealizations and predictions as partially orderable. This is not a very strong assumption. It means only that some, though not necessarily all, idealizations (or predictions) may be comparable with respect to realism, and that those that are comparable are transitively comparable. For example, consider the kinetic model of a gas molecule as two hard spheres connected by a spring, and consider also a model of a gas molecule as a single point source of a short-range repulsive, long-range attractive force. These models are not comparable with respect to realism, since their strong points are with respect to different aspects of molecular behavior: collision versus equipartition of energy. However, the "union" of their best features results in a model more realistic than either: a system of two point sources of short-range repulsive and long-range attractive forces. Similarly, the "intersection" of their worst features results in a model comparably less realistic than either. If the various dimensions of appraisal for the prediction are each quantifiable, then predictions form a stronger structure, namely a lattice. The picture suggested, then, is of a scientific theory as a function that maps partially ordered idealizations into a lattice of predictions.

10. For more details, see Laymon 1978b.

11. This is the second of Lord Kelvin's "clouds over the dynamical theory of heat and light" (Thomson 1901, pp. 493–527).

12. See Rohrlich 1965 for a comprehensive account, and Grünbaum 1976 for a discussion of the runaway solutions.

13. This affects, e.g., the discussion in Thomson and Tait 1912, II, pp. 1–5.

14. For more details on Newton's procedure here, see Laymon 1983.

15. For a detailed analysis of the case of Brownian motion, see Laymon 1978a.

16. Imagine in addition that the function I can be calculated only for some symmetry condition. That is, $\Delta I / \Delta \theta$ cannot be calculated by comparing the calculated differences in I for different inputs of θ.

17. For a detailed analysis of this case, see Laymon 1977.

18. For a review and analysis of this case, see Laymon 1978a.

19. In fact, it is essentially Mill's method of concomitant variation. Some readers will have noticed also that Birkhoff's "plausible intuitive hypothesis" was given in a causal form.

20. This section has been inspired by Buchdahl 1951 and Glymour 1980. See especially pp. 111–116 in the latter.

21. Since incomparable predictions also could be produced, the bootstrapping methodology proposed here must be modified to reflect this possibility.

22. This principle is a close cousin to the "plausible intuitive hypotheses" given by Birkhoff (1960, p. 4). I am unsure as to what its justification could be other than analogy to past experience with mathematics of similar form. One additional benefit to be derived from this principle is that it provides a way of explaining more of the ordinary usage of 'approximately true' than was possible in an earlier section. Say that, as above, T produces a better prediction with I_2 than with I_1. Furthermore, assume that there exist convincing modal auxiliaries showing that if the impediments to real computability were removed then a still better prediction would be produced by the use of idealization I_3. In this case, I_2 would be *approximately true* according to the analysis proposed earlier. Now say that T also produces a better prediction about a new or different phenomenon P' with I_2 than with I_1, and let it be the case that this prediction is correct to within the same or similar degree of accuracy as was produced with respect to P. My thesis about ordinary usage, then, is that I_2 would continue to be called approximately true (now with respect to P') even in the absence of modal auxiliaries for the improvability of predictions about P'. This is because the approbation now would be for the similar degree of accuracy produced.

23. For a critical discussion of the concept of variety of evidence, see Glymour 1980, pp. 139–142. For a discussion of independent testability, see Laymon 1980a.

24. For a list of the ways in which the theory of viscous fluids is still an idealization, see Birkhoff 1960, p. 5.

25. As in the case of the cylindrical pendulum, Stokes (1851, pp. 85–87) gives arguments to show that the discrepancies with observational values could be explained away if the mathematics for some of the disturbing causes could be solved. The bootstrapping methodology also is applied by Stokes within the class of cylindrical pendulum experiments. See pp. 80–81 of Stokes 1851.

References

Birkhoff, G. 1960. *Hydrodynamics: A Study in Logic, Fact, and Similitude.* Princeton University Press.

Brown, R. 1829. "Additional Remarks on Active Molecules." *Edinburgh Journal of Science* 1: 314–320.

Buchdahl, G. 1951. "Science and Logic: Some Thoughts on Newton's Second Law of Motion." *British Journal for the Philosophy of Science* 2: 217–235.

Earman, J., and C. Glymour. 1980. "Relativity and Eclipses: The British Eclipse Expeditions of 1919 and Their Predecessors." *Historical Studies in the Physical Sciences* 11: 49–85.

Glymour, Clark. 1980. *Theory and Evidence.* Princeton University Press.

Grünbaum, Adolf. 1976. "Is Preacceleration of Particles in Dirac's Electrodynamics a Case of Backward Causation? The Myth of Retrocausation in Classical Electrodynamics." *Philosophy of Science* 43: 165–201.

Jeans, Sir James. 1940. *An Introduction to the Kinetic Theory of Gases.* Cambridge University Press.

Kuhn, T. 1961. "The Function of Measurement in Modern Physical Science." *Isis* 52: 161–190.

Laymon, R. 1977. "Newton's Advertised Precision and His Refutation of the Received Laws of Refraction." In *Studies in Perception: Interrelations in History and Philosophy of Science,* ed. P. K. Machamer and R. G. Turnbull. Columbus: Ohio State University Press.

Laymon, R. 1978a. "Feyerabend, Brownian Motion, and the Hiddenness of Refuting Facts." *Philosophy of Science* 44: 225–247.

Laymon, R. 1978b. "Newton's *Experimentum Crucis* and the Logic of Idealization and Theory Refutation." *Studies in History and Philosophy of Science* 9: 51–77.

Laymon, R. 1980a. "Independent Testability: The Michelson-Morley and Kennedy-Thorndike Experiments." *Philosophy of Science* 47: 1–37.

Laymon, R. 1980b. "Idealization, Explanation, and Confirmation." In *PSA 1980* (Proceedings, Philosophy of Science Association), vol. 1, ed. P. Asquith and R. Giere. East Lansing, Mich.: Philosophy of Science Association.

Laymon, R. 1982. "Scientific Realism and the Hierarchical Counterfactual Path from Data to Theory." In *PSA 1982* (Proceedings, Philosophy of Science Association), vol. 1, ed. P. Asquith and T. Nickles. East Lansing, Mich.: Philosophy of Science Association.

Laymon, R. 1983. "Newton's Demonstration of Universal Gravitation and Philosophical Theories of Confirmation." In *Minnesota Studies in the Philosophy of Science* XI, ed. J. Earman. Minneapolis: University of Minnesota Press.

Lorentz, H. A., A. A. Michelson, D. C. Miller, R. J. Kennedy, et al. 1929. "Conference on the Michelson-Morley Experiment." *Astrophysical Journal* 68: 341–402.

Newton, Issac. 1959. *The Correspondence*, vol. I, ed. H. W. Turnbull. Cambridge University Press.

Rohrlich, F. 1965. *Classical Charged Particles: Foundations of Their Theory*. Reading, Mass.: Addison-Wesley.

Smoluchowski, M. 1906. "Zur kinetischen Theorie der Brownschen Molekularbewegung und der Suspensionen." *Annalen der Physik* 21: 756–780.

Stephenson, R. 1960. *Mechanics and Properties of Matter*. New York: Wiley.

Stokes, G. 1851. "On the Effect of the Internal Friction of Fluids on the Motion of Pendulums." *Transactions of the Cambridge Philosophical Society* 8: 8–145. Reprinted with additions in *Mathematical and Physical Papers* (Cambridge University Press, 1901).

Synge, J. L. 1960. *Relativity: The General Theory*. Amsterdam: North-Holland.

Thomson, W. (Lord Kelvin). 1901. *Baltimore Lectures on Molecular Dynamics and the Wave Theory of Light*. London: Clay.

Thomson, W., and P. Tait. 1912. *Treatise on Natural Philosophy*. Cambridge University Press. Reprinted as *Principles of Mechanics of Dynamics* (New York: Dover, 1962).

von Kluber, H. 1960. "The Determination of Einstein's Light-Deflection in the Gravitational Field of the Sun." In *Vistas in Astronomy*, vol. III, ed. Arthur Beer. London: Pergamon.

Weinberg, S. 1972. *Gravitation and Cosmology*. New York: Wiley.

Wimsatt, W. C. 1981. "Robustness, Reliability and Multiple-Determination in Science." In *Knowing and Validating in the Social Sciences: A Tribute to Donald Campbell*, ed. M. Brewer and B. Collins. San Francisco: Jossey-Bass.

7

An Interpretation of Theory and Experiment in Quantum Mechanics

Geoffrey Joseph

The two great physical theories of this century, relativity theory and quantum mechanics, have both been closely associated with philosophical ideas concerning measurement and empirical experimentation. Although there seems to be little doubt that vaguely verificationist ideas played a part in the genesis of relativity theory, there is nowadays also little doubt that such considerations are not essential to the interpretation of the theory. The most widely held interpretation of quantum theory—the Copenhagen interpretation—is, in its orthodox formulation, based on a rather explicit analogy with the early verificationist interpretations of relativity (e.g., Bohr 1951). Realists, from Einstein to the present, have rejected this analogy and have therefore sought to construct an account of this theory that is independent of epistemological concepts. I wish to suggest an interpretation of quantum mechanics that shares certain features with the Copenhagen interpretation but also shares certain features with its "realistic" competitors. I will try to show how this interpretation provides us with an alternative way of viewing both the quantum-mechanical formalism and the experimental results that constitute the empirical confirmation of the theory.

Quantum mechanics presents the puzzling phenomenon of a physical theory whose record of empirical confirmation is virtually flawless but whose theoretical framework has no clearly acceptable physical interpretation. The "problem of interpreting quantum mechanics" is actually a set of interrelated problems—some physical, some metaphysical, some epistemological. The quantum-mechanical states are formally representable by a Hilbert space of square-integrable complex-valued functions. However, the physical correlate, or referent, of the Ψ function has been argued to be, among other things, a single system, an ensemble of similarly prepared systems, or even the epistemic state of a conscious observer. The quantum-mechanical algorithm for calculating probabilities does not generate these probabilities by integration over a phase space, as do classical statistical theories. Indeed, there are well-known proofs (Cohen 1966) that the quantum probabilities cannot be retrieved by such integration over any hypothetical phase space. Yet the confirmation of the theory consists largely in the observation of predicted relative frequencies. This suggests that the quantum-mechanical assertions deal with quantities that, if not identical in all respects to classical probabilities, are nonetheless closely related to them. Finally, there are specific physical phenomena, such as the correlations between spin measurements on paired particles in the singlet state, that have indicated to many that any adequate interpretation of the theory must involve nonlocal forms of explanation.

Despite the diversity of the suggested interpretations that this theory has prompted, they can all be viewed as instances of a kind of conceptual revision describable as the physicalization of a naive, or pre-scientific, concept. A concept can be said to be physicalized when its features are given by a successful physical theory rather than by intuitions about its role in everyday discourse. Since the features present in everyday usage often reflect the essentially Newtonian character of terrestrial, macroscopic experience, the physicalization of a concept usually goes hand in hand with a significant revolution in physical theory. The special and general theories of relativity provide perhaps the clearest and least controversial examples of such physicalization. The special theory replaces the linear time order of ordinary experience with a partial order; events with spacelike separation have no invariant time order. The general theory of relativity, both in its conception of the geometry of space-time as non-Euclidean and in its conception of that geometry as a dynamically determined object, rejects crucial aspects of the Newtonian and Kantian viewpoints. The physicalization of a concept is not the formulation of an "operational definition" or "empirical explication" of it. However, an adequate account of the operations of measurement and experimentation ought to follow from such physicalization, for measurement and experimentation are physical interactions of a specific sort—namely, those involving human observers and their experimental apparatuses.

The guiding ideas of the several philosophical investigations of quantum mechanics are alternative suggestions that the classical world view be revised by the physicalization of one or another of its concepts. The research programs based on these suggestions can be seen as attempts to show that, if one is prepared to amend drastically some classical concept, one can give a coherent explanation of the many problematic aspects of quantum theory. It is futile to criticize any of these suggestions on the grounds that it is unintuitive, for they are all unintuitive. Quantum mechanics is unintuitive. Rather, the appropriate criterion of success for any "interpretation of quantum mechanics" is a systematic criterion. No interpretation that succeeds in giving a coherent explanation of the sources of the conspicuously nonclassical formal and operational features of the theory can be rejected, regardless of the extent to which it may offend our *a priori* preconceptions.

The two most widely studied examples of such approaches to the quantum theory are the Copenhagen interpretation and the quantum-logic interpretation. The Copenhagen interpretation relativizes the attribution of precise values to observables to specific, macroscopically describable, experimental contexts. Measurement, which is formally described by the projection postulate as a discontinuous reduction of the wave packet from a superposition to an eigenstate, is said to effect a transition from an indeterminate condition to the possession of one of the eigenvalues in the spectrum of the relevant operator. Measurement, in this view, does not merely reveal a pre-existent state of affairs; it brings about that state of affairs. Since simultaneous measurement of conjugate observables is not generally possible, the Copenhagen interpretation refrains from attributing joint values to such observables. For this reason it is usually described as an instrumentalist or anti-realist interpretation of the theory. In its adoption of the projection postulate, the Copenhagen interpretation physicalizes the epistemological notion of "measurement" in a very specific manner, a manner not initially forced by the quantum formalism itself. The Copenhagen interpretation's further adherence to a strictly operational interpretation of the concept of an observable possessing a definite value thus leads to a denial of simultaneous values of observables represented by noncommuting operators. It was surely this epistemological motivation that led its early proponents to reject realist interpretations of the quantum theory even before the appearance of the alleged "no hidden variables" theorems.

The quantum-logic interpretation physicalizes the propositional logic underlying the theory. By correlating the rays of Hilbert space with singular atomic propositions attributing eigenvalues to observables and by further correlating the lattice-theoretic relations between subspaces of Hilbert space with logical relations between propositions, this interpretation arrives at the radical position that the logical relations between physical states of affairs form a nondistributive lattice. Some proponents of quantum logic (Putnam 1970) have argued that the adoption of a revised logic is a way of accommodating the existence of simultaneous values for conjugate observables and hence supports a form of quantum-mechanical realism. Other proponents (Finkelstein 1973) take a more operationist attitude toward the significance of the nondistributivity. Such operationist approaches are compatible with the Copenhagen interpretation; indeed, it has been suggested (Heelan 1970) that quantum logic so viewed is essentially just a formalization of Bohr's ideas.

Although an exhaustive survey of these two interpretations is not appropriate here, it is useful to review some the problems they have encountered. Later I shall compare their answers to these difficulties with the answers suggested by the alternative approach, which will be discussed here.

The account of measurement adopted by adherents of the Copenhagen interpretation has been subjected to extensive criticism. One of the several morals to be drawn from the thought experiment of Einstein, Podolsky, and Rosen (1935) is that it is possible to measure, in Einstein's sense of "predict with certainty," the values of certain observables without physically interacting with them. By physically interacting with one member of a pair of correlated particles in the singlet spin state, one can come to know the value of an observable of the distant, unperturbed particle. Surely the projection postulate is false in this case. The unperturbed particle undergoes no change of state; its wave function does not collapse to an eigenfunction.[1] In this case, not only is the projection postulate a superfluous addition to the other postulates of quantum theory; it is also logically inconsistent with them. Even in cases in which measurement involves direct physical interaction with the object being investigated, the resultant state of the object is not always an eigenstate of the observable being measured (Sneed 1965; Teller 1981). In some cases the result of the measurement is the destruction of the object being observed—for example, when a photon is absorbed by a detector. There does not seem to be any simple and uniform relationship between the epistemological concept of measurement and any single physical concept in quantum mechanics, any more than there is in classical physics. In neither case does there appear to be a unique and uniform answer to the question "What physical state must characterize an object once a conscious subject has ascertained the value of a certain one of its observables?"

The quantum-logic program has often been criticized on the ground that a nondistributive propositional logic is, like intuitionist propositional logic, too weak to generate all the mathematics necessary for physics. The usual constructions of Hilbert space, for example, are carried out within classical logic and set theory and cannot obviously be mimicked in a weaker system. Even if one is prepared to adopt the extremity of weakening propositional logic, it is not clear that one will have made any progress toward clarifying the quantum-mechanical puzzles. Although the quantum-logic program is often presented as a realistic alternative to the Copenhagen interpretation, there are several surprising analogies between the two. Consider the quantum-logic account of the two-slit experiment. It is said (e.g., in Putnam 1970) that the conjunction "(particle p arrives at the detector screen) & (p passes through slit #1 .v p passes through slit #2)" is true but that the logically distributed proposition "(p passes through slit #1 and p arrives at the detector .v p passes through slit #2 and p arrives at the detector)" is not true. Yet when we change the experimental situation by placing a counter behind, say, slit #1, we always obtain conclusive evidence of passage through one or another of the slits. Apparently the act of measurement effects a "nonlocal collapse of the disjunction to one of its disjuncts," a process as mysterious as the "nonlocal collapse of the wave packet." Indeed, it appears that the relation between observed (measured) and unobserved systems differs little from its relation in the Copenhagen interpretation, although it is described in different terminology. It is hard to see how either of these interpretations can be termed more or less realistic than the other.

Despite the several differences between the view to be suggested here and the two research programs just described, one important point of emphasis remains the same: Each of the two programs locates the central interpretive issue in the relationship between a superposed state $\sum c_n \Psi_n$ and the (finite or infinite set of) eigenstates $\{\Psi_n\}$ of which it is composed. What physical relation one takes to correspond to this relation of mathematical composition is the key to one's interpretation of the theory. The Copenhagen interpretation finds this relation in the theory of measurement. The quantum-logic approach finds the relation in the logical structure of physical states of affairs. Let us now turn to another possible diagnosis, one that locates the relevant facts in the relation between quantum-mechanically possible ensembles of particles. We will first examine the conceptual structure of this interpretation, and then apply it to an analysis of the possible experimental data available in a quantum-mechanical world.

The viewpoint to be described here, a version of the statistical interpretation of quantum mechanics, associates the state vector not with an individual system but with an ensemble of similarly prepared systems. However, the nonstandard features of quantum probability have led most to doubt the plausibility of such an approach. In order to be viable, the statistical interpretation must, like the other interpretations, physicize one of its central concepts. Quantum mechanics is a probabilistic theory. Probability is a quantative relation between an event, or a kind of events, and a reference class. The character of single point events, microscopic or macroscopic, does not appear to be fundamentally different in quantum mechanics than in classical mechanics. Rather, it is other relata of the probability relation—the notion of "set," or "ensemble"—that must be amended to accord with a quantum mechanically acceptable viewpoint. Recognition of this fact will enable us to construct alternative explanations of certain puzzling features of the theory.

The pre-scientific concept of "aggregate" or "collection" can be made precise in several different ways. How it ought to be made precise, and how the resulting notion is to be formalized, are questions whose answers depend on the particular task at hand. To ask which is the unique correct concept is to overlook the diversity of tasks that the various notions of "aggregate" are called upon to perform in different contexts. Zermelo-Frankel (ZF) set theory, for example, formulates the concept of set appropriate to the formalization of classical (nonconstructive) mathematics. The mereological calculus of individuals formulates the concept of aggregate appropriate to a particular (nominalistic) philosophical program. The theory of recursively enumerable, or Turing-generable, sets specifies a class of sets contained in, but distinct from, the universe of Zermelo-Frankel set theory. Various physically motivated concepts of "aggregate" or "collection," such as "thermodynamically closed system" and "ensemble of particles in thermal equilibrium," are familiar from classical thermodynamics and statistical mechanics. Such concepts involve complicated physical properties and relations of the elements of such aggregates, and are not equivalent to a mere set-theoretic enumeration of these elements.

These diverse notions of aggregate do not have the same properties. The respective domains of objects of the standard models of the theories constitutive of these concepts are, therefore, not generally closed under the same operations. In ZF set theory, the power set of any set is again a set. But the aggregates satisfying the calculus of individuals obviously do not satisfy this axiom. In ZF theory, the axiom of separation (*Aussonderung*) guarantees that any subset of an existent set is again an existent set. The Turing-decidable sets, in contrast, do not obey an analog of *Aussonderung*. The set N of all natural numbers is trivially recursive. However, certain proper subsets of N are demonstrably not recursive; the set of Gödel numbers of truths of first-order arithmetic, for example, is not recursive. Because of the highly specific nature of the concept of recursiveness and the highly specific formalism that characterizes that concept, there is no simple way to characterize the class of recursive sets from the viewpoint of abstract set theory. "Which sets are recursive?" and "Which subsets of a given recursive set are recursive?" are questions that cannot be answered solely with the resources of abstract set theory.

It is possible to study the formal features of these various theories of aggregate from a philosophically neutral point of view. To do so is to view them not as competing answers to a single question but rather as answers to distinct questions. Thus, for example, the mathematical Platonist can view the recursive sets as those answering to certain decidability requirements without viewing them as an exhaustive catalog of the existent universe of abstract objects, nor does the positing of any one of these classes of aggregates commit one to a "nonstandard logic" or a "nonstandard set theory."

Let us now return to quantum mechanics and ask what concept of physical collection or aggregate is appropriate to that theory. A specific concept of this sort is firmly entrenched in its theoretical framework and in its operational procedures. The quantum-mechanical state functions describe ensembles of particles. It is the distribution of values of observables in these ensembles that is predicted by the quantum-mechanical probability algorithm, and it is the empirical observation of these distributions in suitably prepared ensembles that constitutes the empirical confirmation of the theory. In some cases such an ensemble of particles describable by a pure Ψ function can be produced by some physical process, or filtration device, which results in an eigenstate preparation for some given observable. Typical examples of this eigenstate preparation are ensembles produced by passage through a Stern-Gerlach apparatus with a fixed orientation and ensembles of particles passing through a localizable potential barrier such as a one-slit diaphragm. Ensembles described by a superposition of state vectors of a given observable are typically produced in atomic decay phenomena and in scattering cross sections obtained in collision phenomena. Whether the ensemble is described by an eigenstate or by a superposition of eigenstates, the resulting collection of particles can of viewed as the result of subjecting input particles to a uniform set of physical conditions. The successive generation of the members of such ensembles of particles is a temporally extended process. Some physically possible process, whether microscopic or macroscopic, successively produces or generates the individual systems that are elements of the quantum ensemble. The criterion for membership in a quantum ensemble of "similarly prepared systems" or "identical copies of a system" is the sameness of the physical conditions to which the system is subject. Since systems qualifying for membership in an ensemble are subject to the same force fields, they are describable by the same Hamiltonian. As the wave function evolves temporally according to Schrödinger's equation, the statistical properties of the ensemble are calculable at any instant from this evolving wave function.

In our description of these ensembles of particles we must initially maintain a certain systematic ambiguity in the interpretation of their statistical properties. This is so as not to prejudge the issue of "realism." One component of the dispute over quantum-mechanical realism is the question whether these statistics represent actual premeasurement distributions of values of observables or whether they are the result of the interaction of ensembles with detectors. Since the collection of the statistics that confirm the theory must always be mediated by such detectors, the mere fact of the theory's empirical confirmation is neutral between these two interpretations. Although some of the remarks I will make about quantum ensembles are logically compatible with either of these two interpretations of the statistics, I will also argue that some of them constitute reasons for skepticism concerning certain of the "no hidden variable" theorems, which have widely been thought to decide the question of realism in the negative.

Let us now define the notion of aggregate that reflects the features of this concept in quantum mechanics. A collection of particles characterized by the predicate Fx is a *quantum-mechanically generable ensemble* if and only if there exists some physically possible mechanism whose output over time is the set $\{x \mid Fx\}$ of all and only the particles satisfying Fx. This definition is intended to provide a physical analog to the standard notion of definability used in the comprehension axiom of formal semantics and set theory, in which a set is said to be definable if and only if its members and only its members are those satisfying a given predicate. However, the idea of generability referred to here is a distinctively physical one. An ensemble is generable, in this sense, if there is some black box, of unspecified constitution, microscopic or macroscopic, whose successive output over time is precisely the set of elements of the ensemble. In just the same sense, a set of natural numbers is recursively enumerable if and only if there exists some possible Turing machine whose output is all and only the elements of the set.

Having defined the concept of a generable ensemble, we must now ask which ensembles are in fact generable. We must find axioms that are satisfied by exactly the class of generable ensembles. We must find a theory of aggregates or collections of particles for which the class of physically generable ensembles is the standard model. On the physicalized statistical interpretation, quantum mechanics is that theory. Each pure state vector in Hilbert space corresponds to a physically generable ensemble. The statistical distribution of values of observables in each such ensemble is given by the quantum-mechanical algorithm. That algorithm yields numbers that are the nearest quantum-mechanical analog to classical probabilities. Probability, in both quantum and classical physics, is a relation between a reference class and its subclasses. The class of meaningful probability statements is restricted in the quantum case to those dealing with physically possible, and hence experimentally replicable, reference classes. These are the reference classes whose existence is physically possible, that is, the reference classes that are physically generable. Quantum-mechanical probabilities express relative frequencies present in the class of physically generable ensembles. Thus, the formal properties of quantum probability are similar to, but not the same as, those of classical probability. All quantum probabilities are relative frequencies, but not all mathematically possible relative frequencies are quantum-mechanically well-defined probabilities. Quantum probabilities are only those relative frequencies that are expressible as relationships between cardinalities of generable ensembles and certain of their subensembles. The "possibility structure" of quantum theory thus reflects the possibility of uniformly generating certain aggregates of particles. Therefore, the actual quantitative predictions of the theory are derived from the relations between state vectors in Hilbert space rather than from classical measure-theoretic relations between sets of particles and their set-theoretic subsets.

Since the distribution of values of observables in generable ensembles is given by the quantum algorithm, it is governed by the Heisenberg dispersion relation: If A and B are conjugate operators and $[A,B]$ is their commutator, then

$$(\Delta A)_\Psi (\Delta B)_\Psi \geq \frac{h}{2\pi}[A,B] .$$

On the statistical interpretation of the theory, this is the relation between the statistical spread, or dispersion, of observables in an ensemble characterized by a state vector Ψ. From this fact alone it follows that the class of physically generable ensembles has closure properties quite different from the usual closure properties of abstract set theory. Most significant for the problem of understanding the structure of quantum probability, the analog of the set-theoretic axiom of separation is not satisfied; that is, the class of generable ensembles is not closed under the subset operation. Consider a state preparation of an ensemble in an eigenstate Ψ of some observable O_1, and let $\Psi = \Sigma\, c_n \Phi_n$, where the Φ_n are the eigenstates of some observable O_2 incompatible with O_1. (For example, Ψ can be taken to be an eigenstate of spin in a fixed direction, the Φ_n to be eigenstates of spin in some distinct direction, and the state preparation to be effected by means of passage through a Stern-Gerlach apparatus.) The particles emerging from the apparatus constitute a uniformly generated ensemble: $\{x\,|\,x$ is in state Ψ; x has eigenvalue a_1 of observable O_1; x has either eigenvalue b_1, or eigenvalue b_2, or . . . of observable $O_2\}$. The set-theoretic decomposition of this collection into collections of particles each characterized by a fixed eigenvalue of O_1 and a fixed eigenvalue of O_2 yields collections which, according to abstract set theory, each exist. But none of these dispersion-free subcollections is physically generable, for the generation of any one of them would violate the statistical dispersion relation between conjugate observables. Arbitrary subsets of generable ensembles are therefore not themselves generable. The description "is characterized by state vector Ψ" is therefore predicable of an entire generable ensemble. It is meaningless to apply it to the individual elements of the ensemble, and it is false to apply it to the set-theoretic subsets of the ensemble.

Let us apply this analysis to the notorious two-slit or "particle interference" experiment. Consider three experimental arrangements, as illustrated in figure 1. The one-slit diaphragm in part a of the figure generates the ensemble $\{x\,|\,x$ is in position eigenstate Ψ_A and lands on detector screen$\}$; that in part b generates the ensemble $\{x\,|\,x$ is in position eigenstate Ψ_B and lands on detector screen$\}$; that in part c generates

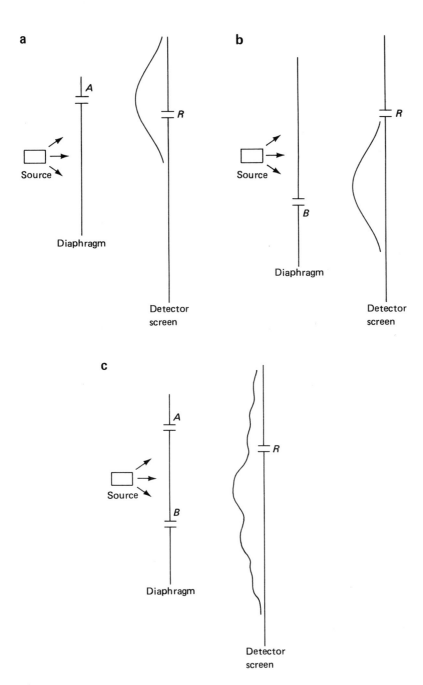

Figure 1

the ensemble $\{x \mid x$ is in superposition $\Psi_A + \Psi_B$ and x lands on detector screen$\}$. Calculation of the temporal evolution of the state vector for each of these ensembles yields the distribution of positions shown on the respective barrier detectors. The first two of these distributions are referred to as evidence of "particle diffraction," and the third as evidence of "particle interference." The appearance of the interference pattern in c is often taken to indicate a causal anomaly, or, worse, to indicate the necessity of a "wave-particle" duality according to which particles are said to be both nonlocalizable and localizable. The alleged calculation that is supposed to lead one to expect in c the sum of the two one-slit patterns instead of the actually observed interference pattern is as follows:

Let $\text{prob}_{\Psi_A + \Psi_B}(R)$ be the probability, relative to superposed state $\Psi_A + \Psi_B$, that a particle lands in region R on the detector screen. Let $\text{prob}_{\Psi_A + \Psi_B}(R \mid A)$ be the probability, relative to $\Psi_A + \Psi_B$, that a particle lands in region R, given that it has passed through slit A. Let $\text{prob}_{\Psi_A + \Psi_B}(R \mid B)$ be the probability, relative to $\Psi_A + \Psi_B$, that a particle lands in region R, given that it has passed through slit B. Let $\text{prob}_{\Psi_A}(R \mid A)$ be the probability, relative to eigenstate Ψ_A, that a particle lands in region R, given that it has passed through slit A. Let $\text{prob}_{\Psi_B}(R \mid B)$ be the probability, relative to eigenstate Ψ_B, that a particle lands in region R, given that it has passed through slit B. Then

$$\text{prob}_{\Psi_A + \Psi_B} R = \text{prob}_{\Psi_A + \Psi_B}(R \mid A) + \text{prob}_{\Psi_A + \Psi_B}(R \mid B)$$
$$= \text{prob}_{\Psi_A}(R \mid A) + \text{prob}_{\Psi_B}(R \mid B).$$

But if probabilities are to be calculated only with respect to generable ensembles, then this calculation is fallacious. Neither $\{x \mid x$ is in $\Psi_A + \Psi_B$ and x passes through slit $A\}$ nor $\{x \mid x$ is in $\Psi_A + \Psi_B$ and x passes through slit $B\}$ is a generable ensemble. Therefore, we cannot meaningfully ask what fraction of a physically possible ensemble meeting either of these descriptions leads to a certain outcome. Quantum mechanics respects this condition, for the two conditional probabilities $\text{prob}_{\Psi_A + \Psi_B}(R \mid A)$ and $\text{prob}_{\Psi_A + \Psi_B}(R \mid B)$ are not defined in the quantum formalism. Since these probabilities are not defined at all, they are *a fortiori* not defined as $\text{prob}_{\Psi_A}(R \mid A)$ and $\text{prob}_{\Psi_B}(R \mid B)$. The use of classical probability theory contains suppressed factual presuppositions about the existence of certain reference classes, presuppositions that are not true in the quantum domain. A particle that is an element of the generable ensemble $\{x \mid x$ is in $\Psi_A + \Psi_B$ and x passes through slit A or slit

B} is not an element of either of the putative ensembles $\{x \mid x$ is in $\Psi_A + \Psi_B$ and x passes through slit $A\}$ or $\{x \mid x$ is in $\Psi_A + \Psi_B$ and x passes through slit $B\}$. The demand that experimental arrangement c in figure 1 should yield the sum of the two one-slit patterns is in effect a demand that a well-defined quantum-mechanical probability be expressed as the sum of two numbers, neither of which is an admissible quantum-mechanical probability. This demand is not reasonable, and its failure to be met does not constitute any kind of anomaly. Nor does this experiment, or any similar experiments relying essentially on the difference between superpositions and mixtures, give any immediate reason for adopting the metaphysics of the "wave-particle duality" or the semantical dualism of "complementarity." This description of the experiment is compatible with the hypothesis of exclusive passage by each particle through one or another of the slits. It is also compatible with the equality between the number of particles in the ensemble described by the superposition that land in region R with the sum of the number of particles that land in region R and pass through slit A plus the number of particles that land in region R and pass through slit B. But this equality does not provide grounds for the expectation that a relative frequency with respect to a generable ensemble should be equal to the sum of relative frequencies with respect to two hypothetical ensembles neither of which is generable, nor does it provide grounds for the expectation that such a frequency should equal the sum of relative frequencies with respect to two generable ensembles (the two one-slit arrangements) the physical conditions necessary for whose generation are each incompatible with those necessary for the generation of the superposed ensemble.

Let us return now to a needed qualification in the concept of a quantum-mechanically generable ensemble. Our discussion has up to this point been "Platonistic" in the sense that it refers to ensembles that are potentially infinite in size and that therefore are not experimentally realizable. How do these considerations bear on the finite ensembles actually produced in experimental situations?

In claiming that dispersion-free ensembles are not physically possible in the sense that they cannot be uniformly produced by a physically possible device, we must understand this as an assertion about a limiting condition that obtains as the size of the ensemble becomes infinite. But this same proviso holds for any of the statistical assertions of the theory. Finite dispersion-free ensembles have a small probability of being generated in a given experimental arrangement and therefore must be considered physically possible. Consider, for example, a beam of particles prepared in an eigenstate of spin in some given direction e and hence in a superposition of spin states in some orthogonal direction d. If the beam is then passed through a Stern-Gerlach filter, which separates it into two beams corresponding to the two possible eigenvalues of spin in the direction d, there is a nonzero probability that, if the initial ensemble contains N particles, all N will be detected in a single one of the split beams. This probability decreases as N increases, and approaches zero as N approaches infinity. Therefore, the classification of ensembles as physically generable or not must refer to those that have a nonzero probability of being generated as N approaches infinity. This same qualification applies to all the statistical assertions of quantum mechanics. In any particular run of an experiment there remains a small finite probability that the results will display no dispersion. There is a small finite probability, for example, that in a particular run of the particle-diffraction or particle-interference experiments all the particles will be detected at a single point on the barrier screen. The occasional appearance of such an experiment is not taken as disconfirmation of the theory, however, for the theory predicts only that such a result will become increasingly unlikely as the size of the ensemble increases.

Although one must not confuse physical questions concerning the interpretation of quantum mechanics with epistemological questions, a connection arises in the present context. Since we have granted a small probability to the existence of finite dispersion-free ensembles, it might appear that we must address the question of the well-definedness of probability distributions with respect to such ensembles. For example, if in the two-slit experiment one recognizes the possibility that a 1,000-member dispersion-free ensemble $\{x \mid x$ is in $\Psi_A + \Psi_B$ and x passes through slit $A\}$ is generated, it appears that one must address the question of the distribution of particles detected on the barrier screen in this case. There is, however, a motivation for not considering this question well defined, for it is physically impossible to answer it. The only way to know whether this finite ensemble has actually been generated in a particular run of the experiment is to place a detector behind slit A. By doing so, one changes the state of the resulting ensemble, and thus one obtains a distribution on the barrier screen different from the one that would have been obtained had one not verified that the dispersion-free ensemble had been generated.

If there is an answer to the question of what distribution results from the dispersion-free ensemble, it is physically inaccessible. However, one need not revert to verificationism to see that it is an illusion to consider the original question well defined. In order that the question of the probability distribution relative to the finite dispersion-free ensemble be meaningful, it is not sufficient that this ensemble occasionally be generated. The resulting distribution on the barrier screen on any single instance of the generation of such an ensemble may be far different from the alleged true distribution. In order that the distribution be well defined, one must be able to generate a (potentially) infinite series of replications of this ensemble and find the distribution that is the limit of the observed data. But this hypothesis has zero probability of being fulfilled. That is, the probability of generating an infinite number of ensembles, each of them dispersion-free and each having only a very small probability of being generated, approaches zero as the number of such ensembles increases. Thus, to expect the well-definedness of the original question is to make a factual presupposition that is quantum-mechanically impossible to satisfy. The initial question, therefore, has no answer. Quantum mechanics again respects this fact, for it yields no well-defined answer to the question. What might have appeared to be an epistemologically inaccessible physical fact is not a physical fact at all.

As noted above, the central problem addressed by both the Copenhagen interpretation and the quantum-logic interpretation is to describe the physical correlate of the mathematical relationship between a superposed state $\sum c_n \Psi_n$ and the eigenvectors $\{\Psi_n\}$. Some reasons for doubt concerning the customary association of this correlate with the measurement operation were mentioned. It is also important to mention in this context a principle similar to the projection postulate, one that Fine (1973) has termed the "eigenvalue-eigenstate link." This alleged principle states that if a system is characterized by a specific eigenvalue of an observable, then it is in the corresponding eigenstate of that observable. Fine points out that, although the converse of this statement is indeed part of the quantum theory and is as well confirmed as the rest of the theory, the eigenvalue-eigenstate link is not part of the theory and has been thought to be plausible only because of the belief in the "collapse of the wave packet." The quantum formalism itself does not warrant any inferences from the value of an observable to the quantum state of the system involved. Neither the projection postulate nor the eigenvalue-eigenstate link need be adopted under the ensemble interpretation discussed here. In fact, it is meaningless under this interpretation to attribute a state vector to a particle *simpliciter*. Here the assignment of state vectors to particles is conditional on their being elements of a generable ensemble characterized by that state vector.

On the view advocated here, the transition from superposition to eigenstate does not concern the properties of a single particle, nor is it essentially connected with a measurement operation. It concerns instead the description within the quantum formalism of a certain set of physical procedures: the filtration procedures by which a generable subensembles characterized by an eigenvector of an observable are extracted from ensembles characterized by a superposition of such eigenvectors. Thus, the transition from superposition to eigenstate is the mathematical description of the physical operation that is the closest physical analog to the operation of subset formation familiar from abstract set theory. A subensemble of a given ensemble is a generable subensemble if and only if there exists a physically possible filtration device that, when applied to the initial ensemble, emits all and only the members of that subensemble. The generable subensemble E of a given ensemble characterized by the fact that all of its members have the relevant eigenvalue a of a given observable O_1 is an ensemble characterized by the eigenvector Ψ_a. This fact follows directly from quantum theory itself. It is a consequence of the quantum formalism that the eigenvector Ψ_a of the observable O_1 is the unique state relative to which the probability of O_1 taking on the eigenvalue a is 1. Hence, the generable ensemble characterized by this eigenvector is the unique generable ensemble all of whose elements are characterized by this eigenvalue of O_1.

In the Copenhagen interpretation the transition from superposition to eigenstate was associated with the measurement process. The non-commutativity of operators representing conjugate obervables, $AB\Psi \neq BA\Psi$, was therefore associated with the doctrine of the non-simultaneous measurability of these observables. On the ensemble view, one can state the significance of this noncommutativity without recourse to the epistemological concept of measurement. On the Copenhagen interpretation, the transition was associated with a certain physical operation: a measurement. On the ensemble view, the transition is also associated with a physical operation: a filtration procedure by which a subensemble characterized by a reduced state vector is extracted from a given ensemble characterized by a superposition.

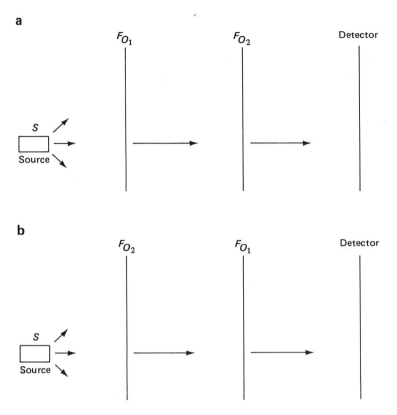

Figure 2

Consider two physically distinct possible arrangements of filtration devices (Stern-Gerlach magnets, diaphragms with slits, etc.), as illustrated in figure 2. S is the source of a generable ensemble in state Ψ; O_1 and O_2 are conjugate observables represented respectively by noncommuting operators A and B; F_{O_1} is a filter that generates an O_1 eigenensemble. The first arrangement involves two successive processes of subensemble formation. The first filter F_{O_1} produces an ensemble characterized by the O_1 eigenvector Ψ_a; the second selects from this an ensemble characterized by the O_2 eigenvector $\Psi_{a'}$. The second arrangement of filters results in an ensemble characterized by the O_1 eigenvector Ψ_a. The two ensembles resulting from these two arrangements of filters are distinct physical aggregates with distinct statistical properties: the ensemble characterized by Ψ_a exhibits dispersion in O_2 and no dispersion in O_1; the ensemble characterized by $\Psi_{a'}$ exhibits dispersion in O_1 and no dispersion in O_2. This ensemble interpretation of the noncommutativity of operators entails nothing about the possibility or impossibility of simultaneous measurements of conjugate observables for a single particle. Because of the change of state upon eigenensemble formation at F_{O_1} and again at F_{O_2}, these respective arrangements of filters are not suitable for measuring the values of observables of any single particle. Since each of the two filters perturbs the values of the conjugate observables, successful passage by a particle through the sequence of filters does not justify the retrodictive inference that it has the same values in the final resultant subensemble as it did in the initial ensemble, nor is there any requirement that this sequence of devices be linked to a macroscopic instrument suitable for inspection by a human observer. Unlike the dubious "disturbance theory of measurement," this "disturbance theory of eigenensemble formation" follows directly from the dispersion relations deducible in the quantum formalism. This interpretation is also neutral on the question of whether a particle can simultaneously be assigned precise values for conjugate observables. One may, after consideration of the several "no hidden variables" theorems, conclude that such simultaneous assignments cannot be made. But it is surely too hasty to draw such a conclusion solely on the basis of the noncommutativity of operators.

Consider now another prominent mathematical fact about the formalism of quantum mechanics: The subspaces of Hilbert space—or, isomorphically, the projection operators each having such a subspace as its range—form a nondistributive lattice. On the Copenhagen interpretation, this nondistributivity is taken as a reflection of the non-Boolean character of the set of possible measurements, or projections onto eigenstates. The quantum-logic interpretation gives this mathematical fact an even more radical meaning: Since the rays of Hilbert space are identified with singular propositions each assigning an eigenvalue to a system, the nondistributivity of the lattice of states is taken to indicate that the laws of classical propositional logic are not satisfied in the quantum domain. Under the ensemble interpretation, the nondistributivity of the lattice reflects a rather straightforward physical fact about the impossibility of the decomposition of generable ensembles. Prepare an ensemble in a state $\Psi_{a'}$ which is the eigenstate corresponding to the eigenvalue a of observable O_1. Let O_2 be a conjugate observable having eigenstates Φ_n with corresponding eigenvalues

$b_1, b_2, \ldots, b_n$. Express $\Psi_a = \sum m_i \Phi_i$. Then the following expression characterizes the prepared ensemble:

$\{x \mid (x$ is in $\Psi_a)$ & $(x$ has eigenvalue a of O_1 & $(x$ is in $\sum m_1\Phi_i)$

    & $(x$ has eigenvalue b_1 of O_2 or x has eigenvalue b_2

    of O_2 or $\ldots$ or x has eigenvalue b_n of $O_2)\}$.

Consider the "distributed" terms $\{x \mid x$ is in Ψ_a .&. x has eigenvalue a of O_1 and x has eigenvalue b_1 of $O_1\}$, etc. The putative ensembles corresponding to each of these terms are not uniformly generable. Although the initial ensemble preparation produces particles having eigenvalue a of observable O_1 and having one of the eigenvalues of O_2, the uniform generation of the initial ensemble is not a generation of any of the "distributed" dispersion-free subensembles. As the number of particles in the ensemble increases, the probability that any one of these subensembles will be generated approaches zero. The nondistributivity of the lattice of states is a reflection of the fact that a uniform generation of an ensemble is a physical process not equivalent to a "conjunction" or "union" of physical processes that constitute uniform generations of its set-theoretical subsets. Because of the dispersion relations governing any ensemble generation, the physical conditions necessary for the generation of such an ensemble are incompatible with those necessary for the generation of the subensembles. Such an interpretation allows one to retain the customary Boolean properties for propositional logic and abstract set theory. In relating the nondistributivity of the lattice of states to empirically ascertainable statistical facts, the ensemble interpretation is in closer accord with the actual application of quantum theory than are some of the more radical interpretations.

Let us now consider another nonclassical feature of the quantum formalism and its explanation according to the physical ensemble interpretation. In general, joint probability distributions are not well defined in the quantum formalism. Consider a state $\Psi = \sum c_i\Psi_i = \sum d_j\Phi_j$, which is expressible as a superposition of eigenstates $\{\Psi_i\}$ of some observable O_1 and is also expressible as a superposition of eigenstates $\{\Phi_j\}$ of some conjugate observable O_2. Let $\text{prob}_\Psi(v(O) = x_i)$ be the probability, relative to state Ψ, that observable O takes on eigenvalue x_i. Then each of the probabilities $\text{prob}_\Psi(v(O_1) = x_i)$ and $\text{prob}_\Psi(v(O_2) = y_j)$ is well defined in the theory and equals the length of the projection of Ψ onto the relevant eigenvector, $|\langle\Psi,\Psi_i\rangle|^2$ and $|\langle\Psi,\Phi_j\rangle|^2$. But none of the joint probabilities $\text{prob}_\Psi(v(O_1) = x_i$ and $v(O_2) = y_j)$ is defined in the quantum formalism.

Consequently, none of the quantities $\mathrm{prob}_\Psi(v(O_1) = x_i)$ can be expressed as the sum of the usual marginal probabilities

$$\mathrm{prob}_\Psi(v(O_1)) = x_i = \sum_j \mathrm{prob}_\Psi(v(O_1) = x_i \mid v(O_2) = y_j).$$

It is instructive to recall the features of the classical viewpoint that lead one to expect that joint probabilities and conditional probabilities ought to be well defined. The classical probability calculus is an application of measure theory to classical set theory. The existence of joint probability distributions of functions of random variables is guaranteed by the existence of Cartesian products of those functions and of well-defined measures, such as cardinality or integral measures, on arbitrary subsets of the product space. The seemingly *a priori* status of the law of marginal probability can be traced to its grounding in the set-theoretic axiom of separation:

A = set of events each of kind α,

B_i = set of events each of kind α and each of kind β_i for some i,

$A = \bigcup_i B_i$.

If every event of kind α is both an event of kind α and an event of kind β_i, and if $A = \bigcup_i B_i$, then the measure of A is the sum (or integral) of the measures of the B_i. But none of these classical presuppositions is true of generable ensembles. If the ensembles $\{z \mid v(O_1) = x_i\}$ are each generable subensembles of an initial ensemble characterized by the superposition $\Psi = \sum_n c_n \Psi_n = \sum_j d_j \Phi_j$, then the dispersion-free ensemble $\{z \mid v(O_1) = x_i \text{ and } v(O_2) = y_j\}$ is not a generable ensemble. Similarly, as noted above, the probability of generating the set A of systems of kind α is not equal to the sum of the probabilities of generating the putative subensembles corresponding to the set-theoretic decompositions of A. If, as suggested above, quantum probability theory has as its admissible reference classes all and only the physically generable ensembles, then it follows that certain classically well-defined probabilities have no analogs in this theory. A physically reasonable theory of probability should provide the mathematical means for dealing with statistical distributions resulting from any physically replicable initial conditions. The quantum-mechanical dispersion relations governing the physical producibility of ensembles allow for meaningful probabilistic assertions with respect to certain ensembles but not with respect to their set-theoretic subsets. There is no reason to expect that one should be able to interpolate probabilities into quantum theory so as to recover the entire structure of classical probability theory when quantum mechanics itself denies the factual presuppositions necessary for the well-definedness of these probabilities. As in the other cases we have considered, restricting the class of admissible aggregates to the generable ensembles provides an explanation of the ill-definedness in the quantum formalism of joint probability distributions and of conditional probabilities. This explanation is independent of assumptions about measurability and neutral toward the question of the existence of simultaneous precise values for conjugate observables of individual systems.

There is one sense in which the interpretation of the quantum-theoretic formalism sketched here is closer in spirit to the Copenhagen interpretation than it is to those interpretations that attempt to find a neoclassical phase-space interpretation of that formalism. If one separates from the Copenhagen interpretation its Kantian overtones, its overly hasty skepticism about the existence of a microworld, and its dubious operational theory of meaning, one is left with a very suggestive claim concerning physical modality. The core insight of that interpretation is that noncommuting projection operators do not represent "simultaneously real" entities; they represent instead the results of physically possible operations, only one of which can be made actual in a given experimental situation. But the Copenhagen interpretation goes awry in epistemologizing this modal structure and identifying these physical operations with macroscopically describable measurements. The approach described here retains the modal constructivism of the Copenhagen interpretation but attempts to implement it within a statistical interpretation of the theory that is, at least initially, compatible with microrealism.[2] The noncommuting projection operators each correspond to a uniformly generable subensemble of a given ensemble. Because of the dispersion relations governing the generation of ensembles, at most one of these subensembles can be actualized; the actualization of one physically precludes the actualization of any other. The assignment of an eigenvector to a particle is, on the Copenhagen interpretation, warranted by the mere detection of the relevant eigenvalue; this "reduction of the wave packet" is inconsistent with the rest of the theory and must therefore be rejected. Eigenvectors are assigned more sparingly under the proposed ensemble interpretation. A particle is describable by an eigenvector only when it is an element of an ensemble relative to which the detection of the relevant eigenvalue has probability 1. The physical conditions necessary for the realization of such eigenensembles corresponding to conjugate variables are alternative, but not simultaneously realizable, possibilities. Which of these conditions is actualized is, as Bohr would have said, dependent on a choice made by the experimenter. As a consequence, both the ensemble interpretation and the Copenhagen interpretation share with all other modal interpretations a denial of the possibility of a neoclassical phase-space reconstruction of the theory. In order for such a phase-space reconstruction of quantum mechanics to be possible, it would be necessary to interpret the theory in such a way that noncommuting observables would represent elements of the physical world that would simultaneously possess actual, and not merely possible, reality. Although the ensemble intepretation and the Copenhagen interpretation differ on their diagnosis of what the appropriate "elements of reality" are, they concur in reading quantum theory as a set of assertions about which of these elements are co-realizable.

Much of the philosophical study of quantum theory has centered on the question of hidden variables. Although the intensive analysis of this question has led to some of the deepest theorems concerning the quantum-mechanical formalism, it may have diverted attention from certain other important issues, such as the interpretation of the Ψ function and of the transition from superposition to eigenstate. The construction of such interpretations, of which the present ensemble interpretation is no doubt only one among many, will be necessary before an adequate philosophical understanding of quantum theory can be achieved. A brief examination of the hidden-variable dispute from the viewpoint of the ensemble interpretation is, however, appropriate at this time.

The several "no hidden variables" theorems all have the following logical structure: A complicated deduction having many logically independent premises is shown to lead to quantitative relations between observables that contradict the predictions of quantum mechanics. Among these logically independent premises have been the assumption of the simultaneous existence of precise values for noncommuting observables, the soundness of classical logic and mathematics, the soundness of classical probability theory, and various locality conditions. Although these theorems have been motivated by the question of hidden variables, the multiplicity of their premises makes it possible to reach divergent diagnoses as to the specific presupposition(s) that in fact conflict with quantum mechanics. This situation is similar to that surrounding the discussions of the paradoxes of classical set theory. It is clear that the conjunction of several premises leads to contradiction, but which premise is the erroneous one (use of classical logic, use of impredicative definitions, etc.) is open to debate. The suggestion to revise the naive notion of ensemble and to restrict the admissible reference classes for quantum probabilities to the physically generable ensembles allows one to view the "no hidden variables" theorems is still another way: They need not be taken as evidence against the simultaneous existence of precise values for all observables of individual systems; they can instead be taken as evidence against the calculation of probabilities with respect to dispersion-free ensembles.

Some of the "no hidden variables" theorems are explicit proofs that the expectation values of observables calculated with respect to any hypothetical dispersion-free states must conflict with the expectation values predicted by quantum theory. Von Neumann's (1955) theorem deduces a contradiction from the existence of what he calls "homogeneous (dispersion-free) ensembles." Bell (1966) proves a similar result but notes that a suppressed and not at all obvious premise of such proofs is the assumption that the operators yielding expectation values for dispersion-free states must display the same linearity as the standard quantum operators. In a study that is generally critical of the "no hidden variables" theorems, Fine and Teller (1974) claim in passing that the existence of dispersion-free states is equivalent to the existence of a valuation function that assigns a precise value to every observable. Fine and Teller claim that certain contradictions with quantum mechanics can be attributed to the use of the sum rule. However, inspection of their theorems shows that they apply this rule to hypothetical dispersion-free ensembles, in violation of quantum-mechanical practice.

Gleason (1957) demonstrates that there are no dispersion-free probability measures definable on the closed subspaces of a n-dimensional Hilbert space ($n > 2$), and furthermore that all measures that are so definable can be expressed by a quantum-mechanical density operator. However, one can interpret these theorems not as results about the definiteness of values of observables of individual systems characterized by state vectors, and hence of individual systems that are elements of ensembles characterized by those state vectors. One can instead isolate (and question) the premise of these arguments, which is to treat as equivalent the assumption of the existence of precise values and the assumption that expectation values calculated with respect to hypothetical dispersion-free states (ensembles) must agree with those predicted by quantum mechanics. Bub (1974) has suggested that Gleason's theorem be interpreted as a kind of completeness result that characterizes all the mathematically possible probability measures definable on the closed subspaces of Hilbert space and shows that the quantum-mechanical density operator formalism is capable of expressing all of them. But it is a more conservative reaction to this theorem to conclude that it is the class of quantum states that is the complete class of physically possible states, and hence that it is the class of (dispersive) quantum ensembles that is the complete class of physically possible ensembles. There are no states, or corresponding ensembles, with the property that, for all observables, the probability of measuring a given value of each observable is 1. But this finding is compatible with the assumption that each generable ensemble characterized by a Ψ function consists entirely of systems each characterized by a precise value for each observable. To interpret these theorems as showing limitations on the class of ensembles rather than on values of observables of the elements of those ensembles is more conservative than the standard "no precise values" interpretation for a simple reason: Measurement always yields precise values for individual observables, but measurements of statistical distributions of such values always reveal dispersive ensembles. There is therefore some plausibility in interpreting these theorems as showing limitations on the class of ensembles rather than on the values of observables of the members of these ensembles.

The Bell-Wigner theorem is directed against local realism, that is, against the conjunction of the proposition that quantum observables all have precise values and the proposition that these values do not depend on events occurring at a spacelike separation from the measurement operation. There are by now many Bell-type theorems and many experiments designed to compare their predictions with those of quantum theory (Clauser and Shimony 1978). In addition, it has become clear that one can formulate several logically distinct "locality" conditions, some of which are clearly inconsistent with the special theory of relativity and some of which are not (Jarret 1981). No single discussion can deal with all the issues surrounding these theorems. The following remarks are intended only as an analysis of the initial derivation of the Bell inequality from the viewpoint of the ensemble interpretation of the theory.

Consider Wigner's (1970) version of the derivation, as simplified by Bub (1974): A source emits pairs of particles in the singlet spin state. Each particle enters one of two adjustable Stern-Gerlach filters, each of which can be set to measure the component of spin in any spatial direction. Let a, b, and c be three spatial directions, and define the sextuples $(a^1_\pm\ b^1_\pm\ c^1_\pm\ a^2_\pm\ b^2_\pm\ c^2_\pm)$ as the number of pairs of particles in the given ensemble such that the component of spin of particle 1 in directions a, b, and c are $+$ $(-)$ and the components of spin of particle 2 in directions a, b, and c are respectively $+$ $(-)$. Thus, each sextuple represents the number pairs of particles in the ensemble that satisfy a conjunction of six atomic propositions. Because of the anticorrelation of spins in any one direction,

$$(a^1_i\ b^1_j\ c^1_k\ a^2_i\ b^2_m\ c^2_n) = O \quad \text{if } i = l \text{ or } j = m \text{ or } k = n.$$

Thus,

$$\text{prob}(a^1 = a^1_+ \ \& \ c^2 = c^2_+) = \sum_{jklm}(a^1_+\ b^1_j\ c^1_k\ a^2_i\ b^2_m\ c^2_+)$$

$$= (a^1_+\ b^1_+\ c^1_-\ a^2_-\ b^2_-\ c^2_+) + (a^1_+\ b^1_-\ c^1_-\ a^2_-\ b^2_+\ c^2_+)$$

$$= w + z,$$

$$\text{prob}(b^1 = b^1_+ \ \& \ c^2 = c^2_+) = (a^1_+\ b^1_+\ c^1_-\ a^2_-\ b^2_-\ c^2_+) + (a^1_-\ b^1_+\ c^1_-\ a^2_+\ b^2_-\ c^2_+)$$

$$= w + x,$$

$$\text{prob}(a^1 = a^1_+ \ \& \ b^2 = b^2_+) = (a^1_+\ b^2_-\ c^1_+\ a^2_-\ b^2_+\ c^2_-) + (a^1_+\ b^1_-\ c^1_-\ a^2_-\ b^2_+\ c^2_+)$$

$$= y + z.$$

Then, since x, y, z, and w are positive real numbers,

$$w + z \leq w + x + y + z.$$

Thus,

$$\text{prob}(a^1 = a^1_+ \ \& \ c^2 = c^2_+)$$

$$\leq \text{prob}(b^1 = b^1_+ \ \& \ c^2 = c^2_+) + \text{prob}(a^1 = a^1_+ \ \& \ b^2 = b^2_+).$$

But this inequality is not satisfied by quantum mechanics. The quantum-mechanical expressions for these probabilities are, respectively, $1/2 \sin^2 1/2\ \theta ac$, $1/2 \sin^2 1/2\ \theta bc$, and $1/2 \sin^2 1/2\ \theta ab$, and the inequality is violated when, for example, b bisects the angle between a and c.

One straightforward deduction from the conflict between Bell's inequality, and the quantum predictions seems unavoidable: Given the assumption that the three probabilities $1/2 \sin^2 1/2\ \theta ac$, $1/2 \sin^2 1/2\ \theta bc$, $1/2 \sin^2 1/2\ \theta ab$ each represent relative frequencies characterizing the initial ensemble of pairs of particles, independent of and temporally prior to any interaction with the Stern-Gerlach apparatuses, all that is required to reach a contradiction are unassailable principles of finite counting such as could be expressed in a Venn diagram. If this assumption is part of what is meant by "quantum-mechanical realism," then the conflict between Bell's inequality and the empirically observed quantum-mechanical predictions is a sufficient reason to reject this sort of "realism" (Healey 1979). Apparently these probabilities must be understood as the result of the interaction between the elements of the initial ensemble and the various settings of the apparatus. Once this concession is made, however, it is far from clear that the failure to obey Bell's inequality is evidence for nonlocality. It is necessary to examine the possibility that we can understand the probabilities as the result of local stochastic interaction between the Stern-Gerlach apparatuses and the individual elements of the pairs of particles. We need to ask whether the derivation of Bell-type inequalities can be justified in this case. Of course, there exist alleged demonstrations that such inequalities are derivable in the stochastic as well as in the deterministic case (Clauser and Shimony 1978), but these theorems all assume the applicability of the laws of classical probability theory, including the classical law of marginal probability. Such laws, as we have seen, are not justifiable under the ensemble interpretation.

Under the ensemble interpretation, the three quantum probabilities $1/2 \sin^2 1/2\, \theta ac$, $1/2 \sin^2 1/2\, \theta bc$, and $1/2 \sin^2 1/2\, \theta ab$ represent the relative cardinalities of three physically distinct subensembles of the initial ensemble of pairs. The physical condition necessary for the production of any one of these subensembles (that is, the choice of an orientation for the Stern-Gerlach filters) is incompatible with the conditions necessary for the production of the other two. Since the operators corresponding to these subensembles are pairwise incompatible, no single generable ensemble can simultaneously realize all three of these relative frequencies. On the ensemble interpretation, each of the three quantum probabilities is the relative cardinality (relative, that is, to the cardinality of the original ensemble) of one of three physically distinct generated ensembles, each of which is described in the product state space by the appropriate tensor product of eigenvectors. The crucial step in the derivation of Bell's inequality was an equation of conditional probabilities

$$(a_+^1 \ \& \ c_+^2)\,|\,b_+^1 \ = \ (b_+^1 \ \& \ c_+^2)\,|\,a_+^1$$

based on the equation of the cardinalities of subsets of the initial ensemble. On the phase-space analysis, this equation was justified by the fact that each of the two expressions referred to the same subset of the initial ensemble. That is, each of these expressions was satisfied by the same set of pairs of the initial ensemble. There is no comparable justification for equating these expressions once one takes the three quantum probabilities to refer to three distinct ensembles. There is no reason to expect that the number of pairs having b_+^1 within the ensemble characterized by the tensor product of eigenstates $\Psi \otimes \Psi$ is equal to the number of pairs having a_+^1 within the ensemble characterized by $\Psi \otimes \Psi$. Nor can one equate these two numbers by interpreting them as the number of pairs having these values before they interact with the Stern-Gerlach apparatuses, for on the dispositional interpretation of these probabilities (according to which they represent the results of interaction of the pairs with the apparatuses) the numbers before and after such interaction need not be the same.

There is a particular approach to the values of observables of individual systems that is consistent with our unwillingness to equate the cardinalities of these set-theoretic subsets of distinct ensembles. This is the stochastic hidden-variable program. Here the various results of measurements of components of spin are stochastically jointly determined by the quantum state of a particle and a particular magnetic field in a Stern-Gerlach apparatus. Each of the quantum probabilities expresses the result of allowing an ensemble to interact with one pair of such magnets, thus producing a subensemble characterized by a product of eigenvectors. To compare the cardinality of subsets in distinct subensembles is to assume that certain counterfactual propositions have determinate truth values. That is, one might attempt to justify the subdivision of the initial ensemble and the equation of the cardinalities of its subsets by claiming that a given pair (x,y) of particles that is an actual element of the subensemble characterized by $\Psi \otimes \Psi$ satisfies exactly one of the counterfactual predicates "(x,y) is a pair of particles that would have assumed values $(b_+^1 \ \& \ c_+^2)$ if it had been passed through magnets oriented in directions b and c" and "(x,y) is a pair of particles that would have assumed values $(b_-^1 \ \& \ c_+^2)$ if it had been passed through magnets oriented in directions b and c." It seems reasonable to assume that (x,y) satisfies the *disjunction* of these two counterfactual predicates, since they exhaust the possibilities. However, part of the meaning of the claim that passage through any of the magnets is only stochastically and not deterministically fixed by the quantum state of the pair appears to be the denial that there is any physical property of the pair that could serve to make one of these disjuncts true rather than the other. Once (x,y) is passed through the Stern-Gerlach magnets oriented in a particular direction and assumes the eigenvalues (a^1_+ & c^2_+) and therefore becomes an element of the ensemble characterized by $\Psi \otimes \Psi$, it becomes physically impossible to allow its initial (singlet) state to react stochastically with any other possible orientation of the magnets. In contrast with the classical case, in which one disposition of an object can be actualized without (in principle) perturbing the object's probability of realizing some distinct disposition, the physical conditions necessary for the realization of such a quantum disposition are incompatible with those conditions necessary for the realization of conjugate quantum dispositions. That a pair of particles that has actualized one such disposition satisfies a disjunction of counterfactually defined properties but neither of the counterfactual disjuncts can be considered a kind of nondistributivity governing individual systems. This nondistributivity does not arise at the level of extensional propositional logic, nor is it essentially connected with the measurement process.

It would be an overstatement to claim that the ensemble interpretation succeeds where other interpretations fail in completely removing the mysterious aspects of the spin correlations of paired particles in quantum mechanics. It seems fair to say that we still lack adequate conceptual tools to deal with the rather baffling fact that the predicted and observed probabilities of responses at each wing of the apparatus are the same in any chosen orientation, while at the same time strict correlations appear between the responses at distant wings. How to understand this nonclassical combination of determinism and indeterminism is currently perhaps the most pressing issue raised by quantum theory. I hope, in future work, to examine the question in greater depth from the viewpoint of the ensemble interpretation.

Acknowledgment

This work was supported by National Science Foundation Grant SES-8007236.

Notes

1. Before the measurement, the pair of correlated particles is in the two-body state $1/\sqrt{2}[(\Psi_+ \oplus \Psi_-) - (\Phi_- \oplus \Psi_+)]$. The state of each particle, obtained by projecting onto the respective component Hilbert spaces, is a mixture of eigenstates. If particle 2 is perturbed by the measurement, particle 1 remains in the mixed state (Jauch 1968).

2. For yet another discussion of the modal structure of the theory, see van Fraassen 1972.

References

Bell, J. S. 1966. "On the Problem of Hidden Variables in Quantum Mechanics." *Reviews of Modern Physics* 38: 447–452.

Bohr, N. 1951. "Discussions with Einstein on Epistemological Issues in Atomic Physics." In *Albert Einstein: Philosopher-Scientist*, ed. P. A. Schilpp. New York: Open Court.

Bub, J. 1974. *The Interpretation of Quantum Mechanics*. Dordrecht: Reidel.

Clauser, J., and A. Shimony. 1978. "Bell's Theorem: Experimental Tests and Implications." *Reports on Progress in Physics* 40: 1883–1927.

Cohen, L. 1966. "Can Quantum Mechanics Be Reformulated as a Classical Probability Theory?" *Philosophy of Science* 33: 317–322.

Einstein, A., B. Podolsky, and N. Rosen. 1935. "Can QM Description of Physical Reality Be Considered Complete?" *Physical Review* 47: 777–780.

Fine, A. 1973. "Probability and the Interpretation of Quantum Mechanics." *British Journal for the Philosophy of Science* 24: 1–37.

Fine, A., and P. Teller. 1974. "Algebraic Constraints on Hidden Variables." *Foundations of Physics* 4: 75–83.

Finkelstein, D. 1973. "The Physics of Logic." In *Paradigms and Paradoxes*, ed. R. Colodny. University of Pittsburgh Press.

Gleason, A. 1957. "Measures on Closed Subspaces of Hilbert Space." *Journal of Mathematics and Mechanics* 6: 885–893.

Healey, R. 1979. "Quantum Realism: Naivete Is No Excuse." *Synthese* 42: 1121–1144.

Heelan, P. 1970. "Complementarity, Context Dependence, and Quantum Logic." *Foundations of Physics* 1: 95–110.

Jarrett, J. 1981. The Case Against Local Realism. Unpublished.

Jauch, J. 1968. *Foundations of Quantum Mechanics*. Boston: Addison-Wesley.

Putnam, H. 1970. "Is Logic Empirical?" *Boston Studies in Philosophy of Science* 5: 216–242.

Sneed, J. 1965. "Von Neumann's Argument for the Projection Postulate." *Philosophy of Science* 32: 22–39.

Teller, P. 1981. The Projection Postulate and Bohr's Interpretation of Quantum Mechanics. Unpublished.

van Fraassen, B. 1972. "A Formal Approach to the Philosophy of Science." In *Paradigms and Paradoxes*, ed. R. Colodny. University of Pittsburgh Press.

von Neumann, J. 1955. *Mathematical Foundations of Quantum Mechanics*. Princeton University Press.

Wigner, E. 1970. "On Hidden Variables and Quantum Mechanical Probabilities." *American Journal of Physics* 38: 1005–1009.

8 The Birth of the Magnetic-Resonance Method

John S. Rigden

The development of the magnetic-resonance method was motivated by the desire to measure properties of the atomic nucleus. In more specific terms, it was the desire to measure the magnetic moments and the spins of atomic nuclei that inspired the creation of a series of experimental techniques that built one upon the other throughout the 1930s. The culmination of this evolutionary development of molecular-beam methods was the magnetic-resonance method.

The gyromagnetic ratio of the atom had been measured. This ratio of the magnetic moment to the mechanical angular momentum (or quantum-mechanical spin) was a consequence of Ampère's postulate of molecular currents so long as the charge carriers had an inertial property. After the discovery of the electron and the measurement of its charge-to-mass ratio, there were many attempts to measure the atom's gyromagnetic ratio (Galison 1982).

The molecular-beam method stands in distinct contrast with the earlier methods of measuring the atomic gyromagnetic ratio. The earlier approaches were macroscopic. Bulk samples in the form of iron rods were suspended in a magnetic field, and the additive effect of many atomic magnetic moments led to an observable result. In contrast, the molecular-beam approach is microscopic. With a collimated beam of atoms, measurements are made on individual, isolated particles. Similarly, the magnetic-resonance method stands in distinct contrast with earlier molecular-beam methods in that it gives results with a precision unattainable by the earlier nonresonant beam experiments. More than this, the magnetic-resonance method is in itself a powerful method, a generic method that gave rise to new fields of research and has brought a rich harvest of physical and chemical results.

The particular form of the magnetic-resonance method as it was discovered and first demonstrated in I. I. Rabi's laboratory during the final weeks of 1937 was a direct consequence of the context in which it was discovered. In the most general sense, it was the method of molecular beams that directly determined both the path to the discovery and the form of the discovery. The history of molecular beams is a subject in its own right and is not detailed here. There are numerous general references to the subject of molecular beams for the interested reader (e.g., Zorn 1964).

There were more specific contextual factors that influenced the way the magnetic-resonance method was first demonstrated. First, there were the specific measurements that were being made. These measurements to determine the signs of nuclear magnetic moments were new, and thus they were the primary focus of experimental activity during the months immediately before the discovery. The apparatus for these experiments was the immediate precursor to the experimental system with which the magnetic-resonance method was first demonstrated. Second, there was a set of ideas that was particularly appealing to Rabi's kinesthetic sense of physical intuition. "One day," says Rabi (1981), "I was walking up the hill on Claremont Avenue and I was thinking of [the sign of a nuclear moment] kinesthetically with my body. . . . the whole resonance method goes back to this." Some of these ideas were drawn from the conceptual legacy of the old quantum theory. Third, there was Rabi's style as an experimentalist. Rabi eschewed tedium. He did not want his experimental results to be the end product of long days spent plotting data, working with statistics, or calculating from the data to the results; rather, he wanted to know his results "at the end of the day." This desire was the mother of invention; ingenious experimental methods were devised that brought a simplicity and a directness to his results. The magnetic-resonance method was the ultimate in directness and simplicity.

The Resonance Concept

Consider an atomic two-level system with the energies given by

$$E_i = f(\alpha,\beta, \ldots)\Phi_i m_i, \quad i = 1,2$$

where $\alpha,\beta, \ldots$ are constants, m_i is a quantum number, and Φ_i is an intrinsic property of the atom. The basic problem of spectroscopy is to measure the energy difference between the two levels. This energy separation can be determined by measuring directly the intrinsic property Φ_i. The precision of such a *nonresonant* method is, at best, limited to the precision with which the quantity Φ itself can be measured and by the uncertainties in the constants $\alpha, \beta, \ldots$.

Alternatively, the atomic system can be bathed with radiation whose frequency, ν, can be varied continuously. As long as the inequality

$$\frac{E_2 - E_1}{h} < \nu < \frac{E_2 - E_1}{h}$$

holds, the probability that a transition will be induced between the two levels is very small. This probability, however, becomes large when the condition

$$\nu \simeq \frac{E_2 - E_1}{h}$$

is realized. If this increased transition rate causes a detectable change in the atomic system, the energy separation is essentially reduced to the measurement of the frequency at which the maximum change occurs. This is the *resonant* method. Since frequencies can be measured with great accuracy, the energy separation can be measured to a high precision.

Consider, for example, the nonresonant measurement of an atomic magnetic moment. In their classic experiment, Gerlach and Stern (1924) passed a beam of silver atoms through an inhomogeneous magnetic field. The ground state of silver is a $^2S_{1/2}$ state, so in a magnetic field there are two magnetic quantum states: one in which the magnetic moment is parallel to the external magnetic field and the other with an antiparallel alignment. A transverse force—perpendicular to the particle velocity—is exerted on each silver atom. The direction of this force is dependent on the two spatial alignments of the magnetic moments; those moments tending to be parallel with the field are pulled into the strong-field region, while those with antiparallel alignments are pushed into the weaker part of the field. As a result, the parent beam is split into two components. By detecting these two components and measuring their spatial separation, the magnetic moment of the silver atom could be determined. However, other detailed information was needed (for example, $\partial H / \partial Z$), and specific assumptions had to be made (for example, the nature of the velocity distribution). The precision by this method was poor. The measured value reported by Gerlach and Stern (1924, p. 699) carried an uncertainty of 10 percent.

The resonance approach to the same measurement is simpler and more precise. In the refocusing molecular-beam method developed by

Rabi, atoms leave the source and pass through two deflecting fields on their way to the detector. The two deflections on any single atom can be adjusted so that they are exactly equal and opposite so long as that atom remains in the same quantum state throughout both deflecting fields; hence, such an atom is refocused into the detector. On the other hand, if an atom can be stimulated to undergo a quantum transition as it traverses the region between the two deflecting magnets, then it will be deflected by a different amount in the second deflecting field and it will not be refocused into the detector. The detector signal will accordingly decrease. This is the basis of the magnetic-resonance method. No assumptions need to be made, and only two physical parameters need to be measured: the strength of a homogeneous field and a frequency. Both can be measured with great accuracy. The experimental uncertainty in the magnetic moments measured by this method is less than 1 percent.

Experimental and Conceptual Roots from the Old Quantum Theory

In the magnetic-resonance method, quantum transitions are induced between energy states that are perturbed by a magnetic field. The idea that a magnetic field can influence the spectral light of atoms goes back at least as far as Faraday. It was Pieter Zeeman, however, who first observed a demonstrable effect. In 1897, Zeeman placed a piece of asbestos soaked with a salt solution in the flame of a Bunsen burner which was positioned between the paraboloidal poles of a Ruhmkorff electromagnet. "The light," he reports, "was analyzed by a Rowland grating, with a radius of 10 ft., and with 14,938 lines per inch. The first spectrum was used, and observed with a micrometer eyepiece with a vertical cross-wire. An accurately adjustable slit is placed near the source of light under the influence of magnetism. . . . the distance between the poles was about 7 mm. If the current was put on, the two D-lines were distinctly widened. If the current was cut off they returned to their original position. . . . If the current was put on again the D-lines were widened, becoming perhaps three or four times their former width." (Zeeman 1897, p. 227)

Lorentz (1897) immediately explained Zeeman's findings in terms of his electron theory. Lorentz's success, however, was short lived. When the anomalous Zeeman effect was discovered, the electron theory was found wanting (Endo and Saito 1967). Furthermore, Lorentz's explanation was soon recognized to be an application of a theorem that Larmor (1897) had proved at about the same time, and Larmor (1900) applied his theorem to the Zeeman effect two years later. Larmor's ideas, as they were later embellished by the concept of space quantization, played a leading role in the evolution of the resonance method.

Larmor considered a charged ion under the influence of two forces: a central force and a magnetic force. In its essence, Larmor's theorem results from a balancing of the Coriolis force and the Lorentz force. The theorem says that superimposed on the ion's orbital motion is a precession of the entire orbit about the direction of the magnetic field. The angular frequency of this precession is given by

$$\omega_L = \frac{e}{2mc} H,$$

where e and m are the ion's charge and mass, respectively, and H is the strength of the magnetic field.

The magnetic field affects all orbits similarly except for the sense of the orbital motion—with one sense the ion's motion is retarded; with the other sense it is enhanced. The net result is that in a magnetic field there are three frequencies: $\nu = \nu_0$ and $\nu_{\pm} = \nu_0 \pm \Delta\nu$, where ν_0 is the ion's frequency in the absence of a magnetic field and $\Delta\nu$ is the magnitude of the enhancement (+) or retardation (−) due to the magnetic field. These three frequencies provided an explanation for the triplet of lines characteristic of the normal Zeeman effect.

As the quantum theory was developed during the first two decades of this century, its tenets came to specify not only the size and the form of atomic orbits (at least in the case of hydrogen) but also the position of the orbits in space. Asserting that Bohr's model of the hydrogen atom could be extended by the inclusion of elliptical orbits, Sommerfeld (1928) was able to specify the size and the shape of the Keplerian orbits by subjecting each of the two degrees of freedom to the basic quantum condition $\int p\,dq = nh$. This quantum condition singled out, from all the motions that were mechanically possible, those discrete motions, those real motions that were possible according to the theory of the quantum.

Space quantization resulted when Sommerfeld (1916) extended his analysis of the hydrogen atom to three dimensions. [The idea of space quantization was proposed independently by Debye (1916).] This phenomenon was physically significant only when a direction in space was given favored status, as when a magnetic field is present. Sommerfeld considered the problem in the limit as the magnetic field goes to zero. This was perfectly acceptable. As the magnetic field is slowly reduced to zero, any alterations in the electronic orbitals due to the magnetic field vanish because such alterations vary continuously with the strength of the field. In contrast, the spatial orientation remains because the orientation is discrete.

In this limiting procedure, Sommerfeld was invoking the adiabatic hypothesis of Ehrenfest (1917). The essence of this hypothesis is that the quantum numbers that fix the state of a system are adiabatic invariants. In the old quantum theory, this hypothesis, along with the correspondence principle, was an important bridge between the ideas of classical physics and those of the quantum theory.

The Larmor precession together with space quantization provided a clear physical picture of an atom in a magnetic field. The orbiting electron gives rise to an angular momentum and a magnetic moment that, as vectors, are normal to the plane of the orbit. Since the orbit precesses with the Larmor frequency, the vectors of angular momentum and magnetic moment do so as well, sweeping out a cone whose axis is the direction defined by the external magnetic field. (The energy associated with each discrete spatial orientation is different, since the energy of a magnetic moment in a magnetic field depends on the angle between their relative directions. The Zeeman effect was interpreted in terms of the energy differences between allowed spatial orientations of the atomic orbitals. See Sommerfeld 1928, p. 299.) The Larmor precession is illustrated in figure 1 for an atom with three space-quantized orientations.

Space quantization was the most surprising feature to emerge from the old quantum theory. Sommerfeld (1928, p. 246) was prompted to comment: "Without doubt this spatial quantising is one of the most surprising results of the quantum theory. When we consider the simplicity with which the positions are derived and how simple is the result, it seems almost like magic." Small wonder that its experimental verification attracted so much attention.

Space quantization provides a direct account of the Zeeman effect; however, the Zeeman effect does not provide a definitive test for space quantization. After all, the normal Zeeman effect could be explained classically. It was the experiment of Gerlach and Stern (1922) that brought empirical justification to the strange idea of space quantization. In this experiment, silver atoms were sent through an inhomogeneous magnetic field that exerted a transverse force on each atom. The mag-

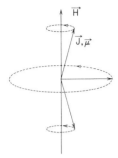

Figure 1
The Larmor precession for an atom with three allowed spatial orientations. The angular momentum vector $\vec{J}$ and the magnetic moment μ (which can be parallel or antiparallel to $\vec{J}$, depending on the sign of μ) precess around the magnetic-field direction $\vec{H}$ with a definite frequency ν.

nitude of the force was a function of the orientation of the silver atom's magnetic moment with respect to the direction of the magnetic field. The parent beam was split into two beamlets, thereby signifying that the silver atom took only two orientations in the field. It was a dramatic result. Later, Max Born (1962, p. 143) pointed to this experiment as providing "the most impressive evidence we have of the fundamental difference between classical and quantum mechanics."

It was the Stern-Gerlach experiment that brought Rabi, as a graduate student, to the recognition that new ideas were required in order to understand the structure of the atom. He thought the whole system of ideas associated with the Bohr atom contrived and silly. "You didn't gain very much," said Rabi in 1981. "You give up the laws of physics and explained the hydrogen atom. I felt that young people like me and my friends would figure this out. But what converted me, more than anything else, was when I heard about the Stern-Gerlach experiment. There was a mystery that was beyond my way of thinking because these atoms had to orient. How did they get oriented? Atoms come out of the hot oven and bounce higgledy-piggledy and there they are—oriented! There was no mechanism. I was great at inventing mechanisms, but I couldn't think of any mechanism. . . ."

The drama associated with the result of the Stern-Gerlach experiment was intensified by the controversy associated with the experiment. Debye did not regard space quantization as having physical reality; he saw it as merely a computational device to account for the additional energy of an atom in a magnetic field. Therefore, Debye did not expect anything unusual to come out of the experiment. Sommerfeld vacillated. At one time he would argue that the magnetic moments of the silver atoms would be distributed relative to the field in a random fashion (Estermann 1975); at another time he would support the hypothesis of spatial quantization and argue that the magnetic moments would take one of three orientations with respect to the field and that the original beam would be split into these components corresponding to $m = \pm 1, 0$. On the other hand, Bohr applied an *ad hoc* argument to rule out the $m = 0$ orientation and looked for the beam to be split into two components. Stern himself did not at that time believe in space quantization. When Pauli was shown the photograph of the two fringes detected in the experiment, he said "This should convert even the nonbeliever Stern" (Estermann 1975, p. 663). As Stern expressed it, the experiment was an open question put to nature.

The Stern-Gerlach experiment confirmed space quantization. The experiment also raised this question: How do the atoms achieve their orientations with respect to the field? This puzzling question was put in a paradoxical form by Einstein and Ehrenfest (1922), who concluded that the time required for an atomic moment to orient itelf via its interaction with a magnetic field would be between 10^9 and 10^{11} seconds, depending on the temperature. Yet in Stern's experiment the silver atoms are in the field for less than 10^{-4} second. Einstein and Ehrenfest were unable to answer this question satisfactorily.

There was no satisfactory resolution to the various puzzles that were born in Frankfurt with the result of the Stern-Gerlach experiment; that is, no resolution until the discovery of electron spin and the development of quantum mechanics.

The Spin-Flip Problem

In 1927, before the full formalism of quantum mechanics was a year old, C. G. Darwin introduced a lengthy paper with the following words: "The matrix and the wave mechanics have both been already developed to great lengths as a calculus of stationary states, but they have not yet got so far in what we may call dynamics, a description of the progress of events. More and more complicated phenomena have been fitted into the same scheme, but not much has been done in making this scheme intuitively understandable." (Darwin 1927, p. 258) The purpose of Darwin's paper was to investigate a number of simple problems from the viewpoint of the new quantum mechanics—particularly from the perspective of wave mechanics, which, as Darwin pointed out, invited the use of superposition. One of the problems that attracted Darwin's attention was the Stern-Gerlach experiment.

Darwin considered a beam of silver atoms that had been split into two components by a Stern-Gerlach-type inhomogeneous magnetic field. He then described a hypothetical experiment: With a slit device, only one of these initial components is allowed to enter the next magnetic field, which is a homogeneous field whose direction is oriented at 90° relative to the first field. The beam of silver atoms then enters a third magnetic field, which is identical to the first inhomogeneous deflecting field. The question Darwin asked about this experiment was: How many beam components are observed at the detector placed behind the third, Stern-Gerlach-type field? With the idea of superposition, Darwin concluded that the new wave mechanics changed the way the above question is answered. Landé (1929) found the same question fascinating and analyzed the Stern-Gerlach-type experiment in greater detail.

Otto Stern, intrigued with the experiment implied by the work of Darwin and Landé, posed an equivalent question (Phipps and Stern 1931): If one component of a beam of silver atoms passes into a second inhomogeneous magnetic field that is oriented at an angle relative to the first field, how is the space quantization expressed in the second field? The answer given by quantum mechanics (and acknowledged by Stern) is that some of the atoms will align themselves parallel and some antiparallel to the new field orientation. For the case where the second field is oriented at an angle of 90° with respect to the first field, Stern acknowledged that the silver atoms would be distributed equally between the parallel and antiparallel orientations.

Theory is one thing, experiment is another. Stern did not believe that the result predicted theoretically could be observed experimentally. Specifically, the theoretical result rests on the assumption that in going from one magnetic field to the next the direction of the field changes in a strictly nonadiabatic manner; that is, the silver atom "sees" a sudden change of field direction. "In reality," wrote Stern (Phipps and Stern 1931, p. 185), "under the conditions that can be obtained experimentally, the situation is just the opposite. . . ." In an implicit fashion, Stern was referring to the fact that the magnetic field at the edge of a deflecting magnet, the fringe field, changes both in magnitude and in direction so that a beam atom "sees" a more gradual change in the orientation of the magnetic field. In other words, an atom's passage from one field to another is, according to Stern, an adiabatic process.

The conceptual context of Stern's thinking in 1932 is interesting. For Stern, space quantization was a proven fact. The magnetic moment of a silver atom can, because of the electron's intrinsic spin of 1/2, take only two orientations with respect to an external magnetic field. The magnetic moment precesses around the field direction in such a fashion that the angle between the field direction and the direction of the magnetic moment is a constant. If the magnetic field varies in direction, how does one describe the magnetic behavior of an atom? What is the process by which the atom orients itself with respect to the changing magnetic field? Or does it?

Stern believed that if the field direction changed gradually, the atomic moment would "follow the turning of the field"; that is, the state of quantization would be the same in both orientations of the external magnetic fields. (Adiabatic process.) On the other hand, if the field direction changed suddenly, the atomic magnetic moment could be flipped to a new orientation. (Nonadiabatic process.) The crucial condition required for an atomic flip to occur was, according to Stern, that the direction of the magnetic field change significantly in the time required for the atom to complete one precessional cycle. It was this requirement that presented the experimental challenge. As Stern showed, the Larmor period is, for a field of 1,000 gauss, 7×10^{-10} sec. If the atom in the beam had a typical velocity of 10^5 cm/sec, it would cover only a distance of 7×10^{-5} cm during one Larmor period. It is, concluded Stern, experimentally impossible to change the field direction by a relevant amount in such a small distance.

The only practical way that Stern could conceive to obtain a nonadiabatic condition would be with a field that, from the atom's reference frame, rotated. Therefore, in between two Stern-Gerlach fields, Phipps and Stern introduced a magnetic field that varied spatially (over a distance on the order of 1 mm) such that an atom passing through this spatial region "saw" a rotating field. Heisenberg had shown Stern that the fraction of atoms flipped would be proportional to $(T_L/T_F)^2$, where T_L is the Larmor period and T_F is the apparent period of the rotating field. At Stern's request, Pauli asked Güttinger (1931) to do an exact calculation that would provide a prediction of the experiment's outcome. With Güttinger's theoretical calculation and their own experimental parameters, Phipps and Stern expected approximately 6 percent of the beam atoms to be flipped. In other words, about 94 percent would retain the same orientation in the second field as they had in the first and 6 percent would have the opposite orientation (assuming spin 1/2). Unfortunately, Phipps, his fellowship at an end, was unable to verify this prediction.

Emilio Segrè, on a Rockefeller Fellowship, came to Hamburg from Rome and joined the Stern group in 1931 (Segrè 1973). Stern suggested that Segrè continue the experiment Phipps had started (Segrè 1982). Phipps had produced his "rotating" magnetic field with three tiny electromagnets. Segrè, inspired by reading Maxwell's *Treatise*, proposed that a current be used in place of the electromagnets. "It was clear to me, qualitatively," said Segrè (1982), "that there would be the flip and that it would do in essence what Phipps and Stern were trying to do." Stern agreed to Segrè's proposal to use a current, and with the assistance of O. R. Frisch he began to set up the experiment.

When Segrè returned to Rome for his Easter vacation, he talked with his friend Ettore Majorana about the experiment and asked him how to predict the fraction of atoms flipped by the field of the current. Majorana (1932) developed a theory that was more directly applicable than Güttinger's theory to the specific conditions of the new version of the Hamburg experiment. In terms of this theory, the early data looked good. In mid August of 1932, Frisch, Phipps, Segrè, and Stern sent to the journal *Nature* a communication that ended with the claim that "the number of re-oriented atoms . . . agreed with the theoretical predictions."

In one sense, the experiment of Frisch and Segrè (1933) was a success. Frisch wrote in 1959: "Almost from the beginning, people had wondered what would happen if atoms passed through a region where the magnetic field rapidly changed direction. Would the atomic magnets follow those changes, however rapid? . . . Experiments in [Stern's] laboratory showed that atoms could indeed be 'shaken off' and made to change their spin orientation. . . ." Frisch and Segrè did successfully discover the phenomenon of spin flips through the nonadiabatic effect of a rapidly changing magnetic field. In the sense of particulars, however, the experiment left a question: The fraction of potassium atoms observed to have undergone a flip did not agree with the fraction predicted by Majorana. At the time, however, Frisch and Segrè concluded that within the accuracy of their experiment there was essential agreement between experiment and theory. Later it was recognized that in the interpretation of their data Frisch and Segrè had omitted a small but fundamental factor. In that omission, they missed the opportunity to record the first beam measurement of a nuclear magnetic moment.

Rabi and Nonresonant Molecular-Beam Methods

In 1927, Rabi left for Europe to learn the new quantum mechanics. After brief stops in Zurich, Munich, and Copenhagen, he arrived in Hamburg, where according to arrangements worked out by Niels Bohr he was to work with Pauli. Otto Stern was also then at Hamburg. Working in his laboratory were two postdoctoral fellows: Ronald Fraser, from Scotland, and the American John Taylor.

With the Stern-Gerlach experiment and the idea of space quantization still making a vivid impression on his thoughts, it was natural for Rabi to pay a visit to the Stern laboratory. However, as the days passed, it was the relaxing opportunity to speak his native English with Fraser and Taylor that made Stern's lab something of a hangout for Rabi. Soon Rabi's theoretical abilities were discovered and he became a resource for Fraser and Taylor. Rabi watched their experimental work and suggested an alternate way of deflecting a beam of atoms that did not require the difficult calibration of an inhomogeneous field. Stern invited Rabi to do the experiment himself. Rabi, regarding such an invitation as a great honor, designed the experiment, and in late 1928 he carried it to a successful completion. The deflecting field that he originated is now called the Rabi field. (See Frisch 1959, p. 6.)

Rabi returned in August 1929 to the United States, where as a new lecturer in the physics department he began what was to become a lifelong career at his alma mater, Columbia University. For almost three years Rabi carried on his research in theoretical physics. Most of this work was a continuation of the solid-state research he had initiated during his days with Heisenberg at Leipzig. He never published any solid-state theory; moreover, he was never really inspired by it. By contrast, Rabi remembered well the charm, the beauty, and the power of the molecular-beam experiments he had watched and contributed to in Stern's laboratory. Taken together, these influences prompted Rabi to begin thinking once again about molecular-beam research. However, the final ingredient that motivated Rabi to start setting up his own molecular-beam laboratory was the work he did in collaboration with Gregory Breit. They showed how the molecular-beam method could be extended to the measurement of nuclear magnetic properties.

Nuclear magnetic moments are smaller than their electronic counterparts by the factor m_e/m_p, where m_e and m_p are the masses of the electron and proton respectively. As the molecular-beam method had been practiced in Stern's laboratory up to that time, this ratio of masses, $1/1,836$, meant one of two things: that if an atom or a molecule possessed a large electronic magnetic moment, the effect of the tiny nuclear magnetic moment would be completely masked out and would not be observed; or that, for those particular cases where the electronic moment was zero (S_0-state atoms or $^1\Sigma$-state molecules), the beam method of Stern would have to be pushed to its limit and, in the process, large experimental uncertainties would compromise the integrity of the results.

The method of Stern employed large magnetic fields, on the order of 10^4 gauss or larger, which acted directly on the magnetic moment of the electron. Breit and Rabi (1931) showed that in weak magnetic fields the electronic and the nuclear magnetic moments are coupled together and that, as a consequence of that coupling, the orientation of the electronic magnetic moment with respect to the external magnetic field is altered. Put another way, the coupling of the electronic and nuclear moments gives rise to $(2J + 1)(2I + 1)$ magnetic quantum states rather than the $(2J + 1)$ states that characterize the strong-field approach used in the Stern-Gerlach-type experiment. (J is the electronic angular momentum in units of $h/2\pi$; I is the nuclear angular momentum in the same units.) It was on the basis of Breit-Rabi theory that Rabi began the design of his molecular-beam experiments.

The molecular-beam method for measuring the magnetic properties of atomic particles, as it was practiced when Rabi began setting up his laboratory, was based on the direct deflection of beam particles. An inhomogeneous magnetic field exerted a force on a beam particle at right angles to its motion. From the magnitude of the resulting deflection, magnetic properties could be determined. Unfortunately, there were many complications. Fast particles were deflected less than slow particles, so a detector would record a distribution of deflections. Each beam component was smeared. This distribution of deflections was a consequence of the way velocities were distributed among the beam particles. The data were analyzed in terms of a most probable velocity, which required a knowledge of the beam temperature. Furthermore, the analysis of data required a knowledge of the magnetic field and its inhomogeneity, which meant that a difficult and error-prone calibration was necessary. Finally, no nuclear magnetic properties had been measured.

From the earliest results that came out of Rabi's laboratory, a stylistic element graced his molecular-beam experiments. In his first major experiment, Rabi, together with Cohen, determined the nuclear spin of sodium. This result was determined directly from their primary data; they merely counted the number of intensity peaks (the number of beam components) observed at the surface-ionization detector (Rabi and Cohen 1933). This direct and simple way of determining the desired result was a consequence of the clever ways they developed to manipulate the beam of sodium atoms. Three inhomogeneous magnetic fields—two strong and one weak—were employed, as well as a selector slit that allowed only a portion of the parent beam to pass. The net result of these three deflecting fields was a focusing effect; that is, each beam component was brought back to the approximate position of the undeflected beam. For sodium, $J = 1/2$; therefore the number of beam components arriving at the detector is given by $(2J + 1)(2I + 1) = 4I + 2$. However, the location of the selector slit was such that half of these components was eliminated; therefore, there were only $2I + 1$ components.

Four distinct intensity peaks were observed (figure 2), which gave the spin of the sodium nucleus as $I = 3/2$ (Rabi and Cohen 1934). When Rabi counted those peaks, he was not only learning the nuclear spin of sodium, he was also reaffirming the existence of space quantization. "Just the fact that it existed," said Rabi (1981), "is entirely out of this world. The awful mystery of quantum mechanics became borne in on me more and more. After all, there are those peaks, four peaks . . . well, it's tremendous." Rabi's excitement never waned.

The elegance of getting a basic result merely by counting intensity maxima never lost its appeal. The zero-moment method (Cohen 1934) was another peak-counting approach. For specific strengths of the deflecting magnetic field, the effective magnetic moment of the beam particle can be zero; in that case it passes through the apparatus undeflected. This is true for both fast and slow particles, so this method is independent of the beam particle's velocity. Accordingly, signal intensities at the detector are enhanced. Thus, by slowly varying the strength of the deflecting field, beam components are slowly swept across the detector. The number of zero-moment intensity peaks coupled with the relative values of the magnetic-field strength at which these peaks are observed enables the nuclear spin to be determined. The method is particularly well suited for cases where I is greater than $3/2$.

While the Cohen-Rabi apparatus was deflecting atoms of sodium, a

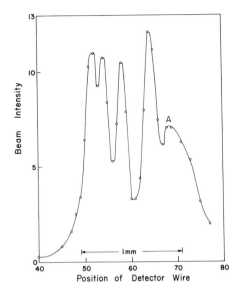

Figure 2
Beam intensity as a function of detector position for a beam of sodium atoms. Four distinct peaks were observed, indicating a spin of 3/2 for the sodium nucleus. (The apparent peak labeled *A* is not part of the deflection pattern; it arises from the fast atoms in the original beam.)

second molecular-beam laboratory was being set up by two postdoctoral fellows, J. M. B. Kellogg and Jerrold Zacharias. In this laboratory, a series of experiments on the hydrogens was initiated in 1933. Coming as they did so soon after the discovery of the neutron, these experiments were of crucial significance in those early days of post-neutron nuclear physics. They began by verifying the anomalous value of the proton's magnetic moment (it was three times larger than predicted), and they ended with the major discovery of the deuteron's quadrupole moment (a discovery that necessitated a fresh approach to the theory of nuclear forces).

The experiments on the hydrogens is in itself a fascinating story. (See Rigden 1983.) In the present context, however, these experiments were important because the experimental techniques that were created formed important stepping stones toward the magnetic-resonance method. A novel feature of the hydrogen experiments was the way the inhomogeneous deflecting fields were produced. In the Stern-Gerlach apparatus, the inhomogeneous magnetic fields were produced by electromagnets whose iron pole pieces were shaped so as to produce a large field gradient. This approach gave rise to one of the largest experimental uncertainties in these measurements: the uncertainties associated with the calibration of magnetic field strength at and over the location of the beam. In the Kellogg-Zacharias-Rabi apparatus, the inhomogeneous magnetic field was produced by parallel current-carrying conductors held in a precision-machined jig. The geometry of these conductors was known with sufficient accuracy that the field strength could be calculated directly (Rabi, Kellogg, and Zacharias 1934). The two-wire fields were adequate for the weak deflecting fields consistent with Breit-Rabi theory; however, the advantages of this field were so great that, when stronger fields were needed, iron pole pieces were designed in the shape of the equipotentials of the two-wire field (Millman, Rabi, and Zacharias 1938).

By the middle of the 1930s, Rabi, his postdoctoral associates, and his graduate students had developed the facility to maneuver and to control beams of atoms or molecules in clever, predetermined ways—"they could play the courses with beams" (Rabi 1981). In the development of the magnetic-resonance method, the next changes that were made in their apparatus were most significant. One of these, the refocusing method, was a natural outgrowth of the ever-increasing ease with which they could manipulate the beam particles. The second change in the apparatus was inspired by Rabi's theoretical analysis of the spin-flip experiments of Stern, Phipps, Frisch, and Segrè.

Some time after the Frisch-Segrè paper (1933) was published, Rabi wrote to Segrè asking for some of the details associated with their experiment. Still later, Rabi wrote again to Segrè. "Do you know why you didn't succeed in flipping all the potassium atoms?," Segrè (1982) remembers Rabi writing. "You have measured the magnetic moment of the potassium nucleus." Then, Segrè said, "Rabi showed that our data could be analyzed in that way and used to measure the magnetic moment of the nucleus. I never would have dreamt that, you see."[1]

Rabi's analysis of the Frisch-Segrè experiment was published in 1936 under essentially the same title as the earlier papers on the spin-flip experiments published from Hamburg. The "rotating" field used in Stern's laboratory was very weak. This, Rabi pointed out, meant that in this field the electronic and nuclear magnetic moments would be coupled together. When this coupling was taken into account, the data from the Frisch-Segrè experiment—that is, the number of atoms undergoing a reorientation (a flip)—were in agreement with Majorana's theory.

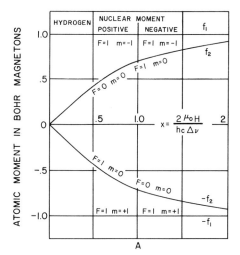

Figure 3
Variation of the effective atomic moments, $\mu_{eff} = f_m\mu_0$, as a function of $x = 2\mu_0 H/hc\Delta\nu$, where H is the magnetic-field strength. The quantum numbers labeling each of these effective moments differ for positive moments (second column) and negative moments (third column); however, the form of these curves and the resulting deflection pattern are independent of the signs.

Rabi's 1936 paper contained more than an analysis of the Frisch-Segrè experiment. Up to this time, there had been no way to measure or to deduce the sign of a magnetic moment from molecular-beam data. The reason for the difficulty can be seen from figure 3, which shows, for $I = 1/2$, the effective moments as a function of magnetic-field strength for the states $F = 1$, $m_F = \pm 1,0$ and $F = 0$, $m_F = 0$. As can be seen from this figure, the sign of the magnetic moment affects only the labeling of these effective moment diagrams; it leaves their form unaltered. Since the deflection of a beam atom is proportional to its effective moment, figure 3 also exhibits the deflection pattern for the case $I = 1/2$. The deflection patterns are identical for both positive and negative magnetic moments. Nonetheless, in the last section of his 1936 paper, Rabi was able to show, in a qualitative way, how the sign of the magnetic moments of nuclei could, in principle, be determined.

Rabi's method employed a "rotating" field to induce nonadiabatic transitions. A beam of hydrogen atoms in a weak deflecting field is, as shown in figure 3, symmetrically split into four components where each component is populated by atoms with the same quantum numbers

F and m_F. Rabi's proposed method to determine the signs of nuclear magnetic moments was based on the fact that a "rotating" field could, in principle, induce nonadiabatic transitions from the $F = 1$, $m_F = 0$ state, but could not induce such transitions from the $F = 0$, $m_F = 0$ state. Therefore, with the weak magnetic-field strength adjusted to give four well-separated components, a selector slit allows only one of the inner components to enter the "rotating" field, which is followed in turn by a strong analyzing field. If the selected component is labeled by $F = 0$, $m_F = 0$, it passes unchanged through both the "rotating" field and the analyzing field. One component is detected. On the other hand, if the selected component is labeled by $F = 1$, $m_F = 0$, the "rotating" field can induce transitions to the states $F = 1$, $m_F = \pm 1$ and atoms in all m_F states of $F = 1$ enter the strong analyzing field. From figure 3 it can be seen that in the limit of large fields the four components merge into two components; therefore, if the selected component is labeled by $F = 1$, $m_F = 0$, two components are detected. One component if the moment is positive, two components if negative.

The signs of nuclear magnetic moments are important in and of themselves; however, in 1935 there was another thing that added significance to their determination. The magnetic moments of both the proton[2] and the "deuton"[3] had been measured.[4] These results, together with the assumption that magnetic moments were additive, allowed the magnetic moment of the newly discovered neutron to have any one of four different values. If the signs of the moments (the proton's and the deuteron's) could be determined, the neutron's magnetic moment could be assigned a specific value.[5]

With these ideas and with these hopes, Kellogg and Zacharias began to redesign and rebuild their molecular-beam apparatus in 1935. As already mentioned, the new design incorporated two important changes. First, the refocusing method was introduced. In this method, the beam particles passed through two deflecting fields in succession. The fields were arranged, however, to deflect the particles in opposite directions, and with the proper current through the parallel wires, the magnitudes of the deflections could be equalized. Under this condition, all atoms, fast and slow, were refocused onto the detector. This was an important advance. The earlier simple deflection experiments carried uncertainties of 10 percent, which were due in part to the necessity of estimating the temperature of the beam in order to establish the velocity distribution of the atoms in the beam. With the refocusing method, no assumptions had to be made about either the velocity distribution or the temperature

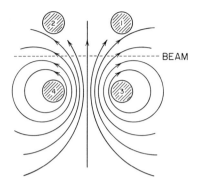

Figure 4.
The T field. Beam particles passing through the T field "saw" the equivalent of a rotating field. With a proper velocity through this field, a nonadiabatic transition from one spatial orientation to another could be induced.

of the beam. (The same was true with the zero-moment method; however, that method could not be applied to spin-1/2 particles.) The refocusing technique was a significant advance in molecular-beam methodology. Furthermore, the signs of magnetic moments could be determined more directly by the refocusing method than by the method originally proposed by Rabi in his theoretical paper of 1936. The second change in the apparatus was the introduction of a static magnetic field, called the T field, whose purpose was to flip atoms in the beam from one space-quantized orientation to another. This field, shown in figure 4, was produced by four current-carrying wires placed between the two deflecting fields. Atoms passing through the T field saw the equivalent of a rotating field; the faster they moved through it, the greater was the frequency of the apparent rotation. If the frequency of the rotating field was approximately equal to the Larmor precession frequency of the atom, the probability of a reorientation of the magnetic moment was significant and the atom could emerge from the T field in a different spatially quantized state. Only those atoms that remained in the same quantum state through both deflecting fields would be refocused into the detector. If, on the other hand, the T field changed the quantum state of an atom, then that atom would not reach the detector and the beam intensity at the detector would decrease accordingly. It was this change in the beam intensity (or lack thereof) that identified the sign of the nuclear moment.

Rabi, Kellogg, and Zacharias (1936) were able to reduce the uncertainties in their measured results from 10 percent to 5 percent, to determine that the magnetic moments of both the proton and the deuteron were positive, and to establish that the magnetic moment of the neutron had the anomalous value of about -2.0 nuclear magnetons. These results were obtained with the refocusing method together with the T field, the apparatus that represented the acme of nonresonant molecular-beam systems.

The dynamic interplay between theoretical suggestion and experimental implementation is the gist of day-to-day science. The Rabi laboratory of the 1930s provides a superb example of this interplay. The essence of Rabi's proposed method for determining the signs of the magnetic moments was implemented, but not in the specific fashion he proposed. The theory in the 1936 paper was indeed crucial, but the experimental virtuosity of Rabi, Kellogg, and Zacharias in that same year enabled them to conceive of the refocusing method, which opened a more direct and smoother path to the sign of the proton's magnetic moment. In turn, the refocusing method (together with the T field) was in itself suggestive. The magnetic-resonance method was only a short step away, and after another theoretical paper was published by Rabi in 1937 that step was soon taken.

The Birth of the Molecular-Beam Magnetic-Resonance Method

The behavior of magnetic moments in various magnetic-field configurations has attracted the attention of many physicists over the years. This problem was, and in some sense continues to be, of interest on two levels. First, starting with Darwin's paper, this problem has played an important role in the interpretation of quantum mechanics.[6] Second, beginning with the papers by Güttinger and Majorana, there was a problem associated with the quantum-mechanical description of spin systems in external fields. In 1977, Julian Schwinger wrote a paper for a Rabi *Festschrift* in which he called attention to the importance that the Majorana formula had had in the development of his own attitude toward angular-momentum theory. In this paper (p. 170) Schwinger writes that "the original Majorana paper was baffling and it was obligatory to find a quantum mechanical derivation of the formula, in particular." The first extensions of Güttinger's and Majorana's work were the papers by Rabi (1936) and Motz and Rose (1936). Schwinger's first treatment of the problem came in 1937 when he and Rabi wrote separate papers on this subject that were published back to back in *Physical Review*.

It was Rabi's 1937 paper, however, that influenced most directly the conceptual context within which the magnetic-resonance method was born. In this paper, Rabi extended his theoretical analysis of the previous year and showed in a more general way how a rotating magnetic field could be used to determine the signs of nuclear magnetic moments. This was the explicitly stated purpose of his paper. However, Rabi's 1937 paper did more than this; it provided the fundamental theoretical basis for the molecular-beam magnetic-resonance method. For this reason, as well as for others, this paper is still cited today.[7]

Rabi derived an expression giving the transition probability for a system with angular momentum $J = 1/2$ to undergo a nonadiabatic transition between the magnetic states $m = 1/2$ and $m = -1/2$. He solved the Schrödinger equation exactly and found the chance that the system in state $m = 1/2$ will, at some later time, be found in the state $m = -1/2$. He expressed his result as

$$P_{(1/2, -1/2)} = \frac{\sin^2\theta}{1 + q^2 - 2q\cos\theta} \sin^2 \frac{2\pi\nu t}{2} (1 + q^2 - 2q\cos\theta)^{1/2}, \qquad (1)$$

where q is the ratio ν_L/ν. (ν_L is the Larmor precession frequency of a magnetic moment in a magnetic field H, and ν is the frequency of the rotating magnetic field.) The ratio q is considered positive if the rotation of the magnetic field is in the same sense as the Larmor frequency. If, on the other hand, the rotational direction of the magnetic field is reversed and the ratio q becomes negative, the result above becomes

$$P_{(1/2, -1/2)} = \frac{\sin^2\theta}{1 + 3\cos^2\theta} \sin^2 \frac{2\pi\nu t}{2} (1 + 3\cos^2\theta)^{1/2}. \qquad (2)$$

The transition probability expressed by equation 1 is much larger than equation 2; therefore, as Rabi (1937, p. 654) concluded, "It is clear . . . that from this qualitative difference and a knowledge of the sense of rotation one can infer the sign of the moment."

Rabi's famous 1937 paper was sent to the editors of *Physical Review* at the end of February. The purpose of the paper was a direct outgrowth of the context in which it was written; members of the Rabi laboratory were applying the new refocusing-rotating-field method to the determination of the signs of nuclear magnetic moments. Since this method was adequate for the nuclear systems being investigated, the more general procedure explicitly described in the 1937 paper was never applied experimentally.

Its adequacy notwithstanding, there were inherent limitations with the T-field method. It was only a qualitative method. The T field was a static field that varied spatially in both magnitude and direction. The apparent field rotation was a consequence of the atom's motion. Specifically, the rotational frequency of the T field, as seen by a beam particle, depended on the transit time through it: The larger the velocity of a passing atom, the larger was the apparent frequency. Since the velocities of the beam particles are distributed according to the Maxwell distribution law, relatively few atoms would have such velocity that their Larmor frequency matched the rotational frequency of the field. Furthermore, each atom saw only half a cycle of the rotating field. Taken together, these factors lessened the magnitude of the effect. Finally, since the velocities of individual particles were unknown, the rotational frequency as seen by these particles was also unknown. Hence the qualitative nature of the method.

The way to transform the T-field method into the quantitatively precise resonance method is implicit in Rabi's 1937 paper. The probability expression of equation 1 (above) was derived for a gyrating magnetic field; that is, a magnetic field H with Cartesian coordinates

$H_x = H \sin\theta \cos\omega t,$
$H_y = H \sin\theta \sin\omega t,$
$H_z = H \cos\theta.$

In other words, the magnetic field gyrates around the z axis with frequency $\nu = \omega/2\pi$.

Suppose we now regard the gyrating field H as the vector sum of two magnetic fields: a constant field $\vec{H}_0$ directed along the z axis and a field $\vec{H}_1$ that rotates in the xy plane with a frequency ν. Considered this way, the gyrating field has a magnitude of $(H_0^2 + H_1^2)^{1/2}$. If the experimental conditions are arranged so that $H_1 \ll H_0$, the angle θ in equation 1 is very small. Under this condition, the small-angle approximation [$\sin\theta \approx \theta$, $\cos\theta \approx 1 - 1/2\theta^2$] allows equation 1 to be rewritten as

$$P_{(1/2, -1/2)} = \frac{\theta^2}{(1 - q)^2 + q\theta^2} \sin^2 \pi\nu\tau \left[(1 - q)^2 + q\theta^2\right]^{1/2}. \tag{3}$$

Equation 3 is the Rabi form of the resonance equation: When $q = 1$ (or, since $q = \nu_L/\nu$, when $\nu = \nu_L$), the probability $P_{(1/2, -1/2)}$ goes through a maximum. The resonance condition is therefore realized when the two frequencies ν and ν_L are equal.

The theory in the 1937 paper became the basis for the powerful, far-reaching magnetic-resonance method. This theory was complete in mid February, and the equipment modifications required to implement the method were quite simple: The static "rotating" field had to be replaced with an oscillating field embedded in a homogeneous magnetic field. Yet it was not until the fall, over seven months later, that equipment was shut down, beam apparatus was dismantled, and the necessary modifications were begun. Several factors contributed to this delay. First, there was, in a sense, a backlog of work to finish. The magnetic moments of various nuclei had been determined by the older techniques, and the T-field method was being used to round out the job by determining the signs of these moments. As mentioned above, the T-field method was adequate for that task. Second, the method itself appealed to Rabi—he enjoyed flipping atoms with the T field. Finally, there was no reason to rush into the next stage of the development. By 1937, as Rabi (1982) has said, "We had the field to ourselves. We had no competition." There was no reason to interrupt a productive line of research just then. This changed that September.

Rabi's laboratory was a natural stopping-off place for European physicists visiting the United States. During September 1937, C. J. Gorter, from the University of Groningen, was such a visitor. In the previous year, Gorter (1936) had tried and failed to detect transitions between nuclear magnetic energy levels. In his experiment, the sample was placed in a strong magnetic field H_0 and an oscillating field of frequency ν was applied perpendicular to H_0. With the sample in a calorimeter and the frequency of the oscillating field fixed, the strong, steady field was slowly varied. At some particular value of the steady field, Gorter hoped to detect absorption as indicated by heating effects. Gorter's failure to discover nuclear magnetic resonance was due to the combined effect of the purity of his samples and the radio-frequency power level (Van Vleck 1970). In short, the populations of the upper and lower levels were essentially equal and the line was saturated. This work was behind Gorter when he visited Columbia in September. As Rabi (1982) remembers their discussion, Gorter said "Why aren't you doing it this way?" By this he meant "Why aren't you using an oscillator?"

Gorter's visit accomplished two things. First, it served to emphasize and to sharpen similar ideas Rabi had had since his 1937 theoretical paper. Zacharias remembers sitting in Rabi's ninth-floor office in Pupin Hall discussing this very idea. According to Zacharias (1982), Rabi said: "You know, I think you can describe the flop [the reorientation of a magnetic moment by the T field] in terms of a frequency. And, if you apply a radio frequency, you can make it flop." The step to apply the radio-frequency field had not been taken, for reasons already discussed. Among these reasons, the satisfaction that the T-field method brought to Rabi cannot be overemphasized. "I liked what we were doing," said Rabi (1982). "It satisfied something deep within me." The second consequence of Gorter's visit was the implication that he might return to Holland and do the experiment himself. "When I saw that he might go after it and we might get some competition," said Rabi (1982), "I said, 'Well, let's do it.'" Gorter's visit was on a Saturday. On Monday, two days later, the Rabi team shut down their vacuum pumps in preparation for the modification of their apparatus.

What was the influence of Gorter's visit on the evolution of the magnetic-resonance method? It should come as no surprise that Gorter (1967) felt his contribution to the discovery of the magnetic-resonance method was "undervalued." Gorter's disappointment was certainly enhanced by the fact that he came close to making the discovery himself and the fact that it had such a provocative influence on the physical sciences. (Not only did Rabi win the Nobel Prize, but others, either for their extension of the magnetic-resonance method or for their application of it, did as well.) Gorter's disappointment is understandable. However, his visit merely hastened the fruition of ideas that had already been given theoretical form in Rabi's 1937 paper and had been discussed during the intervening months. The delay in implementing the ideas behind the magnetic-resonance method was, as much as anything, a consequence of the intense creativity of Rabi and his associates. The idea for new experimental procedures came so rapidly during the mid 1930s that it became necessary to deliberately pause in order to do some nuclear physics. Gorter's visit ended such a pause.

As Sidney Millman (1977) recounts, he, Zacharias, and Polycarp Kusch turned their "immediate attention to the design and construction of the apparatus as required for molecular resonance." The molecular-beam magnetic-resonance system that resulted from their design and construction was a direct descendant of the earlier beam systems, and as such it included many of the experimental features described above. In fact, the only necessary change in their apparatus was the replacement of the static T field with pole pieces to produce a homogeneous magnetic field and with the means to produce an oscillating field; however, the design of the deflecting magnets was also changed.

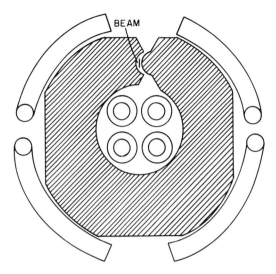

Figure 5
Cross section of the deflecting magnet used in the first magnetic-resonance experiments. The inhomogeneous magnetic field, at the beam location, could be calculated directly.

The new deflecting-magnet design, introduced earlier (Millman, Rabi, and Zacharias 1938), combined the advantages of the strong fields that are possible with iron with the experimental simplicity of the two-wire system. The cross section of the new deflecting magnet is shown in figure 5. The pole pieces forming the magnet gap were designed so as to form equipotential surfaces corresponding to the magnetic field produced by parallel wires, placed 6.2 mm apart, carrying current in opposite directions. With this design, the magnetic-field strength could be calculated analytically. In between these two inhomogeneous magnetic fields was the heart of the magnetic-resonance system. A uniform magnetic field, the C field, was supplied by a magnet of Armco iron. The pole faces, separated by a gap of 1/4 inch, were 10 cm long and 4 cm high. In between the pole faces was a 1/8-inch copper tube bent into the shape of a hairpin (figure 6). A high-frequency current, supplied by a Hartley oscillator, was the source of the oscillating magnetic field. The hairpin was oriented so that the oscillating field was at right angles to the strong homogeneous field. The beam traveled along a path between the legs of the hairpin.

A schematic diagram of the molecular-beam magnetic-resonance method is shown in figure 7. Molecules effuse from the source slit O

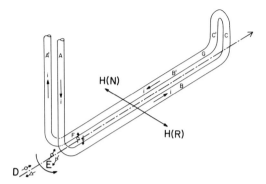

Figure 6
The hairpin-shaped wire that served as the source of the oscillating magnetic field in the first magnetic-resonance experiments. This wire was located in the homogeneous C field.

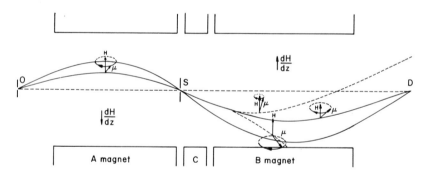

Figure 7
Schematic diagram of the molecular-beam magnetic-resonance system. The magnets A and B produce strong inhomogeneous deflecting fields. Together, the A and B fields refocus beam atoms into the detector D so long as an atom remains in the same quantum state (see solid trajectories). The C-field region is where quantum states can be changed by an oscillating field (not shown). If an atom emerges from the C field in a different quantum state, it is not focused into the detector (see dashed trajectories).

into the evacuated chamber. In the absence of any magnetic deflecting field, the particles making up the beam traverse a straight-line path (dashed line *OSD* in figure 7). With the deflecting fields activated, a downward force is exerted by magnet *A* and an upward force by magnet *B*. The magnitudes of the deflecting forces are such that the deflections themselves are equal in magnitude as long as the particles remain in the same magnetic quantum state through both the *A* and the *B* field. The equal deflections bring the particles converging onto the detector *D* and establish a signal of given intensity.

The change of signal intensity correlated with the magnitude of the *C* field is the datum of significance. The procedure used was to hold the frequency of the oscillating magnetic field constant and to slowly vary the strength, H_0, of the homogeneous *C* field. In this magnetic field, a nuclear magnetic moment μ_I precesses with a frequency given by

$$\omega_L = \frac{\mu_I}{hI} H_o,$$

where *I* is the spin of the nucleus. When the magnetic field H_0 brings the Larmor frequency ω_L into coincidence with the frequency of the oscillating field, ω, the probability for a reorientation is a maximum (equation 3). When a magnetic moment is reoriented, it enters the *B* field in a different quantum state, it is not refocused, and the signal intensity decreases. This is the essence of the magnetic-resonance method.

Four months after Rabi's group began modifying the equipment to implement the magnetic-resonance method, they sent their first brief communication to *Physical Review* (Rabi et al. 1938). The first nuclear-magnetic-resonance curve, taken from this paper, is shown in figure 8. (The abscissa is the current in the *C* magnet and the ordinate is the beam intensity.) The beam was LiCl; the first resonance curve was due to transitions between nuclear magnetic quantum states of Li.

The magnetic-resonance method had distinct advantages over all the previous methods that had been used to measure magnetic properties of nuclei. First, it was direct. In the nonresonant beam methods, the objective of the experiments was to measure the separation between the two hyperfine states that correspond to the parallel and antiparallel configurations of the electronic and nuclear spins. With a value for this hyperfine separation, semiempirical formulas could be used to calculate the magnetic moment; however, these semiempirical formulas also re-

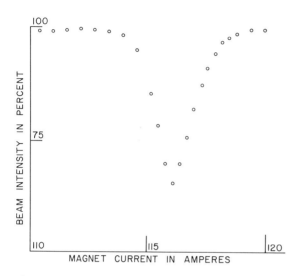

Figure 8
The first magnetic-resonance curve. This curve represents a decrease in beam intensity as a function of the strength of the C field. The atom that gave rise to this curve was lithium.

quired a knowledge of the Schrödinger wave function at the nucleus. Except for hydrogen, these wave functions could only be approximated. With the resonance method, by contrast, the magnetic moments were obtained directly from directly measured quantities. Second, those quantities could be measured with great accuracy. The frequency of the oscillating field, measured by means of a heterodyne frequency meter, was determined with a precision of less than 0.03 percent. The strength of the homogeneous magnetic field, the C field, was measured with a flip coil and a ballistic galvanometer. Uncertainties in this measurement were on the order of 0.2 percent. These two parameters, a frequency and the strength of a homogeneous magnetic field, gave the gyromagnetic ratio of a nucleus with a precision of less than 1 percent— a great improvement over the precisions of approximately 5 percent that were achieved with the nonresonant methods.[8] Third, the magnetic-resonance method was more widely applicable. Many more nuclei could be studied with it than with the nonresonant method. In themselves, these advantages made the magnetic-resonance method a development of great significance to experimental physics.

The oscillating field used in the resonance apparatus had one drawback in comparison with the rotating-field method: An oscillating field is effectively equivalent to two magnetic-field components rotating in opposite directions. Only the component that is rotating in the same sense as the precessing moment is able to influence the moment's orientation in the magnetic field; however, it is impossible to know which component is responsible for inducing a transition. Therefore, in a purely oscillatory field it is impossible to determine the sign of the magnetic moment. Fortunately, the oscillating field in Rabi's apparatus was not purely oscillatory.

From the beginning, Rabi, Millman, and others noticed that the shapes of the resonance curves were not quite symmetrical. One night, while riding home to Brooklyn on the subway after a long day in the laboratory, Millman recognized the reason for the misshapen resonance curves: The oscillating field changed in direction over the full length of the hairpin-shaped conductors used to produce it. In short, there were end effects. Millman (1939) showed how these end effects not only explained the asymmetries in the line shape but also opened the way to determining the signs of nuclear magnetic moments.

From the time the first resonance curve was plotted, it was obvious that the new method was greatly superior to anything that had preceded it; consequently, the magnetic-resonance method became the basis for further experimentation. The molecular-beam system that had been used for the hydrogen experiments was converted into a magnetic-resonance system, and new hydrogen experiments were started. The Rabi laboratory was such a busy place that the first full-length paper was not submitted for over a year after the discovery was announced. In that first full-length paper (Rabi et al. 1939), the new method is described in detail.

The magnetic-resonance method was the culmination of a series of molecular-beam experiments that were brilliant for the physical results that were obtained and were renowned for the experimental methods that were developed. The magnetic-resonance method itself was, and remains, truly epoch-making. When you consider the many fields of research that have been spawned by this method, when you take account of the practical applications that have been made with this method, and when you regard the increasing way that it is serving the welfare of humankind, you are likely to conclude that the magnetic-resonance method ranks as one of the greatest discoveries in experimental science.

Acknowledgments

A large part of this chapter was written while I was a Visiting Scholar in the Department of Physics at Harvard University. I gratefully acknowledge many insights gained from my host, Professor Norman F. Ramsey, who was a student of I. I. Rabi during the time when the magnetic-resonance method was born. I also thank Professor E. M. Purcell for many stimulating conversations.

Many of Rabi's former students and post-doctoral fellows have been generous in their willingness to talk with me about the "good old days." Among them were Polykarp Kusch, Sidney Millman, Julian Schwinger, and Jerrold Zacharias. I thank each of them. I also appreciate the helpful discussion I had with Emilio Segrè.

Most of all I acknowledge my indebtedness to I. I. Rabi. He has talked with me on numerous occasions about the early days of quantum mechanics, about his own physics, and about his laboratory.

Finally, I thank the secretaries Carol Davis and Adele Wright (Harvard) and Sylvia Stephens (University of Missouri–St. Louis).

Notes

1. This correspondence is no longer extant.

2. I. Estermann, R. Frisch, and O. Stern, "Magnetic Moment of the Proton," *Nature* 132: 169–170 (1933); R. Frisch and O. Stern, "Uber die Magnetische Ablenkung von Wasserstoffmolekulen und das magnetische Moment des Protons," *Zeitschrift für Physik* 85: 4–15 (1933); I. Estermann and O. Stern, "Uber die Magnetische Ablenkung von Wasserstoffmolekulen und das magnetische Moment des Protons," *Zeitschrift für Physik* 85: 17–24 (1933); I. I. Rabi, J. M. B. Kellogg, and J. R. Zacharias, "The Magnetic Moment of the Proton," *Physical Review* 46: 157–163 (1934).

3. I. Estermann and O. Stern, "Magnetic Moment of the Deuton," *Physical Review* 45: 761 (1934) and *Nature* 133: 911 (1934); I. I. Rabi, J. M. B. Kellogg, and J. R. Zacharias, "The Magnetic Moment of the Deuton," *Physical Review* 46: 163–165 (1934).

4. For a description of these experiments see Rigden 1983.

5. A specific value for the neutron's magnetic moment could be assigned as long a magnetic moments in the nucleus were assumed to be additive.

6. See, for example, David Bohm, *Quantum Theory* (Englewood Cliffs, N.J.: Prentice-Hall, 1951); Richard P. Feynman, Robert B. Leighton, and Matthew Sands, *The Feynman Lectures on Physics*, vol. III (1965); Eugene P. Wigner, "The Problem of Measurement," *American Journal of Physics* 31: 6–15 (1963).

7. The Rabi "flopping formula" familiar to laser physicists comes from this 1937 paper. For example, see Murray Sargent III, Marlan O. Scully, and Willis E. Lamb, Jr., *Laser Physics* (Reading, Mass.: Addison-Wesley, 1974), pp. 25–27.

8. In some cases the zero-moment method could give results with a precision better than 5 percent.

References

Born, M. 1962. *Atomic Physics,* seventh edition. New York: Hafner.

Breit, G., and I. I. Rabi. 1931. "Measurement of Nuclear Spin." *Physical Review* 38: 2082–2083.

Cohen, V. W. 1934. "The Nuclear Spin of Caesium." *Physical Review* 46: 713–717.

Darwin, C. G. 1927. "Free Motion in the Wave Mechanics." *Proceedings of the Royal Society* A 117: 258–293.

Debye, P. 1916. "Quantenhypothese und Zeeman-Effekt." *Physikalische Zeitschrift* 17: 491–507.

Ehrenfest, P. 1917. "Adiabatic Invariants and the Theory of Quanta." *Philosophical Magazine* 33: 500–513.

Einstein, A., and P. Ehrenfest. 1922. "Quantetheoretische Bemerkungen zum Experiment von Stern und Gerlach." *Zeitschrift für Physik* 11: 31–34.

Endo, S., and S. Saito. 1967. "Zeeman Effect and the Theory of the Electron of H. A. Lorentz." *Japanese Studies in History of Science* 6: 1–18.

Estermann, I. 1975. "History of Molecular Beam Research: Personal Reminiscences of the Important Evolutionary Period 1919–1933." *American Journal of Physics* 43: 661–671.

Frisch, O. R. 1959. "Molecular Beams." *Contemporary Physics* 1: 3–16.

Frisch, R., and E. Segrè. 1933. "Über die Einstellung der Richtungsquantelung II." *Zeitschrift für Physik* 80: 610–616.

Frisch, R., T. E. Phipps, E. Segrè, and O. Stern. 1932. "Process of Space Quantization." *Nature* 130: 892–893.

Galison, P. 1982. "Theoretical Predispositions in Experimental Physics: Einstein and the Gyromagnetic Experiments, 1915–1925." *Historical Studies in the Physical Sciences* 12(2): 285–323.

Gerlach, W., and O. Stern. 1922. "Der experimentalle Nachweis der Richtungsquantelung in Magnetfeld." *Zeitschrift für Physik* 9: 349–352.

Gerlach, W., and O. Stern. 1924. "Über die Richtungsquantelung in Magnetfeld." *Annalen der Physik* 74: 673–699.

Gorter, C. J. 1936. "Paramagnetic Relaxation." *Physica* 3: 503–513.

Gorter, C. J. 1936. "Negative Results of an Attempt to Detect Nuclear Magnetic Spins." *Physica* 3: 995–998.

Gorter, C. J. 1967. "Bad Luck in Attempts to Make Scientific Discoveries." *Physics Today* 20: 76–81.

Güttinger, P. 1931. "Das Verhalten von Atomen in magnetischen Drehfeld." *Zeitschrift für Physik* 73: 169–184.

Kellogg, J. M. B., I. I. Rabi, and J. R. Zacharias. 1936. "The Gyromagnetic Properties of the Hydrogens." *Physical Review* 50: 472–481.

Landé, A. 1929. "Polarization von Materiellen." *Naturwissenschaften* 17: 634–637.

Larmor, J. 1897. "On the Theory of the Magnetic Influence on Spectra; and on the Radiation from Moving Ions." *Philosophical Magazine* 44: 503–512.

Larmor, J. 1900. "On the Dynamics of a System of Electrons or Ions; and on the Influence of a Magnetic Field on Optical Phenomena." *Transactions of the Cambridge Philosophical Society* 18: 380–407.

Lorentz, H. A. 1897. "Uber den Einfluss magnetischer Krafte auf die Emission des Lichtes." *Wiedmannsche Annalen der Physik* 63: 278–284.

Majorana, E. 1932. "Atomi Orientati in Campo Magnetico Variabile." *Nuovo Cimento* 9: 43–50.

Millman, S. 1939. "On the Determination of the Signs of Nuclear Magnetic Moments by the Molecular Beam Method of Magnetic Resonance." *Physical Review* 55: 628–630.

Millman, A. 1977. "Recollections of a Rabi Student of the Early Years in the Molecular Beam Laboratory." *Transactions of the New York Academy of Sciences* 38: 87–105.

Millman, S., I. I. Rabi, and J. R. Zacharias. 1938. "On the Nuclear Moments of Indium." *Physical Review* 53: 384–391.

Motz, L., and M. E. Rose. 1936. "On Space Quantization in Time Varying Magnetic Fields." *Physical Review* 50: 348–355.

Phipps, T. E., and O. Stern. 1931. "Uber die Einstellung der Richtungsquantelung." *Zeitschrift für Physik* 73: 185–191.

Rabi, I. I. 1936. "On the Process of Space Quantization." *Physical Review* 49: 324–328.

Rabi, I. I. 1937. "Space Quantization in a Gyrating Field." *Physical Review* 51: 652–654.

Rabi, I. I. 1981. Interviews with the author, September 10–11, New York.

Rabi, I. I. 1982. Interview with the author, July 22, New York.

Rabi, I. I., and V. W. Cohen. 1933. "The Nuclear Spin of Sodium." *Physical Review* 43: 582–583.

Rabi, I. I., and V. W. Cohen. 1934."Measurements of Nuclear Spin by the Method of Molecular Beams." *Physical Review* 46: 707–712.

Rabi, I. I., J. M. B. Kellogg, and J. R. Zacharias. 1934. "The Magnetic Moment of the Proton." *Physical Review* 46: 157–163.

Rabi, I. I., J. R. Zacharias, S. Millman, and P. Kusch. 1938. "A New Method of Measuring Nuclear Magnetic Moment." *Physical Review* 53: 318.

Rabi, I. I., S. Millman, P. Kusch, and J. R. Zacharias. 1939. "The Molecular Beam Resonance Method for Measuring Nuclear Magnetic Moments." *Physical Review* 55: 526–535.

Rigden, J. S. 1983. "Molecular Beam Experiments on the Hydrogens During the 1930s." *Historical Studies in the Physical Sciences* 13(2): 335–373.

Schwinger, J. 1937. "On Nonadiabatic Processes in Inhomogeneous Fields." *Physical Review* 51: 648–651.

Schwinger, J. 1977. "The Majorana Formula." *Transactions of the New York Academy of Sciences* 38: 170–184.

Segrè, E. 1973. "Otto Stern, 1888–1969." *Biographical Memoirs* 43: 215–236.

Segrè, E. 1982. Interview with author, November 8, Berkeley, California.

Sommerfeld, A. 1916. "Zur Quantetheorie der Spectrallinien." *Annalen der Physik* 51: 1–94.

Sommerfeld, A. 1928. *Atomic Structure and Spectral Lines*, second edition. London: Methuen.

Van Vleck, J. H. 1970. "A Third of a Century of Paramagnetic Relaxation and Resonance." In *Magnetic Resonance*, ed. C. K. Coogan et al. New York: Plenum.

Zacharias, J. R. 1982. Interview with author, March 19, Newton, Massachusetts.

Zeeman, P. 1897. "On the Influence of Magnetism on the Nature of the Light Emitted by a Substance." *Philosophical Magazine* 43: 226–239.

Zorn, J. C. 1964. "Resource Letter MB-1 on Experiments with Molecular Beams." *American Journal of Physics* 32: 721–732.

9   Artificial Disintegration and the Cambridge-Vienna Controversy

Roger H. Stuewer

In April 1919, Ernest Rutherford reported an extraordinary new discovery: The nucleus of nitrogen could be artificially disintegrated by bombarding it with RaC α particles.[1] Reflecting on this discovery a decade later, Rutherford placed it in a broad historical context:

> The idea of the artificial disintegration or transmutation of an element is one which has persisted since the Middle Ages. In the times of the alchemists the search for the "philosopher's stone," by the help of which one form of matter could be converted into another, was pursued with confidence and hope under the direct patronage of rulers and princes, who expected in this way to restore their finances and to repay the debts of the state. . . . The failures were many and the natural disappointment of the patron usually vented itself on the person of the alchemist; the search sometimes ended on a gibbet gilt with tinsel.[2]

Whatever the merits of this capsule history, Rutherford, the modern alchemist, identified in it the principal characteristics of science: observation, experiment, and theory, all supported by an institution in a position to reward or punish the scientist. By the time Rutherford wrote these words, no one was more familiar than he with each of these factors, and their delicate interrelationships, in the field of artificial disintegration. He and his younger colleague at the Cavendish Laboratory, James Chadwick, had just emerged from what K. K. Darrow termed "one of the most famous controversies of modern physics"[3]—a controversy with Hans Pettersson and Gerhard Kirsch at the Institut für Radiumforschung in Vienna. I shall examine the origins, evolution, and outcome of that controversy.

Rutherford's Discovery and Interpretation of Artificial Disintegration, 1919–1922

Rutherford's discovery of artificial disintegration was the culmination of years of intermittent wartime research carried out at the University of Manchester with the help of his laboratory assistant, William Kay. It rested upon the microscopic observation of scintillations—tiny flashes of light produced when charged particles strike a "scintillation screen," a thin glass plate covered with zinc sulfide crystals containing a slight metallic impurity, such as copper.[4] This method of observation had been employed continuously by Rutherford since 1908, when he and Hans Geiger proved that a single charged particle (in this case an α particle) produces a single scintillation.[5] Rutherford also knew that the intrinsic brightness of a scintillation depends only upon the energy or residual range of the particle producing it, while the observed brightness depends upon the optical properties of the microscope employed as well. For good results, the microscope had to have an objective lens of high light-gathering power (as measured by its numerical aperture), an overall magnification of about $40\times$, and a field of view of several square millimeters. It was also advantageous to illuminate the field of view faintly to prevent the eye from straying.

The observations themselves were tedious and delicate. The most accurate counts were obtained when about 40 scintillations were observed per minute; more than about 80 or fewer than about 10 made the counting "troublesome and uncertain."[6] At least two people were required, one to vary the experimental conditions and the other to do the counting. Rutherford described the procedure as follows: "Before beginning to count, the observer rests his eyes for half an hour in a dark room and should not expose his eyes to any but a weak light during the whole time of counting.... It was found convenient in practice to count for 1 minute and then rest for an equal interval, the times and data being recorded by the assistant. As a rule, the eye becomes fatigued after an hour's counting and the results become erratic and unreliable. It is not desirable to count for more than 1 hour per day, and preferably only a few times per week."[7] Even then, the counters generally improved with practice. As Rutherford later explained, "the superior efficiency of an experienced observer appears to be due to greater concentration, to control of spontaneous movements of the eye, and to practice in using the excentral portions of the retina, thereby avoiding the insensitive fovea-centralis."[8] In other words, the observation of scintillations, simple in principle, actually was a highly complex process that depended on the optical system employed and the training, experience, physical health, and psychological state of the observer. Thus, when Rutherford, in reporting his discovery of artificial disintegration in April 1919, thanked William Kay "for his invaluable assistance in counting scintillations,"[9] he was not engaging in polite banter; he was accurately acknowledging Kay's contributions. Kay probably was at the microscope when the "surprising effect" was observed that ultimately led Rutherford to conclude that RaC α particles could disintegrate the nitrogen nucleus, liberating, as he wrote, "probably atoms of hydrogen or atoms of mass 2."[10] If they indeed were hydrogen nuclei, as he strongly suspected, it meant that they had "formed a constituent part of the nitrogen nucleus."

In other words, hydrogen nuclei, or protons (as Rutherford soon called them), had to join α particles and electrons as fundamental constituents of nuclei.[11] But Rutherford went further: He boldly proposed a theoretical interpretation of the disintegration process based upon a definite model of the nitrogen nucleus. Observing that the atomic weight of nitrogen, at 14 units, is of the form $4n + 2$, where n is an integer, Rutherford suggested that the nitrogen nucleus consists of a central core composed of three tightly bound α particles surrounded by two loosely bound hydrogen "outriders" (which, of course, were what the incident α particles were dislodging).[12] Thus, from the outset, theory and experiment were coupled together in Rutherford's mind. He was indeed a "crypto-theoretician," as Maurice Goldhaber recently characterized him.[13]

Rutherford's discovery of artificial disintegration crowned his career at the University of Manchester. A few months after he reported it he transferred to Cambridge to become J. J. Thomson's successor as Cavendish Professor of Experimental Physics. He had hoped to take Kay with him to Cambridge, but at the last minute Kay decided to remain in Manchester for family reasons. One who did accompany Rutherford, however, was James Chadwick. Born in 1891 in Bollington, south of Manchester, Chadwick obtained his M.S. degree at Manchester in 1913.[14] On Rutherford's recommendation he then received an 1851 Exhibition Senior Research Studentship for further study in Hans Geiger's laboratory in the Physikalisch-Technische Reichsanstalt in Berlin-Charlottenburg, where he carried out important researches on β decay and mastered the German language. A year later, however, with the declaration of war on August 4, 1914, Chadwick was trapped in Berlin. He was arrested and imprisoned for ten days, and then interned in racehorse stables near Spandau, on the western outskirts of Berlin. Although he soon met C. D. Ellis and was able to pursue some research with the help of Heinrich Rubens and Walther Nernst (Geiger had been called to active duty), Chadwick suffered severely during the winter famine of 1916–17, which left his digestion permanently impaired. The tall, slender young man, by nature "shy and reserved" and "severe and forbidding" in appearance,[15] returned to Manchester after the armistice (November 11, 1918) at the age of 27 with no particular fondness for Germans in his heart. Half a year later, he went with Rutherford to Cambridge, where he first served as an expert scintillation counter. He was, however, a gifted experimental physicist, and his responsibilities soon increased. By 1921 he was collaborating fully with Rutherford, and a year or so later he was appointed assistant director of the Cavendish Laboratory.

Rutherford extended his researches on artificial disintegration immediately after his arrival at the Cavendish. By June 1920, when he delivered his second Bakerian lecture,[16] he had proved conclusively that the particles from nitrogen were indeed hydrogen nuclei. He also had found that no hydrogen nuclei are expelled by RaC α particles "in appreciable numbers from carbon, silicon, or oxygen."[17] But the result to which he seemed to attach the most importance was his finding that RaC α particles apparently could expel an entirely new particle of charge 2 and mass 3 from nitrogen and oxygen nuclei.[18] Rutherford took this new particle, which he symbolized by X_3^{++}, to be yet another new fundamental nuclear constituent, a tightly bound compound of three hydrogen nuclei and one nuclear electron. It served as the point of departure for his famous speculation on the existence of the neutron,[19] a tightly bound compound of one hydrogen nucleus and one nuclear electron—a speculation that, over the next decade, became firmly implanted in Chadwick's mind.[20] It also led Rutherford to propose "purely illustrative" models of three isotopes of lithium, and of carbon, nitrogen, and oxygen nuclei, [21] as shown in figures 1 and 2. In these models the minus signs represent nuclear electrons, and the circles with mass numbers 1, 3, and 4 inside and plus signs outside represent the proton, the X_3^{++} particle, and the α particle, respectively. Note that the lithium, carbon, and oxygen nuclei contain no protons ready for expulsion, while the nitrogen nucleus contains two. Rutherford pointed out explicitly that when a proton is expelled from nitrogen the "residual nucleus should thus have a nuclear charge 6 and mass 13, and should be an isotope of carbon."[22] This conclusion followed directly from Rutherford's model of the nitrogen nucleus and his picture of the disintegration process; we will see later how he came to abandon it.

Rutherford soon published two further papers on artificial disintegration, both with Chadwick as co-author.[23] By continually improving

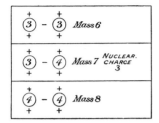

Figure 1
Rutherford's models of three isotopes of lithium. Source: "Nuclear Constitution of Atoms," *Proc. Roy. Soc. A* 97 (1920), p. 398. Reproduced by permission of the Royal Society.

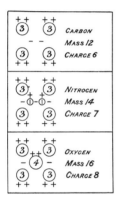

Figure 2
Rutherford's models of carbon, nitrogen, and oxygen nuclei. Source: "Nuclear Constitution of Atoms," *Proc. Roy. Soc. A* 97 (1920), p. 399. Reproduced by permission of the Royal Society.

their optical system to detect weak scintillations produced by particles near the end of their range, Rutherford and Chadwick established that RaC α particles incident on nitrogen expel protons of maximum range 40 cm in air, and that when incident on hydrogen they expel "natural" protons of maximum range 29 cm in air. Thus, if mica absorbers equivalent to, say, 32 cm of air were inserted in front of the scintillation screen, no "natural" protons from possible hydrogen contaminants would be able to reach it. By means of this technique Rutherford and Chadwick found that, in addition to nitrogen, RaC α particles could expel protons of range greater than 32 cm from five, and only five, light elements between atomic numbers 3 and 16: boron (proton range ca. 45 cm), fluorine (over 40 cm), sodium (ca. 42 cm), aluminum (90 cm), and phosphorus (ca. 65 cm). The remaining light elements they tested (Li, Be, C, O, Mg, Si, S), and certain elements of higher atomic number (Cl, K, Ca, Ti, Mn, Fe, Cu, Ag, Sn, Au), yielded no detectable disintegration protons. In other experiments, Rutherford and Chadwick found that the maximum range of the protons from nitrogen and aluminum decreased as the range of the incident α particles decreased. They also found that the direction of emission of the protons from aluminum was largely independent of the direction of incidence of the bombarding α particles, although the maximum proton range in the backward direction (67 cm) was smaller than that in the forward direction (90 cm). They established that such a difference in range existed for the protons expelled from each of the six disintegrable elements.[24]

These experimental results blended into and consequently reinforced Rutherford's theoretical interpretation of the disintegration process. Thus, each of the six disintegrable elements ($^{11}_5$B, $^{14}_7$N, $^{19}_9$F, $^{23}_{11}$Na, $^{27}_{13}$Al, $^{31}_{15}$P) was of odd atomic number, and the atomic weight of each was of the form $4n + 3$ or (for nitrogen) $4n + 2$. This indicated to Rutherford that each nucleus consisted of a central core of α particles surrounded by either two or three loosely bound proton outriders—"satellites," as Rutherford now called them. Since both core and satellites are positively charged, Rutherford also noted that this model "implicitly assumes that positively charged bodies attract one another at the very small distances involved."[25] An incident α particle, therefore, would be subjected first to a repulsive force and then to an attractive force when penetrating into these nuclei.

Still more could be inferred from the experimental results. Since the maximum range of the protons from nitrogen and aluminum decreased as the range of the incident α particles decreased (that is, since there

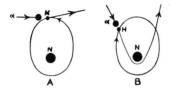

Figure 3
Rutherford's model of artificial disintegration. An α particle strikes a proton satellite at different points in its orbit about the central nuclear core. Source: "The Artificial Disintegration of Light Elements," *Phil. Mag.* 42 (1921), p. 822. Reproduced by permission of Taylor & Francis, Ltd.

was a correlation between the two ranges), Rutherford argued that the disintegration process could not be attributed to an "atomic explosion"— the intensity of an explosion does not depend upon how strongly the detonator is depressed. Rather, this correlation followed directly from the collision process envisioned by Rutherford. In addition, by assuming that the incident α particle can strike the proton satellite at different points in its orbit about the central core, as shown in figure 3, it was understandable that the protons are expelled in all possible directions. In sum, Rutherford's satellite model of the nucleus provided an internally consistent interpretation of all the artificial-disintegration results; theory and experiment meshed harmoniously.

The Challenge from Vienna, 1923–24

For more than three years after mid 1919, Rutherford and Chadwick had the field of artificial disintegration to themselves; new experimental results and deeper theoretical understanding emerged solely from the Cavendish Laboratory. This state of affairs changed dramatically in 1923. Publications began to appear in increasing numbers from researchers working in another laboratory, the Institut für Radiumforschung in Vienna. This institute had been established by funds provided to the Vienna Academy of Sciences in August 1908 by a private donor for the purpose of exploiting the radioactive ores of the Austro-Hungarian Empire—"the cradle of the radioactive substances"—for purely scientific purposes.[26] The model of Marie and Pierre Curie in Paris, who had received large quantities of ore at nominal cost from Joachimstal in Bohemia, indicated what might be achieved in Vienna. Franz Exner was named director of the Vienna institute. However, Exner immediately turned the planning and supervision of construction over to one of his assistants, Stefan Meyer.[27] A descendant of an old and honored Viennese family, Meyer had become an authority in the field of radioactivity after receiving his Ph.D. degree in Vienna in 1896 and while serving as an assistant to Ludwig Boltzmann until Boltzmann's suicide in 1906. Meyer therefore became the actual leader of the Institut für Radiumforschung from the beginning. He saw that the entire four-story structure, including its furnishings and apparatus, was finished and in place on the Boltzmanngasse in only two years. The institute was officially opened in October 1910. When Exner retired in 1920, Meyer was officially named its director.

Meyer, born in 1892, was less than a year younger than Rutherford, and the two became good friends. In 1907, Meyer arranged a loan of about 300 mg of $RaBr_2$ from the Vienna Academy of Sciences to be used jointly by Rutherford and William Ramsay, and when jurisdictional difficulties developed between them Meyer arranged a second loan in 1908 of about 400 mg of $RaCl_2$ to Rutherford alone.[28] This loan enabled Rutherford to establish his research program in Manchester, and Rutherford never forgot Meyer's generosity. In 1910 the two became co-workers when the International Radium Standard Commission was established in Brussels, and Rutherford was elected president, Meyer secretary. The following year, Rutherford was elected a corresponding member of the Vienna Academy, undoubtedly on Meyer's initiative. In 1916, Meyer and Egon R. v. Schweidler published their extensive treatise *Radioaktivität*,[29] which became the standard reference work in the German language, just as Rutherford's *Radioactive Substances and Their Radiations*[30] (1913) became the standard reference work in the English language.

Meyer was beloved by all. During the Great War, he went far out of his way to protect the English physicist R. W. Lawson from internment, and he provided Lawson with research facilities at the institute, personal loans without interest, and gifts on holidays and birthdays—a very different treatment than Chadwick received in Berlin.[31] After the war, in 1921, when informed of the rampant inflation and desperate living conditions in Vienna, Rutherford prevented the radium lent to him earlier by Meyer from being confiscated by the British government as enemy property, and he arranged to purchase 20 mg of it for £540 in hard English currency. In 1927, Rutherford made arrangements to purchase 250 mg more for £3,000 over six years.[32] Both purchases contributed substantially to saving Meyer's institute from financial ruin, and they strengthened the bond of friendship and respect between the two directors. Meyer regarded Rutherford, the outspoken New Zealander, as the undisputed leader in the field of radioactivity. Rutherford had a high regard for Meyer as a scientist and as a man of never-failing kindness. Rutherford directed his laboratory with a firm hand. Meyer was a tolerant and benevolent leader, a mild critic of those who worked under him.[33] Both envisioned their laboratories as contributing cooperatively to the advancement of knowledge in nuclear research.

In late 1921, Meyer received a letter dated November 28 in Monaco from a 33-year-old Swedish scientist from Göteborg, Hans Pettersson.[34] Son of a famous oceanographer, Pettersson had earned first and second degrees at the University of Uppsala in 1909 and 1911, had studied further in London under William Ramsay, Rutherford's old rival, and had received his doctorate at the University of Stockholm under K. Ångstrom in 1914.[35] By the end of 1921, Pettersson had almost thirty publications to his name in the fields of oceanography and geophysics, many dealing with the development of new instrumentation. He found himself in Monaco at that time, by invitation of Prince Albert I, measuring the radioactivity of deep-sea sediments. Pettersson's request to Meyer was simple: He wished to work at the Institut für Radiumforschung making related physical measurements. Meyer, who had a long record of providing facilities and hospitality to foreign scientists, cordially invited Pettersson to come to Vienna.

In April 1922, about a month after arriving in Vienna, Pettersson opened up a collaboration with Gerhard Kirsch. Born in Vienna in 1890, son of the professor of technical mechanics in Vienna's Technische Hochschule, Kirsch had entered the University of Vienna in 1911, had had his education interrupted by the Great War, and had studied further in Vienna and Uppsala after the war. He received his doctorate under Meyer in 1920 and then became one of Meyer's assistants at the Institut für Radiumforschung, working on his *Habilitationsschrift*. In 1922, the 31-year-old Kirsch was two years younger than Pettersson and had only had a few publications to his name. Not surprisingly, Pettersson immediately established himself as the leader of the pair (although this was not generally known outside Vienna). This arrangement suited Pettersson's personality. Pettersson was an energetic, charming, aggressive man who could be autocratic, domineering, and hot-tempered.[36] He was also a gifted fund raiser. As time went on, he secured increasing amounts of financial support from a variety of Swedish foundations, and at least one private Swedish donor, for the construction of new apparatus and the furtherance of his and Kirsch's work.[37] From the beginning, that work departed from Pettersson's original plans. Pettersson had recognized that the most challenging and exciting branch of physics in 1922 was the study of artificial nuclear disintegration.

Pettersson's research in Vienna was interrupted regularly by his return trips to Sweden, where since 1914 he had held an endowed lectureship in oceanography at the Göteborgs Högskola. Two such interruptions had already occurred between April 1922 and July 1923, and since a third was imminent he and Kirsch decided to publish a report of their progress to date. They were required by the statutes of the Institut für Radiumforschung to present their results to the Vienna Academy of Sciences for publication in its *Sitzungsberichte*. This they did on July 12. At the same time, they sent a summary to *Nature* and a full report to the *Philosophical Magazine*,[38] establishing a pattern of simultaneous publication in German and English periodicals that they were to repeat frequently. Their summary in *Nature* appeared in print first, on September 15, 1923.

From the outset, Pettersson displayed the ingenuity in the design and construction of experimental apparatus that would mark all of his work. His first achievement in this respect, which stimulated his subsequent work on artificial disintegration, was the development of new sources of RaC α particles of high and relatively constant intensity.[39] These consisted of long, thin capillary tubes filled with radium emanation and dry oxygen. Half the length of each tube was lined with the element to be bombarded; the other half was to be used in control experiments.

Ultimately, Pettersson and Kirsch decided to make their capillary tubes of "pure fused silica," which led to their first surprising result: They concluded that protons were being ejected from their capillary tubes, or, as they put it, that "silicon atoms under intense bombardment with α-particles break down in a manner similar to that of the elements disintegrated in the experiments of Rutherford and Chadwick."[40] This result was "unexpected" because Rutherford and Chadwick had not detected any disintegration protons from silicon. Pettersson and Kirsch therefore were moved to pursue a broader investigation, to examine "some of the lighter elements found 'non-active' in the experiments of Rutherford and Chadwick."[41] They invented another new α-particle source and employed three or four observers alternately, generally at night, for one or two minutes. They reported that beryllium, magnesium, silicon, and lithium were all disintegrable, yielding protons with maximum ranges of about 18 cm, 13 cm, 12 cm, and 10 cm, respectively.

Thanking Meyer for his support and his "kind interest" in their research,[42] Pettersson and Kirsch concluded, inevitably, that "the hydrogen nucleus is a more common constituent of the lighter atoms than one has hitherto been inclined to believe."[43]

Pettersson and Kirsch's letter to *Nature* provoked an immediate response from Cambridge. L. F. Bates and J. Stanley Rogers, two of Rutherford's research students at the Cavendish Laboratory, published a reply in the very next issue (September 22, 1923).[44] They noted that for the past six months they had been investigating the α particles emitted by radium active deposit and had observed scintillations that "strongly" suggested that, in addition to the usual α particles with a range of about 7 cm, RaC emits very small numbers of α particles of longer ranges: 9.3 cm, 11.1 cm, and 13.2 cm. Thus, it was likely, they concluded, that Pettersson and Kirsch's 10-cm, 12-cm, and 13-cm particles were not disintegration protons at all but rather these long-range α particles being emitted directly from the source.

Pettersson and Kirsch replied three weeks later: "The difference in brightness between the scintillations from α-particles and from H-particles viewed under identical conditions is so conspicuous, that no mistake is possible. Comparing the former to stars of the first magnitude, the latter would be of about the third magnitude; that is, a ratio in luminosity of about 6 to 1."[45] Pettersson had demonstrated this difference in brightness the previous summer before the physical section of the Skandinaviska Naturforskaremötet in Göteborg. In their present experiments, Pettersson and Kirsch admitted, they had observed "a small number of scintillations of α-type"; however, they assumed them to be produced by the "particles found by Sir Ernest Rutherford to be expelled from oxygen," which Rutherford had earlier taken to be X_3^{++} particles, and "now, apparently, α-particles of abnormally long range."[46] Thus, they were forced to uphold their "former view."

That conclusion prompted a mild, private intervention by Rutherford.[47] Writing to Meyer on November 24, 1923, he asked if Pettersson and Kirsch were working under Meyer's direction. He concluded that their research was a "valiant piece of work," but he added that in view of Bates and Rogers's results he was "afraid that one or two of the conclusions are doubtful." Three weeks later (December 15), Bates and Rogers themselves replied, in print. Using a new method "devised by Sir Ernest Rutherford," they had satisfied themselves "that the long range particles from radium active deposit are α-rays."[48] Moreover they had found that thorium active deposit emits small numbers of long-range α particles of ranges 11.5, 15.0, and 18.4 cm, and that actinium active deposit, and even the radioactive element polonium, seemed to fit this pattern as well.[49] Ten days earlier, Rutherford had communicated the full report of Bates and Rogers to the *Proceedings of the Royal Society.*[50]

A few months later, on April 7, 1924, a new voice was heard from Vienna. Dagmar Pettersson, Hans's wife, had sent RaC α particles through gold and copper absorbing foils and had found that the number that source emitted of range greater than 9.2 cm "was nil."[51] Simultaneously, Elisabeth Kara-Michailova and Hans Pettersson, using a special comparison microscope they had developed, reported that the brightness of scintillations produced by protons expelled from hydrogen or paraffin was about one-third that of α particles emitted by polonium.[52] Hence, proton and α-particle scintillations could be distinguished easily from each other by their brightness. Both reports appeared in *Nature* after full descriptions had been submitted to the *Sitzungsberichte* of the Vienna Academy.

Up to this point, the dispute had involved purely observational or experimental questions. Hans Pettersson, however, was not one to leave the theoretical ramparts unassailed. In November 1923 he sent a long paper to the Physical Society of London, which evidently was read to the society by A. W. Porter, F.R.S. It appeared in the society's *Proceedings* in April 1924.[53] In this paper Pettersson launched a direct attack upon Rutherford's satellite model of the nucleus as an interpretation of artificial disintegration. In the words of the abstract, Pettersson would offer "arguments for an alternative hypothesis which assumes that the α-particle communicates its energy to the nucleus as a whole, precipating an explosion."[54]

Pettersson proceeded to undercut Rutherford's "main arguments" for his satellite hypothesis one by one. He claimed that, by detecting relatively low-range protons from beryllium, magnesium, silicon, and lithium he and Kirsch had demonstrated that disintegrability was not an "exceptional quality" confined to nuclei with protons in "an exposed position"; it was "a general property, common to the nuclei of *all* atoms," as required by his explosion hypothesis. Similarly, the Vienna researchers had found relatively large numbers of disintegration protons, which argued against the small probability of a "satellite hit" implied by Rutherford's model. Finally, the fact that the proton range decreased as the α-particle range decreased, instead of arguing for Rutherford's satellite model, actually supported Pettersson's explosion hypothesis:

A "system of elastic forces under high tension," Pettersson asserted, "will undoubtedly react in proportion to the intensity of the shock [received]."[55]

Pettersson continued his demolition of Rutherford's satellite model by carrying out a detailed calculation from which he concluded that the observed ranges of the disintegration protons were consistent with his explosion hypothesis, whereas no such result was "suggested by the satellite hypothesis." There was also an aesthetic consideration: Rutherford's model, in which the positive central core attracted the positive satellites, implied that a positive α particle penetrating it would experience first a repulsive force and then an attractive force—the sign of the nuclear force (Coulomb's law) would alternate. "But on the whole," Pettersson claimed loftily, "the tendency of recent years has been against this view."[56] In general, the complexity of Rutherford's satellite model compared unfavorably with the simplicity of Pettersson's explosion hypothesis. Experiment, of course, would have to "decide between the alternative views," and Pettersson closed by listing several possible tests. "The best line of advance," he declared, "undoubtedly lies along the road of direct experiment . . . , which has been inaugurated by Sir Ernest Rutherford and his collaborators."[57]

Perhaps it was Pettersson's scientific arguments, but more likely it was the aggressive tone of his presentation that drew an immediate reaction from E. N. da C. Andrade, one of Rutherford's greatest admirers:

I have listened with great interest to Dr. Pettersson's Paper, but I think it would have been of greater value if some of the experiments fore-shadowed at the end had been performed before the theory was pro-pounded. . . . [The] distinction between the satellite and explosion theories is mainly a question of words at the present stage. . . . The only experimental evidence put forward [by Pettersson] . . . is not very striking. . . . The position is much as if a man having measured up a box and guessed from shaking it that it contained pieces of metal were to start speculating on the dates of the coins inside it. . . .[58]

Pettersson had a chance to reply by letter before Andrade's comments were published. "If," he wrote, "I may use the descriptive metaphor adopted by Prof. Andrade, even if we cannot hope to ascertain the date of the coins within the box, our only chance of getting to know anything at all about them seems to lie in shaking the box as thoroughly as possible, both by experiments and by speculation."[59] The debate clearly was heating up a bit.

Rutherford was not in the least inclined to change his views. He had lectured on his satellite model at the Royal Institution in the summer of 1923,[60] and again in Liverpool in September as president of the British Association.[61] Furthermore, he and Chadwick were making good experimental progress. At the end of March 1924 they sent off a preliminary report of new experiments to *Nature*,[62] and in August they submitted a full report to the *Proceedings of the Physical Society of London*.[63] Both journals were obviously chosen with Pettersson and Kirsch in mind.

The new experiments of Rutherford and Chadwick exploited their earlier observation that the disintegration protons are emitted in all directions with respect to the direction of the incident α particles. This meant that by choosing an angle of observation of 90° they could automatically eliminate any protons from possible hydrogenous contaminants, since such protons would be driven forward at lesser angles by the incident α particles. This "simple method" enabled Rutherford and Chadwick to detect "with certainty" disintegration protons of range down to only 7 cm in air. Their results were surprising. They found that, besides their original six elements (B, N, F, Na, Al, and P), which yielded protons with ranges of 40–90 cm, seven additional elements (Ne, Mg, Si, S, Cl, A, and K) yielded protons with ranges of 16–30 cm, and one other (Be) gave a "small effect," although the latter was probably due to the presence of a fluorine impurity. No definite results were found for any elements between calcium and iron (atomic numbers 20–26) or for twelve elements of higher atomic number. "The other light elements, hydrogen, helium, lithium, carbon, and oxygen, . . . [gave] no detectable effect beyond 7 cm."[64] Rutherford and Chadwick summarized their results for the light elements between boron and potassium in the bar graph shown in figure 4. Note that elements of odd atomic number generally yielded protons of much greater range than those of even atomic number, and note especially the complete absence of protons from carbon and oxygen.

Four days after these results were summarized in *Nature* (March 29, 1924), Pettersson and Kirsch dashed off a response.[65] They could not contain their delight that Rutherford and Chadwick had "confirmed [their] results" for magnesium and silicon, with "less decisive" results for beryllium. They went on to claim that Rutherford and Chadwick's new right-angle method of observation "appears to be in many respects similar to one we have been using for some time"—one they had first described at a meeting of the Deutsche Physikalische Gesellschaft in Vienna six weeks earlier, on February 25. Finally, Pettersson and Kirsch reported entirely new experimental results. They had found recently,

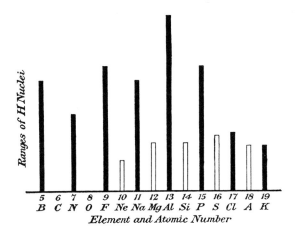

Figure 4
Rutherford and Chadwick's bar graph of the disintegrable elements. The heights of
the bars are proportional to the maximum ranges in air of the expelled protons.
Source: "Further Experiments on the Artificial Disintegration of Elements," *Proc. Phys.
Soc. Lond.* 36 (1924), p. 421. Reproduced by permission of The Institute of Physics.

they wrote, "that carbon, examined as paraffin, as very pure graphite,
and finally as diamond powder, gives off H-particles of about 6 cm.
range," with a yield "of the order [of] 200 per 10^7 of the [incident] α-
particles." Moreover, "Other experiments . . . seem to prove that oxygen
is also disintegrable and gives off α-particles of 9 cm. range in the
forward direction." This was "the first example of α-particles as a product
of artificial disintegration," and it accounted for the 9.3-cm particles
of Bates and Rogers. In general, it was "of interest to observe" that
these results "lend support to . . . the 'explosion hypothesis'
. . . propounded by one of us."[66]
    Meyer, too, as he informed Rutherford on May 13, 1924, was glad
that Rutherford and Chadwick had confirmed "some of the results"
of Pettersson and Kirsch.[67] Neither that message from Vienna nor Pettersson and Kirsch's paper pleased Rutherford and Chadwick. They
squeezed in a reply before their full report went to press. They stated
bluntly that Pettersson and Kirsch's high proton yields from beryllium,
magnesium, and silicon simply "cannot be reconciled with ours, and
the probable explanation . . . is that the particles they observed were
the long-range α-particles emitted by the source."[68] But there was an
even more serious discrepancy: Rutherford and Chadwick had looked
for disintegration protons down to a range of 2.6 cm of air from carbon

in the form of a "sheet of Acheson graphite" and had found no evidence for their existence whatsoever. "This result," they said, "is in complete disagreement with the experiments of Kirsch and Pettersson . . . , who found a very large number of H-particles of 6 cm. range."[69]

Rutherford and Chadwick also stood firm on the theoretical issue. Indeed, they reported new evidence favoring Rutherford's satellite model. Thus, the change from a repulsive to an attractive force within the nucleus implied the existence of a critical potential surface where the forces balanced, and this in turn had two implications: that the energy of the incident α particles would have to exceed a certain minimum value before they would be able to penetrate the target nucleus, and that the disintegration protons would possess a certain minimum energy or range. Both predictions seemed to be verified for the aluminum nucleus; it appeared that the incident α particles had to have an energy of about 3 MeV before penetration became possible, and that the expelled protons had a minimum range of about 13–14 cm. The "fate" of the incident α particles was unknown—it was "possible" that they became "in some way attached to the residual nucleus." At any rate, Rutherford's satellite model, which had suggested the existence of this critical potential surface within the nucleus, had "the great merit of simplicity."[70]

Rutherford and Chadwick immediately carried their experiments forward to settle conclusively what was by this time an old question: By a variety of different means, they established to their complete satisfaction that RaC does indeed emit long-range α particles with ranges of 9.3 and 11.2 cm, as Bates and Rogers had claimed. At the same time, they found no evidence for α particles with a range of 13 cm—these particles, instead, were probably protons expelled from the mica absorbing foils Bates and Rogers had used in their experiments.[71] These results were a mixed blessing from Rutherford's point of view. On the one hand, they meant that Pettersson and Kirsch were wrong in believing that the 9.3-cm range particles were disintegration protons. On the other hand, they meant that Rutherford was wrong in believing that these particles were new X_3^{++} particles from oxygen and nitrogen. Therefore, the nuclear models Rutherford had presented in his Bakerian lecture of 1920 were definitely incorrect. Rutherford and Chadwick also found that the slightest imperfections in thin copper foils greatly influenced their absorbing power. This, they said, explained the "failure of Fr. Dagmar Pettersson to detect the long-range [α] particles" from RaC when using such absorbing foils in her experiments.[72]

Rutherford and Chadwick submitted these results to the *Philosophical Magazine* in July 1924. This was an unusually busy period for Rutherford. Not only was he concluding the above experiments; he had just returned from London, where he had received the Franklin Medal at the Foreign Office, and he was struggling with preparations to preside over the meeting of the British Association in Toronto later that month, to lecture at the centenary celebrations of the Franklin Institute in Philadelphia, and to lecture at various Canadian and American universities. In the midst of this rush, on July 17, 1924, he received an extremely upsetting letter from Pettersson.[73] Pettersson informed Rutherford that in April he had visited Rutherford's esteemed friend Niels Bohr in Copenhagen, and that Bohr had expressed keen interest in the Vienna experiments. Furthermore, as Pettersson was now about to leave Vienna for Göteborg for the summer, he was enclosing a copy of a new letter to *Nature* by himself and Kirsch that he said he hoped might "interest" Rutherford. Pettersson closed his letter by conveying Meyer's "kindest regards" to Rutherford.

Rutherford knew he had to act. The very next day, July 18, 1924, Rutherford wrote a long letter to Bohr that left Bohr in no doubt whatsoever about Rutherford's views on Pettersson and his work:

I gather from Pettersen [sic] that you have been in touch with him on his work in Vienna. He seems a clever and ingenious fellow, but with a terrible capacity for getting hold of the wrong end of the stick. From our experiments, Chadwick and I are convinced that nearly all his work published hitherto is either demonstrably wrong or wrongly interpreted. For example, he claims to get a large number of particles from carbon. We found practically none under the same conditions and we consider that there is no evidence at all of the disintegration of carbon. We have equally failed to observe any effect in lithium or oxygen, and only a slight trace from beryllium possibly due to impurity in the form of fluorine. I am writing to Pettersen privately giving him my views of the situation. It is a very great pity that he and his collaborators are making such a mess of things, for it is only making confusion in the subject.

All the experiments look easy, when they are really very difficult and full of pitfalls for the inexperienced. So much is this so that I have decided not to get any other work done except under my personal eye.

I am sorry that Pettersen has made such a mess but it looks to me as if he has not done nearly enough experiments on broad experimental lines to make sure of his points, but jumps precipitately to conclusions from rough evidence.

I thought I would tell you about this matter which *please regard as private*, as you might be wondering what was the real situation.[74]

True to his word, Rutherford immediately wrote to Pettersson as well.[75] He opened graciously, telling Pettersson that he was "very pleased" that the "Radium Institute under my old friend Prof. Stefan Meyer was attacking the problem of artificial disintegration." This work, as no one recognized more than he, was filled with "many difficulties and pitfalls . . . , apart from the difficulties inherent in the counting of weak scintillations." He appreciated Pettersson's "care and ingenuity" in devising new experimental methods. At the same time, he had been "considerably surprised not only at the attitude taken but also at some of the results." Thus, it was "desirable at this stage to be quite frank." He and Chadwick were of the "definite opinion" that Pettersson's "work and conclusions" were "demonstrably wrong" and "tended to confuse the whole subject." To set matters right, Rutherford therefore listed, in detail, Bates and Rogers's and his and Chadwick's "very strong" evidence on the long-range α particles emitted by RaC, and on the nondisintegrability of carbon and oxygen, wondering if Pettersson's conflicting results might not arise from contaminants in Pettersson's apparatus. In general, Rutherford hoped that Pettersson would regard this as "a friendly if critical letter." He appreciated Pettersson's "many good original ideas on this problem, but 'experientia docet.' " The subject was "full of pitfalls for the beginner and even for the veteran, as I may claim to be." Chadwick, if Pettersson could manage to come to Cambridge, would be happy to discuss everything further while Rutherford was away. "It is better to discuss these divergences of view in private than in print. Workers in this field are too few and too select to misunderstand one another."

Rutherford told Pettersson that he was sending a copy of his letter to Meyer as well. This he did, along with a covering letter that must have shocked his friend:

Two days ago I received from Dr Pettersson an advance copy of a long letter he proposes to publish in *Nature*. Chadwick and I have definite evidence that many of the results in it are either wrong or wrongly interpreted. So I have written him a long letter giving him my reasons and suggesting further consideration before publication. . . . You know me well enough to appreciate that I would not interfere unless I thought the situation was serious. I do not know Pettersson personally or his co-workers, but you do. He seems to me a man of originality and ingenious in his arrangements but I should judge he jumps to conclusions on insecure evidence. The subject of artificial disintegration is full of difficulties and pitfalls and wants investigators who are very careful in experiment and with good judgment. I am sorry to bother you over the matter but you will quite appreciate how important it is to Dr Pettersson and to your laboratory and also to the subject not to go along wrong lines.[76]

Let us pause for a moment to appreciate the situation. Here was Rutherford—holder of the Rumford Medal of the Royal Society awarded in 1905, Nobel laureate of 1908, knighted in 1914, recipient of the Copley Medal (the highest award of the Royal Society) in 1922, president of the British Association in 1923, and slated to become president of the Royal Society (1925–1930) and recipient of the most coveted award of all, the Order of Merit (1925)—here was Rutherford, by far the most distinguished nuclear physicist of the period, placing Pettersson, the novice, on notice, and not only Pettersson but Meyer and his Institut für Radiumforschung as well. It would be enough to stop almost anyone.

Yet Meyer was unconvinced. He replied to Rutherford from his summer residence in Bad Ischl on July 24, even though he had no idea when and where Rutherford would receive his letter.[77] Meyer appreciated Rutherford's "kind intention most heartily," but, he said, he knew Pettersson and his co-workers to be "very capable trustworthy solid researchers" whose "numerous measurements" were made "with every care." Both they and he were "rather sure" that Bates and Rogers's experiments were "not convincing," for neither in Vienna nor, apparently, in Berlin (researchers in Otto Hahn and Lise Meitner's laboratory) could the long-range particles from RaC, ThC, or Po be detected. Perhaps, Meyer suggested, the Cambridge and Vienna experimenters were both right: Rutherford and his co-workers might be using a metallic RaC source and observing α particles emitted directly from it, while the source employed by Pettersson and his co-workers might be somewhat oxidized, so that the emitted α particles might be expelling H nuclei from oxygen close to the RaC nucleus. This suggestion might be "totally wrong," but Meyer thought it would be "very useful to clear up this point." In any case, Meyer agreed that the "divergences" should be resolved through "personal communications."

Pettersson himself was completely unwilling to yield. He was in the ascendent and was supremely confident of his position. And he had more at stake than Rutherford was aware of. He and Kirsch had just published a long review article on artificial disintegration in *Die Naturwissenschaften* (June 20, 1924).[78] Moreover, only three days before writing to Rutherford, on July 10, 1924, he and Kirsch, singly and jointly, had presented to the Vienna Academy four papers that would soon appear in its *Sitzungsberichte*. The first, by Gustav Ortner and Pettersson,[79] described an improved RaC source; the second, by Kirsch and Pettersson,[80] attempted to establish their priority in developing the right-angle method of observation; the third, by Pettersson alone,[81] described his experiments proving (he maintained) that protons of range 16 cm in the forward direction and of range 8 cm at 90° are expelled from carbon; and the fourth, by Kirsch alone,[82] discussed his and Pettersson's observations of protons expelled from nitrogen and oxygen nuclei. The last two of these papers, each comprising 16 printed pages, also argued at length that Pettersson's explosion hypothesis was preferable to Rutherford's satellite model as a theoretical interpretation of all the existing experimental data on artificial disintegration both from Vienna and from Cambridge.

Pettersson revealed his position directly some weeks later when he drafted a response to Rutherford's letter, which had been forwarded to him in Sweden:

Allow me to say how much I appreciate the spirit of friendly criticism in which your letter is written! I have followed your suggestion and written to the editor of Nature asking him to defer the publication of our last letter, submitted to you in M.S. I feel that we owe that also to Professor Meyer who has generously helped us with all facilities for our work, both materially and with most valuable advice and with many suggestions. Although he has practically allowed us a free hand with regard to publishing our results, we cannot claim his authority for the views we have put forward. The mistakes we may have made are our own, and I feel certain that Prof. Meyer will agree with you, as I do myself, that the less of eventually erronous [sic] views there appear in print the better.[83]

However, Pettersson trusted that Rutherford would allow him to "state frankly" his "personal view" of the points Rutherford had made. Apparently Rutherford now agreed that Bates and Rogers's 13-cm-range particles from RaC "do not exist," and that Bates and Rogers were wrong in claiming that the particles found in Vienna a year earlier from beryllium, magnesium, and silicon were long-range α particles. Moreover, Pettersson supposed that Rutherford had seen in Die Naturwissenschaften that "Dr. Philip working in the laboratory of Frau Lise Meitner" had failed to find particles from ThC of range greater than 11.3 cm. At the same time, Pettersson conceded that Rutherford's evidence for the 9.3-cm particles from RaC was "no doubt strong." But that was as far as Pettersson would go. He listed five reasons for his belief that protons with a range of about 9.3 cm are expelled from oxygen and that they are "distinctly more numerous" than those with a range of about 11 cm from nitrogen. "I must confess," he stated, "that I cannot at present see any explanation for the discrepancy between these results and those found in your laboratory."

The question of carbon was even more serious. Pettersson declared that he was "most decidedly" of the opinion that the H particles found in Vienna from carbon "cannot be due to any impurities, as the substances we have examined have been very pure indeed (dry carbonic acid, powdered white diamonds, Siberian graphite of the purest kind, and, lately, Acheson 98% graphite)." The presence of radioactive contaminants was also "most unlikely," because of the various methods they had developed for their elimination. Furthermore, wrote Pettersson,

... our newest microscope with the scintillation-screen directly attached to the front lens ... is so superior, with regard to brilliancy of the scintillations viewed through it, ... that we feel much more confident now, not only in differentiating between scintillations from H- and from α-particles, but also in not overlooking the former even when the particles are relatively near the end of their range. For this reason alone I regard a confusion between H-particles and contamination α-particles as improbable.

Of course, Pettersson would "gladly avail" himself of the opportunity to "confer with Dr. Chadwick," but he was not certain that he could come to England in the fall. He would, however, send copies of his letter to Chadwick, as well as to Kirsch and Meyer. He begged Rutherford's "indulgence" for stating his views "at such length." "Anyhow," he concluded, "I hope that I have not forfeited the privilege of communicating our results to you and of asking your opinion on questions belonging to this most interesting field of research, where beginners can have no higher ambition than that of working along the lines laid out by its veteran."

Pettersson wrote to Meyer that same day, July 27.[84] It was, he told Meyer, "very advantageous to be in direct correspondence" with Rutherford, and because Rutherford "considers it important" he has withdrawn his and Kirsch's letter to *Nature*, although there would be nothing to fear even if it were published without change. To do so, however, would certainly prompt an answer from Cambridge, a sharp polemic, which might not help their continued work in Vienna, and which, if he was not mistaken, Rutherford also wished to avoid because he then would have to "partially retract" the Cambridge experimental results in print and perhaps his satellite hypothesis as well.

Pettersson enclosed a copy of the draft of his answer to Rutherford, and after receiving Meyer's approval of it, he sent it off on August 8, as well as a copy and covering letter to Chadwick. Meyer also had sent Pettersson a copy of his own reply to Rutherford, and when Pettersson wrote again to Meyer on August 8, he thanked Meyer warmly "for backing us up so generously."[85] He also thanked Meyer for calling his attention to a report, published in *Engineering*,[86] of a lecture that Rutherford had given at the Royal Institution in London on April 4, 1924, summarizing the Cambridge experiments. Pettersson analyzed the disintegration data presented in this report in detail for Meyer, showing in particular how it supported his explosion hypothesis. Finally, Pettersson told Meyer that the latter's suggestion concerning the possible difference between the RaC sources used in Cambridge and in Vienna would certainly give Rutherford "food for thought."

Rutherford was abroad when Pettersson's letters arrived in Cambridge in August 1924, and hence Chadwick was the first to learn of his intransigence. Chadwick reacted by writing to Rutherford in September while vacationing in the Scottish Highlands.[87] He first brought Rutherford up to date on work at the Cavendish. Of particular importance was the cloud-chamber evidence that P. M. S. Blackett had just obtained indicating that an α particle, when disintegrating a nitrogen nucleus and expelling a proton, was actually captured by the nitrogen nucleus. This was "a very fine addition to the evidence for the attractive field," and it fitted in "very well with our expectations," wrote Chadwick. Referring to other experiments, he remarked "I think we shall have to make a real search for the neutron"—a suggestion that Chadwick pursued unsuccessfully for about the next two years.[88] Finally, Chadwick turned to Pettersson: "I suppose," Chadwick wrote, "you have received Pettersson's letter. He sent me a copy and also an additional short letter to which I made a very mild and conciliatory reply in the hope of inducing him to visit us. I don't think he will come, however."

Chadwick's intuitions were correct. Pettersson returned directly to Vienna from Göteborg in the fall of 1924 to resume his researches with Kirsch and their collaborators, all with the support of Meyer. Meyer—there can be no doubt—was strongly attracted to the dynamic visitor from Sweden, who was inspiring everyone around him with his evident talent and determination. In fact, all hope for reconciliation would have to evaporate before a resolution of the controversy could take place.

Stalemate, 1924–1927

Observational, experimental, and theoretical issues had become intimately intertwined for researchers at both the Cavendish Laboratory and the Institut für Radiumforschung. Neither Rutherford and Chadwick nor Pettersson and Kirsch could modify one aspect of their position without affecting others. As their positions hardened, each team became less prepared psychologically to see the other's point of view and more determined to reinforce its own.

Rutherford had become increasingly convinced of the validity of his satellite model of the nucleus as a theoretical interpretation of artificial disintegration after 1919. Far from being a caricature of the narrow experimentalist, he fully appreciated the necessity of theory. On September 12, 1923, he made his general philosophy of science clear in his inaugural lecture as president of the British Association. "Experiment without imagination, or imagination without recourse to experiment," he told his audience, "can accomplish little, but, for effective progress, a happy blend of these two powers is necessary."[89] Rutherford followed his own dictum: In 1924 he saw that not only could his satellite model of the nucleus serve as an interpretation of the artificial disintegration of light elements, it could also be extended to serve as an interpretation of the natural disintegration of heavy radioactive elements.

Rutherford revealed the direction his thoughts were taking as early as April 1924, in the lecture at the Royal Institution noted above.[90] By September, when he delivered a major address at the Franklin Institute in Philadelphia, he had developed them further,[91] and in succeeding years he clarified and refined them to the point where he published a quantitative theory of radioactivity in the fall of 1927.[92]

The principal stimulus to Rutherford's thought consisted of a puzzle he identified in the behavior of uranium. Rutherford knew that when uranium I ($^{238}_{92}U$) decays radioactively it emits an α particle of about 4.25 MeV in energy. Assuming that it acquires this energy by being repelled by the inverse-square field of force surrounding the uranium nucleus, he could readily calculate that it had to originate at a distance of about 7×10^{-12} cm from its center. Yet, as he and Chadwick carried out more and more α-scattering experiments, seeking further evidence for the existence of a critical potential surface (first within light nuclei such as aluminum and later within heavy nuclei), they found that the much more energetic α particles from RaC ($^{214}_{83}Bi$) failed to penetrate the uranium nucleus and were simply scattered away after approaching to a distance of about 3.2×10^{-12} cm from the center of the uranium nucleus. How, then, could relatively low-energy α particles be emitted radioactively from a point 7×10^{-12} cm from the center of the uranium nucleus by means of a nuclear disruption, whereas relatively high-energy α particles could penetrate to a distance of 3.2×10^{-12} cm without causing a nuclear disruption?[93]

Rutherford found the resolution of this puzzle in his satellite model of the nucleus. He reasoned that the 4.25-MeV α particles emitted spontaneously during the radioactive decay of the uranium nucleus had existed previously as satellites orbiting at a distance of about 7×10^{-12} cm from its center. At first he assumed these α satellites to be ordinary α particles, but in 1926 P. Debye and W. Hardmeier proved that neutral particles could be held in stable orbits by electrical polarization forces.[94] This led Rutherford to conclude that the α satellites consisted of ordinary α particles that had become neutralized by acquiring two electrons. Radioactive α decay, then, occurred when these two electrons somehow became detached from the neutral α satellites and fell into the central core, leaving a doubly charged α particle to be repelled outward. In α-scattering experiments, however, the more energetic RaC α particles incident upon the uranium nucleus from the outside would pass unimpeded through the ring of neutral α satellites and penetrate to a distance of 3.2×10^{-12} cm from its center before being scattered away by the inner core.

Rutherford's theory of radioactivity also encompassed β decay and the emission of γ radiation: β decay occurred when nuclear electrons, orbiting about the nuclear core inside the ring of neutral α satellites, were expelled; γ rays were emitted when the neutral α satellites dropped from higher to lower Bohr-type nuclear-energy levels.[95] Rutherford's theory therefore was a general theory of radioactivity based upon an imaginative extension of his satellite model of the nucleus, and as he transformed it from a qualitative idea into a quantitative theory between 1924 and 1927 his confidence in that model continually increased.

But we now must return to October 1924, when Rutherford returned from Canada and the United States to Cambridge. He was soon brought down to earth: In the November issue of the *Physikalische Zeitschrift* appeared two long papers, one by Pettersson and the other by Kirsch,[96] that left no doubt that Rutherford's efforts in July had been to no avail. Both papers had been presented in August at the 88th meeting of the Deutsche Naturforscher und Ärzte in Innsbruck, and both asserted Pettersson and Kirsch's claims in experiment and theory. That was the initial volley. On November 23 and 30, Pettersson sent two further papers, written in English, to the *Arkiv för Matematik, Astronomi och Fysik*.[97] These constituted another direct attack on Rutherford's satellite model of the nucleus. On December 4 and 11 he presented the same two papers in German to the Vienna Academy.[98] In the first he argued that his evidence for the disintegration of carbon and numerous other light and heavy elements, including copper and nickel, contradicted Rutherford's satellite model and supported his own explosion hypothesis. In the second he supported his claims by reporting further observations utilizing an improved optical system and a new method for observing low-range disintegration protons at an angle of approximately 150°—a "retrograde method," as he termed it. His new microscope, Pettersson declared, "necessarily" made the scintillations appear much brighter than those observed by Rutherford and Chadwick—"a fact which no doubt explains why these authors find much smaller numbers of particles than we from the same elements and why they have so far failed to observe the H-particles from carbon."[99] Pettersson took the offensive still further by speculating that, consistent with his explosion hypothesis, his new results indicated that "the α-particle may penetrate into the nucleus itself and eventually remain attached to it for a shorter time or definitely," forming a "system of nucleus plus an α-particle." This view, he added in a footnote, now apparently was "accepted as plausible also by Sir Ernest Rutherford . . . , although it is not clear on what experimental data hitherto published his conjecture is based."[100]

That was a gratuitous remark, because Rutherford had already suggested the possibility "that the α-particle is in some way attached to the residual nucleus"[101] on the basis of his and Chadwick's evidence for a critical potential surface in the aluminum nucleus. Moreover, Rutherford and Chadwick regarded that evidence as so significant that they decided to pursue it thoroughly after Rutherford's return to Cambridge.[102] They designed a new apparatus to permit observations at 135°, and by varying the incident α-particle energy by means of absorbing foils they found, by mid 1925, "a very striking departure from the usual laws of scattering."[103] They found that as the energy of the incident α particles increased, the number scattered through 135° by both aluminum and magnesium first dropped and then rose. They accounted for this behavior by proposing that these nuclei consist of a "charged central core surrounded by a satellite distribution of positive and negative charges,"[104] as shown schematically in figure 5. Low-energy α particles would not be able to penetrate the outer positive ring, E_3, and would be scattered away normally; more energetic ones would be able to reach the negative ring, $-E_2$, where some would be captured, reducing the number scattered away; still more energetic ones

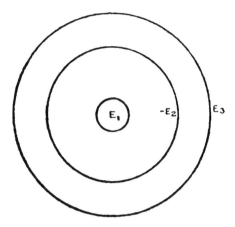

Figure 5
Rutherford and Chadwick's schematic model of aluminum and magnesium nuclei.
Positive (E_3) and negative ($-E_2$) rings of satellites surround the positive central core
(E_1). Source: "Scattering of α-Particles by Atomic Nuclei and the Law of Force," *Phil.
Mag.* 50 (1925), p. 911. Reproduced by permission of Taylor & Francis, Ltd.

would be able to reach the strong positive central core, E_1, and would
again be scattered away normally. As already mentioned, it was scattering experiments similar to these with uranium that stimulated Rutherford to extend his satellite model to the heavy radioactive nuclei.

The idea that the incident α particles are actually captured by the
target nucleus in the disintegration process was confirmed in a striking
way by Blackett's cloud-chamber photographs of the disintegration of
nitrogen,[105] which Rutherford examined after his return to Cambridge.
Blackett had taken some 23,000 photographs containing over 400,000
tracks, exactly 8 of which showed the incident α particle, the disintegration proton, and only one other particle: the residual nucleus. The
latter, therefore, had to have a mass of 17 units and a charge of 8; it
had to be an isotope of oxygen. Thus, Blackett's photographs proved
beyond doubt that Rutherford's earliest interpretation of the disintegration process—that the incident α particle simply collides with a
proton satellite, ejecting it and leaving an isotope of carbon—was incorrect. At the same time, as Chadwick had written to Rutherford, the
photographs confirmed the existence of a critical potential surface within
the nitrogen nucleus. Rutherford pointed out the significance of Blackett's photographs explicitly on February 4, 1925, when he delivered
the eighth Trueman Wood lecture before the Royal Society of Arts.[106]

They indicated, he said, "that the α-particle is captured by the nitrogen nucleus while the latter loses one of its constituent protons." "It is not unlikely," he added, "that a similar result may hold in the case of aluminum. . . ."[107]

It seems that, shortly after speaking these words, Rutherford began to have nagging thoughts that a similar interpretation of the disintegration process had been advanced months earlier by Pettersson, in the July 1924 letter to *Nature* that Rutherford had forced Pettersson to withdraw from publication. To settle this question in his own mind, Rutherford wrote to Pettersson on February 20, 1925,[108] asking Pettersson to send him once again a copy of the withdrawn article. Pettersson responded by return mail, on February 27.[109] In his covering letter Pettersson pointed out that the conclusion he had drawn from his experiments at that time, "viz. that the α-particle may remain attached to the nucleus (a conclusion subsequently confirmed through numerous experiments in this institute and quite recently by the results found by Blackett)," was "independent of the results . . . to which you took exception." Pettersson had put the matter as follows in the withdrawn article:

The excess in numbers of the H- over the α-particles . . . would seem to indicate either that the explosion of each disintegrated nucleus is more radical than one has hitherto been inclined to believe, or that the α-particle . . . remains attached to the nucleus it hits, while an H-particle is expelled as the immediate effect of the impact. The residual nucleus should in that case have a nuclear charge one unit higher and a mass three units higher than the atomic nucleus before the collision.[110]

Reading these words, Rutherford realized their true implications: Not only did Pettersson believe that Rutherford had deprived him of priority in this particular interpretation of the disintegration process, he also assumed that Rutherford had passed that interpretation on to Blackett.

Determined to set the record straight, Rutherford drafted a reply to Pettersson on March 5.[111] He was greatly agitated; the draft of his letter contains several false starts and is nearly illegible, even to those who have become accustomed to Rutherford's notoriously difficult-to-read handwriting. Rutherford first thanked Pettersson politely for his *Nature* manuscript, which he once again returned. He told Pettersson that when he first received it he was "mainly interested in the radium C question for the possibility of capture of the α-particle was not a new idea to me." Jean Perrin had proposed it as early as April 1921 while commenting on Rutherford's paper at the third Solvay Conference

(Rapports, p. 68), and again in his 1924 book *Les Atomes* (p. 318). Rutherford also told Pettersson that he had been intending for some time to write a historical note to *Nature* on this subject, and he would take that opportunity to refer to Pettersson's recent article in the *Physikalische Zeitschrift*, "which contains an almost identical statement of your views." As for Blackett, "I need hardly say that he was quite ignorant of the contents of your original letter which, of course, I regarded as a private communication." In the future, Rutherford commented, he "would prefer not to know" of Pettersson's results "until actual publication in the normal way," so as to avoid the "unnecessary difficulties" of referring to unpublished documents. However, he would "be glad to receive copies" of Pettersson's papers "as soon as they have been printed." Rutherford closed by extending his greetings to Stefan Meyer, and by informing Pettersson that in four months, in July 1925, he would be leaving on a six-months trip to Australia and New Zealand. Rutherford's historical note duly appeared in the April 4, 1925, issue of *Nature*.[112]

Rutherford needed a long break. He was overburdened by a variety of professional duties. We also know from an interview that Charles Weiner conducted with Chadwick in 1969 that Rutherford was not well at this time, and that as a consequence he and Chadwick had a falling out over how the Cavendish Laboratory should be run.[113] Rutherford's long and nostalgic trip, during which he was accorded a hero's welcome as he lectured in numerous cities in Australia and New Zealand, was just the tonic he required.

But Rutherford's letter to Pettersson and his long absence from Cambridge did not earn him and Chadwick a respite from their controversy with Pettersson and Kirsch. In April 1925, Kirsch sent off an article to the *Physikalische Zeitschrift* criticizing the Cavendish work,[114] and in early July, Pettersson and Kirsch reported to the Vienna Academy new experiments[115] on intermediate and heavy nuclei (iron, selenium, tin, gold) from which they concluded that "probably all elements are disintegrable."[116]

The most provocative article, however, was one published by Kirsch in the July 1925 *Physikalische Zeitschrift*.[117] Kirsch argued at length that the capture of the incident α particles contradicted Rutherford's "unnecessarily complicated and specialized satellite hypothesis"[118] and supported Pettersson's simple explosion hypothesis. In the final section of this nine-page, double-column paper, Kirsch explained how Pettersson's explosion hypothesis was consistent with the view that nuclei consist of alternating "shells" of "quasi H-particles and free electrons."[119] That was too much for Chadwick. As soon as he had read Kirsch's article, he wrote to Rutherford (who by this time had sailed from England), giving full vent to his exasperation:

Our friend Kirsch has now let himself loose in the Physikalische Zeitschrift. His tone is really impudent to put it very mildly. He takes our old experiments the very first results and proceeds to show what fools we are to talk of satellites and then tells what clever fellows he & Pettersson are to think of a nucleus which is simply chuck full of satellites rings and rings of them. Kirsch & Pettersson seem to be rather above themselves. A good kick from behind would do them a lot of good. The name on the paper is that of Kirsch but the voice is the familiar bleat of Pettersson. I don't know which is the boss but as Mr. Johnson said there is no settling a point of precedence between a louse and a flea.[120]

Chadwick added that he had just received an offer from Arthur Compton to go to Chicago in October after A. A. Michelson's retirement, but there was "no use thinking about it under the circumstances."

The circumstances, in fact, were becoming even worse from Rutherford's and Chadwick's point of view. Pettersson and Kirsch, so far as the outside world could judge, were establishing a solid reputation as leading investigators in the field of artificial disintegration. R. W. Lawson, then at the University of Sheffield, made this point directly in a letter of December 22, 1924, to his old friend and protector Meyer.[121] Lawson remarked that he was particularly pleased with the great success of Pettersson and Kirsch, who appeared to be holding their own with Cambridge. We also know, for example, that Merle Tuve at the Carnegie Institution in Washington, D.C., was monitoring the Vienna and Cambridge results closely.[122] Indeed, Pettersson was raising the Institut für Radiumforschung to prominence in this field, and as a result was attracting an increasing number of young researchers to Vienna to work under him: Elisabeth Kara-Michailova and G. Ortner had co-authored papers with Pettersson in 1924,[123] E. A. W. Schmidt, Georg Stetter, and Marietta Blau published supporting papers in 1925,[124] and Elisabeth Rona and Bertha Karlik soon joined the group.[125] Moreover, on July 31, 1925, Pettersson and Kirsch signed the preface to a 247-page monograph entitled *Atomzertrümmerung: Verwandlung der Elemente durch Bestrahlung mit α-Teilchen*[126] [Atomic Disintegration: Transmutation of Elements through Radiation by α Particles], in which they took every opportunity to assert their claims. For example, they included a table of "elements that with certainty yield H-particles"—a total of 27 between lithium and iodine (atomic numbers 3–53), of which they claimed priority for first disintegrating 16, leaving only 11 to Rutherford and Chadwick.[127] Just to make certain that no reader missed the point, they closed their monograph with a chronological overview of all work on artificial disintegration carried out in both laboratories.[128] Lawson soon wrote a laudatory review of this work.[129] Nor was that all. By 1926 Pettersson's and Kirsch's reputations had grown to the point that they were invited to co-author a long review article on "Atomzertrümmerung" for the Handbuch der Physik[130] and Kirsch was invited to contribute a second review article under the same title for the Ergebnisse der exakten Naturwissenschaften.[131] All in all, Pettersson and Kirsch were reaching German-speaking audiences at least as effectively as Rutherford and Chadwick were reaching English-speaking ones. In three years, 1923–1926, Pettersson and Kirsch had established the Institut für Radiumforschung as a serious competitor to the Cavendish Laboratory in the field of artificial disintegration.

Rutherford and Chadwick recognized the increasing seriousness of the situation. During Rutherford's absence from Cambridge, Chadwick attempted to meet the challenge in two ways. First, in a most insightful step, he calculated the energy balance in artificial disintegration reactions under the assumption that the incident α particle is captured by the target nucleus when a proton is expelled from it.[132] "Unfortunately" however, Chadwick wrote, "the atomic masses of magnesium, aluminum, and silicon have not been determined with sufficient accuracy to afford any test of the mass relations,"[133] and the same was true for all other nuclei. Thus, this crucial theoretical tool could not be employed to ascertain whether carbon, oxygen, and other light nuclei are in principle disintegrable by α particles emitted by RaC or other naturally radioactive elements. Therefore, the only avenue open to Chadwick was the second: direct experiment. On November 21, 1925, he informed Rutherford by letter that "I can still find nothing from Be or C or O or gold."[134]

Thus, when Rutherford returned to Cambridge two months later, he was more convinced than ever that there was something seriously amiss in Vienna. On February 8, 1926, he stated one possible reason in a letter to Bohr: "While I have been away Chadwick has been hard at work going over some of the results claimed by Kirsch and Pettersen [sic]. We are convinced that they are radically wrong for some reason. The idea that you can discriminate between slow α particles and H particles by the intensity of the scintillation is probably the cause of their going wrong. Under the normal conditions of the experiments such a discrimination by eye is terribly dangerous."[135]

Still, what could be done? As we know from his comment in his July 19, 1924 letter to Pettersson, and from the interview that Weiner conducted with Chadwick in 1969, Rutherford was adamantly opposed throughout his career to airing scientific disputes in the open literature; he believed they should be ironed out in private.[136] Yet this avenue had failed with Pettersson, and even with Meyer, who was unwilling to restrain Pettersson and his co-workers. Thus, the task of presenting the Cavendish position in public fell to Chadwick. In November 1926 he published a long article in the *Philosophical Magazine*[137] that was the most thorough attempt of the Cavendish researchers to date to clarify and analyze the experimental and observational differences existing in Cambridge and Vienna.

Chadwick began his article by making those differences explicit. He pointed out that by means of the right-angle method of observation he and Rutherford had been able to search for disintegration protons down to a range of 7 cm in air, and even to lower ranges "with special precautions." They had found that RaC α particles (whose range is 7 cm in air) can disintegrate only the "light elements up to and including potassium . . . with the exceptions of helium, lithium, beryllium, carbon, and oxygen."[138] This was in sharp contrast to the results of Pettersson and Kirsch, who had found that they could disintegrate "almost every element they [had] been able to examine." In particular, Pettersson and Kirsch had found that carbon could be disintegrated and that it yielded protons of maximum range 16 cm in the forward direction and 8 cm at 90°. "These particles," continued Chadwick, "should have been recorded in the experiments of Sir E. Rutherford and myself, and it is a very surprising fact that we have been unable to detect them. Our failure to do so is attributed by Kirsch and Pettersson to the low numerical aperture [or low light-gathering power] of the microscopes we use for observing scintillations. This explanation . . . is advanced . . . as the general reason why our observations give no evidence of disintegration in so many cases where they have obtained positive results."[139]

The first issue to settle, therefore, concerned the properties of Rutherford and Chadwick's optical system. Years of experience had taught them that the observing microscope should have an overall magnification of greater than 30× or "some counters [would] fail to observe a fairly large fraction of the scintillations."[140] Its field of view at the same time should be greater than about 10 mm², to encompass "convenient numbers for counting" (at least 20 scintillations per minute). These conditions were satisfied by Rutherford and Chadwick's microscope, which had a Watson holoscopic objective lens of 0.45 numerical aperture and a "specially constructed wide-angle eyepiece" that together produced an overall magnification of about 35 × and a field of view of 50 mm². Nevertheless, they commissioned Adam Hilger and Co. to construct a new hemispherically shaped objective lens of 1.06 numerical aperture, which together with the eyepiece produced an overall magnification of 32 × and a (projected) field of view of 80 mm². In this system ZnS crystals could be deposited directly upon the objective lens. The result was that the scintillations appeared "much brighter," and hence the counting was "much easier." Comparative tests showed, however, that with either of the two microscopes "about the same number of scintillations was observed." Therefore, wrote Chadwick, "we concluded . . . that in general our observations of scintillations were trustworthy and that we were not failing to record the existence of particles owing to any weakness in our optical arrangement."[141]

The second issue to settle concerned the "personal powers" of the Cambridge observers. "We have had a long and varied experience in counting scintillations," Chadwick pointed out, "and our assistants have been carefully trained. . . . [We] have no evidence that one counter can observe scintillations of a type which are invisible to another."[142] Chadwick had instituted a strict drill for counting scintillations, permitting the Cavendish observers to count only for one minute at a time, followed by a period of rest, and so on.[143] Furthermore, during the past two years the actual efficiencies of the Cambridge counters had been tested systematically by a method developed by Geiger and Werner.[144] In this method, two observers view the scintillation screen simultaneously through two different microscopes, and each records the scintillations observed by tapping an electrical contact key connected to the moving tape of a chronograph. If, then, λ_1 is the efficiency of one observer, λ_2 that of the other, and N the total number of scintillations actually present, the first observer would record $N_1 = \lambda_1 N$ scintillations and the second $N_2 = \lambda_2 N$, and $C = \lambda_1 \lambda_2 N$ would be the number both would record simultaneously. Since the numbers N_1, N_2, and C could be determined directly from the tape, the efficiencies of the observers, λ_1 and λ_2, could be calculated.

By testing some thirty different observers, Rutherford and Chadwick found that "in a short time" their average efficiency increased from about 80 percent to about 95 percent when counting α particles with ranges of a few cm, and from about 60–70 percent to more than 90 percent when counting α particles with ranges of a few mm. For protons with a range of about 7 cm, the efficiency of an untrained observer was about 60 percent; the efficiencies of two experienced observers were 80 percent and 88 percent. Other tests proved that the Cambridge observers could see scintillations produced by protons and α particles with ranges of only a few tenths of a millimeter. For various reasons, a number of these tests had been conducted under "much less favorable" observing conditions than those prevailing in their ordinary disintegration experiments. Therefore, Chadwick felt justified in concluding that the Cavendish observers were "trustworthy" and that "no large fraction of the scintillations present on the screen had been missed."[145]

That left the actual experimental differences to be discussed. Chadwick reported that his most recent observations with the Hilger microscope showed that no protons of range greater than 4 cm in air were expelled from beryllium, carbon, or oxygen, and none of range greater than 4.3 cm in air from lithium. These results, he noted, conflicted directly with those of Pettersson and Kirsch, who had found large numbers of protons from beryllium and carbon with maximum ranges of 9 cm and 8 cm in air, respectively. Moreover, Chadwick pointed out that even where the Cambridge and Vienna results agreed the agreement was "more apparent than real." For example, he and Rutherford had found protons from silicon with a maximum range of 24 cm in air; Pettersson and Kirsch had found them to have a maximum range of only 10–11 cm in air. If we "cannot observe weak scintillations," wrote Chadwick, "we should [have found] smaller maximum ranges. . . ." Therefore, Pettersson and Kirsch were wrong in believing that Rutherford and Chadwick's failure to observe the particles from carbon and beryllium was "due to inefficiency of counting."[146]

As for the possibility that the disintegration protons from beryllium and carbon have a range of less than 4 cm in air (even though Pettersson and Kirsch had claimed that their maximum ranges were 9 cm and 8 cm, respectively), Chadwick pointed out that any disintegration protons of such small ranges would be mixed in with α particles from the source that had been scattered through 90° and therefore had comparable ranges. "To distinguish separately" these short-range disintegration protons and scattered α particles by the brightness of their scintillations, Chadwick stated, "is, as far as our experience goes, impossible." He continued: "The Vienna workers, on the other hand, claim to be able to distinguish [these scintillations] . . . , and, relying on this, they have pursued the search for H-particles to the smallest absorptions, well within the range of the scattered α-particles. This difference in procedure is partly responsible for the difference between their results and ours."[147] As one example "out of many," Chadwick noted that E. A. W. Schmidt in Vienna had found protons with a range of less than 4 cm from aluminum, with no indication that they possessed a minimum range, whereas Rutherford and Chadwick had found that the protons from aluminum possessed a minimum range of 10–12 cm in air. Whatever view was taken of the structure of the aluminum nucleus, Chadwick argued, these very different figures corresponded to very different estimates of the radius of the aluminum nucleus: By a simple calculation, they corresponded to radii of 20×10^{-13} cm and 8×10^{-13} cm, respectively. The former figure, wrote Chadwick, seemed "improbable," since his and Rutherford's α-scattering experiments indicated that the inverse-square law held down to about $10–12 \times 10^{-13}$ cm.

"In my opinion," Chadwick declared, "the difference between our results and those of Schmidt is chiefly due to his procedure in counting within the range of the scattered α-particles, and in judging from the brightness of the scintillation whether it was due to an α-particle or to a H-particle. As I have stated previously, we have never, in this laboratory, felt confident of our ability to pick out the scintillations of H-particles from those of a heterogeneous beam of [scattered] α-particles, although we have had considerable experience in counting."[148] In sum, the Cambridge experiments indicated that disintegrability is not "a common property of all elements," as maintained by Pettersson and Kirsch. Hence, the main points at issue remained "without explanation."[149]

Roughly at the time Chadwick's paper was going to press, in November 1926, Hans Thirring, Stefan Meyer's colleague in Vienna, visited Cambridge briefly. Rutherford took that opportunity to ask Thirring to extend his greetings to Meyer and to reiterate his invitation to Pettersson to visit Cambridge. Meyer responded to Rutherford on December 17,[150] offering holiday greetings and renewed thanks for the invitation to Pettersson. However, he told Rutherford that it was unlikely that Pettersson could come for "at least a couple of months," as he had "to complete some urgent work for Sweden." Perhaps, Meyer suggested, Rutherford could visit Vienna; if that was impossible, perhaps Chadwick could come. "We should be most pleased," Meyer wrote, "to show him all we have and no doubt the outstanding differences between the results found here and his own might be cleared up without difficulties." Meyer noted that Pettersson could not take his instruments to Cambridge, but in Vienna he could show Chadwick "what he considers to be positive evidence at his disintegration experiments."

Rutherford replied on December 23, telling Meyer that he was "glad to hear . . . about the prospects of an interchange of visits . . . to get at the bottom of the reasons of the differences in results obtained in the two Institutions."[151] He was too occupied with his duties as president of the Royal Society to consider coming. Unfortunately, it also was unlikely that Chadwick could come for some time, as he was "expecting an interesting domestic event in February." Nevertheless, Rutherford supported the idea that Chadwick should visit Vienna and "see for himself what is the cause of the divergences." "I quite agree with you," he told Meyer, "that it is highly important that this question should be amicably settled for I myself feel the whole subject of nuclear disintegration must remain in confusion pending a comparative investigation."

The question of an exchange of visits between the two laboratories, therefore, was again raised by their directors at the end of 1926 and was again set aside indefinitely as not practicable. Chadwick's article had raised the level of interest significantly, but no definite arrangements were forthcoming. Quite likely neither Rutherford nor Meyer imagined that the degree of confusion could be increased still further before an exchange of visits could occur, yet they soon discovered otherwise.

At the end of February 1927, Pettersson and Kirsch submitted a long and detailed rebuttal to Chadwick's article, in German to the Zeitschrift für Physik[152] and in English to the Arkiv för Matematik, Astronomi och Fysik.[153] Two months later they presented essentially the same paper to the Vienna Academy for publication in its Sitzungsberichte.[154] This triple publication left no doubt in the mind of any reader that Pettersson and Kirsch regarded Chadwick's analysis as flawed throughout.

Pettersson and Kirsch began by agreeing that the discrepancies between the Cambridge and Vienna observations were "of a much too serious nature to be attributed to ordinary experimental errors." The "quality of the eyes of the observer, his training, his subjective state of fatigue or strain" had to be considered. They also argued that the observations depended upon the α-particle source used: RaC, a relatively high-intensity source, decays rapidly (its half-life is about 20 minutes), so that the background illumination produced by the γ rays it emits also diminishes rapidly. The "faintest scintillations" then become "apparent against the darker background like the fainter stars coming out against a darkening evening sky." This effect could be eliminated by using a relatively low-intensity polonium source, which has a much longer half-life (about 138 days) and emits no γ rays. The background then remains "practically dark in the whole course of the experiment," making the "conditions of visibility . . . constant and optimal." The latter source had been introduced in Vienna about two years earlier when Schmidt found that the relatively low-energy α particles from polonium (range 3.9 cm in air) could produce artificial disintegration just as well as the relatively high-energy ones from RaC (range 7 cm in air), the source used in Cambridge.[155]

To Pettersson and Kirsch, however, a much more significant difference between the methods employed in Cambridge and Vienna concerned the microscopes. Pettersson and Kirsch agreed with Chadwick that for "dependable counts" it is necessary to observe about 20 scintillations per minute, but they pointed out that in improving their microscopes to achieve this counting rate the Cambridge and Vienna researchers had proceeded in opposite directions. Between 1921 and 1924 Rutherford and Chadwick had held the light-gathering power of their objective lens constant by keeping its numerical aperture at 0.45, but by successively employing eyepieces of "special construction" they had increased the field of view of their microscopes from 8 to 25 to 50 mm². By contrast, between 1923 and 1925 Pettersson and Kirsch had increased the light-gathering power of their objective lens by increasing its numerical aperture from 0.45 to 0.70 to 0.80, which, together with the eyepieces they had employed, had reduced the field of view of their microscopes from 20 to 10 to 3.2 mm². The result, wrote Pettersson and Kirsch, was that "the higher light-gathering power of the microscopes we have used and also the smaller intensity of the sources we employ, producing less background illumination of the screen, have allowed the weak scintillations from slow H-particles to be counted here, whereas they may to a large extent have been overlooked under the conditions of the Cambridge experiments."[156] In fact, Rutherford, in his 1920 Bakerian lecture, had "convinced himself" that aluminum yields "no sensible number" of protons, but the very next year he and Chadwick, after introducing "the large-aperture counting microscope from Watson," had found that aluminum yields protons "in larger numbers and of greater range even than nitrogen."

The opportunity for a critical comparison of the Cambridge and Vienna microscopes had presented itself less than a year earlier, Pettersson and Kirsch continued, when L. R. Hasche, a "perfectly unbiased" young American physicist, "endowed with very good eyes and having little or no previous experience of counting scintillations,"[157] arrived in Vienna. Consider the microscopes described in table 1. The "subjective field" is defined as the product of the "objective field" and the square of the linear magnification (call it M). D is the diameter of the exit pupil, given by

$$D = \frac{(n.a.) \times 2 \times 250 \text{ mm}}{M},$$

where n.a. is the numerical aperture of the objective lens and 250 mm the "distance of ordinary vision." The quantity I.L., in arbitrary units, is the "integrated light" from a scintillation, which is proportional to the square of the numerical aperture of the objective lens. The microscopes numbered 1–3 are those used in Vienna between 1923 and 1925, number 4 is one of "special construction" used in Vienna, number 5 is one that was designed in Vienna as a first attempt to replicate Rutherford and Chadwick's 1924 model, number 6 is a second attempt, and number 7 is Rutherford and Chadwick's improved Hilger model. Since the optical properties of the last two had been published only recently, Hasche's tests were confined necessarily to the first five only.

Hasche's results were telling. He found that microscopes 3 and 4 displayed the highest number of visible proton scintillations per unit area, that number 2 was almost as good, and that numbers 1 and 5 were distinctly "less satisfactory"—number 1 because its objective field was "too large to be watched efficiently" and number 5 because of its "low magnifying power." From these results, Pettersson and Kirsch could now draw certain conclusions about numbers 6 and 7, Rutherford and Chadwick's most recent models. They argued that number 6 must be inferior to number 1 for observing weak scintillations, owing to its larger subjective field of view and its smaller magnifying power; it might be comparable to number 5, whose differences in these two respects tended to balance each other out. As for number 7, the "extraordinary size" of its subjective field of view necessarily would prevent the observer from seeing a considerable number of weak scintillations. Moreover, the diameter of its exit pupil, 16.5 mm, was more than twice that of the dark-adapted pupil, which is about 7.5 mm, so that under even the best conditions only $(7.5/16.5)^2$, or 21 percent, of the scin-

Table 1
Pettersson and Kirsch's specifications of the microscopes used in Vienna (numbers 1–4) and Cambridge (numbers 5–7). Source: H. Pettersson and G. Kirsch, "The Artificial Disintegration of Elements," *Ark. Mat., Astr. Fys.* 20A, no. 16 (1927–28): 12.

No.	Designer	f	n.a.	magnification	Obj. field	Subj. field	D.	I.L.
1	Watson I	16 mm	0,45	× 45	21 sq.mm	430 sq.cm	5,0 mm	0,20
2	Watson II	12 "	0,70	× 70	9 "	200 "	7,5 "	0,49
3	Watson III	16 "	0,80	× 80	3,2 "	155 "	5,7 "	0,64
4	Zeiss	20 "	0,65	× 90	2,3 "	180 "	3,6 "	0,42
5	Watson-Reichert	16 "	0,45	× 18	44 "	154 "	12,5 "	0,20
6	Rutherf.-Chadw.	16 "	0,45	× 35	50 "	610 "	6,5 "	0,20
7	Hilger	"	1,06	× 32	105 "	820 "	(16,5) "	(1,12)
							7,5 mm	0,23

tillations could be observed. At the same time, if one assumed the diameter of the exit pupil to be 7.5 mm instead of the "superhuman" dimension of 16.5 mm and then calculated the integrated light, the result for microscope 7 (0.23) would be very close to that for number 6 (0.20). Hence, it was "not so surprising" that Rutherford and Chadwick obtained the same counts with both. In any case, the entire chain of reasoning proved that Rutherford and Chadwick's most recent microscopes, numbers 6 and 7, were inferior to Pettersson and Kirsch's number 1, and hence greatly inferior to Pettersson and Kirsch's best microscopes, numbers 3 and 4. Thus, Pettersson and Kirsch concluded that "the preceding examination of the properties of the optical resources applied here and in Cambridge . . . supports our views . . . regarding the probable explanation for the discrepancies found by the two sets of workers."[158] In other words, the optical properties of the microscopes used in Vienna permitted the observation of very weak proton scintillations, whereas this was not true for the microscopes used in Cambridge—Chadwick's words to the contrary notwithstanding.

Nor was Chadwick's analysis of the efficiencies of the Cambridge observers meaningful, for, wrote Pettersson and Kirsch, "the method of Geiger and Werner, as its authors expressly state, is only applicable to counts of scintillations of a high and constant brightness, so that all of them should be easily visible to both observers."[159] Chadwick's observers, to the contrary, had viewed protons varying in range from 7 to 0 cm in air, so that the scintillations they produced varied in brightness from some maximum value down to zero. Moreover, wrote Pettersson and Kirsch, "according to all our experience" the efficiencies of different observers converge "towards zero at different rates" as the scintillations diminish in brightness, and hence all would not be visible to both observers at the same time. Therefore, Chadwick's claim that the Cambridge observers had counted "all or nearly all such particles" consistently was "no proof" that they had observed all of the particles actually present. Indeed, "the majority of the particles of short range [that were] present [might have been] overlooked. . . ." This possibility was more plausible still in light of the deficiencies of the Cambridge microscopes, for, even under "the most favourable conditions" prevailing in Vienna, Elisabeth Kara-Michailova had found that more than 70 percent of the protons with a residual range of 7 mm were overlooked. "We cannot, therefore," Pettersson and Kirsch summarized, "accept the conclusions Chadwick infers from the [Cambridge] counting tests with H-particles. . . ."[160]

In accordance with Chadwick's order of presentation, the next item on Pettersson and Kirsch's agenda concerned the actual experimental discrepancies. First, they would clear up a "misunderstanding": Chadwick was wrong in claiming that even where agreement existed, as in the case of magnesium and silicon, it was "more apparent than real." It was not surprising that Rutherford and Chadwick had found protons from these elements of much longer ranges than Pettersson and Kirsch had found, for if only a few penetrated relatively thick absorbers one would conclude that their range was high, and "experimental conditions in Cambridge [had] been deliberately chosen for detecting such very rare particles: Large size of the screen area observed, high intensity of the source . . . [whereas] in Vienna . . . they [were] necessarily less favourable . . . , viz. small area of the screen observed, weak sources and also shorter periods of counting [20–30 seconds versus 1 minute]."[161]

Thus, to Pettersson and Kirsch the "discrepancy noted by Chadwick" was "quite in line" with the differences in the optical resources used in Cambridge and Vienna, and to them this was still the crux of the matter. The disintegration protons of range $\geq$ 4 cm in air from beryllium, carbon, and oxygen "ought to have been visible" in the Cavendish Laboratory. "If the failure of the Cambridge workers to observe these atomic fragments is not due to the properties of their counting microscopes . . . and to the background illumination of the screen, caused by the intense RaC sources which they use . . . , we confess ourselves to be unable to suggest any explanation."[162] Under these circumstances they could only describe their own experimental methods and results.

Most of the Vienna observations during the preceding two years had been made with the "retrograde method," involving an angle of observation of about 150°. The main advantage of this method was that both the number and the range of any α particles from the source scattered through such a large angle would be greatly reduced and, in comparison with any disintegration protons present, "would only have the magnitude of a correction term." "In both respects," Pettersson and Kirsch wrote, "the advantages of the retrograde method as compared to the 90° method of observation are manifest, and we much regret that it has so far not been tried in Cambridge for disintegration work. . . ."[163]

The main point at issue, however—one that Pettersson and Kirsch said had "repeatedly been suggested from the side of the Cambridge physicists"—was that the Vienna observers had confused scattered α particles with disintegration protons in attempting to distinguish between the two by the brightness of the scintillations they produced. Pettersson and Kirsch addressed this criticism by describing various experimental techniques designed to avoid confusing the two particles. They asserted that they drew "no definite conclusions" except when "a considerable number of scintillations [were] visible" beyond the theoretical range of the scattered α particles or when inside that range the number of observed particles was "greatly in excess" of the calculated number of scattered α particles. They concluded flatly:

> ... we have taken every precaution against a confusion of atomic frag-
> ments [protons] with scattered α-particles and, in cases where such a
> confusion appeared theoretically possible, we have desisted from draw-
> ing any inferences from our results regarding the disintegration of new
> elements. The fact emphasized by Chadwick, that we have pursued
> the search for H-particles well within the range of the scattered α-
> particles cannot, therefore, be used as an explanation for the discrep-
> ancies considered.[164]

With respect to the elements that Pettersson and Kirsch had found to be disintegrable, they would only note that "some forty different sets of experiments with carbon in its purest form and under conditions which precluded all complications from scattered α-particles or from natural H-particles" yielded "numerous weak scintillations . . . at ab- sorptions between a few mm and 4 to 6 cm." They also had found disintegration protons from beryllium, oxygen, and ten heavy elements between titanium and iodine. Therefore, they said, "At present we . . . must . . . uphold our contention, that . . . [all these] ele- ments . . . are disintegrable . . . , much as we regret that this effect has until now not been confirmed in Cambridge."[165] Disintegrability was a "common phenomenon," supporting Pettersson's explosion hypoth- esis over Rutherford's satellite interpretation.

In spite of their success with the scintillation method, Pettersson and Kirsch pointed out that they had been pursuing other methods of ob- servation in Vienna. They disagreed with Chadwick's feeling that the scintillation method was "dependable" and "well adapted to give quantitative results of high accuracy." They had been devoting a "large amount of labour" to the development of three other "less laborious and more accurate" methods of observation: the Wilson cloud-chamber method, the photographic method, and H. Greinacher's electric-acoustic method. The first was being pursued by R. Holoubek, who had con- firmed the disintegration of beryllium, carbon, oxygen, and other ele- ments with it. The second was being used by Marietta Blau, who had found "confirmatory evidence of considerable weight" for the disintegration of carbon. The third was being developed by G. Ortner and G. Stetter, who to date had found only that the impact of α particles and protons could be made "audible as cracks in a loud speaker, distinctly differing from each other."[166]

Finally, again following Chadwick's lead, Pettersson and Kirsch discussed the case of aluminum. Chadwick had noted that Schmidt in Vienna had found a proton yield about 15 times greater than his and Rutherford's value. Pettersson and Kirsch turned the tables by calculating that Blackett's cloud-chamber photographs established a proton yield from nitrogen—which "for general reasons" should be smaller than that from aluminum—that actually approached Schmidt's value. Similarly, Chadwick had reduced his estimate of the minimum range of the α particles required to disintegrate the aluminum nucleus from 4.9 to about 3 cm, "which is at all events an approach towards the results obtained in Vienna, where Schmidt has found α-particles of 1.3 cm range and less to be still able to release short-range protons from aluminum."[167] The most significant discrepancy with aluminum, however, concerned the minimum range of the expelled protons, which Rutherford and Chadwick had first estimated at 13–14 cm and Chadwick now gave as 9–10 cm, while Schmidt had found no minimum range down to 1 cm. Pettersson and Kirsch again turned the tables. Granted, the Cambridge value corresponded to a radius of 8×10^{-13} cm for the aluminum nucleus, while the Vienna value corresponded to 20×10^{-13} cm, but was this not analogous to the two different values Rutherford and Chadwick themselves had found, "to their surprise," for the radius of the uranium nucleus? "As a way out of this dilemma," Pettersson and Kirsch wrote, Rutherford had proposed a "special [satellite] model for the uranium nucleus." Assuming, "for the sake of argument, this model . . . to be correct," they asserted, "one may ask why the aluminum nucleus might not also be assumed to have a similar structure. . . ."[168] And was it not absurd to contend, as Chadwick had, that one should not look for disintegration protons within the range of the scattered α particles? The "same argumentation" would say that, because α particles of range 7 cm are still scattered normally by the uranium nucleus, one should not look for α particles of lesser range emitted radioactively—and thereby everyone would have missed discovering those of 2.7 cm range. It "appears advisable," Pettersson and Kirsch concluded, that in "a field of research so imperfectly explored as that of the artificial disintegration of elements" one should not "limit the scope of one's investigations by theoretical reasoning." "We are convinced," they summarized, "that if other experimenters will apply the methods we have developed . . . , they cannot fail to observe the phenomena we have seen."[169]

Those other experimenters who were following the controversy must have been more uncertain than ever as to which side to believe after reading Pettersson and Kirsch's rebuttal of Chadwick's article. Nor did Pettersson stop at that point. In succeeding months he buttressed the Vienna position still further. At the end of March 1927 he sent off a long article to the *Zeitschrift für Physik* in which he presented a plethora of experimental evidence for the disintegration of carbon,[170] and the following month he presented essentially the same paper to the Vienna Academy for publication in its *Sitzungsberichte*.[171] He knew, however, that if he was to convince others of the correctness of his and his co-workers' observations, he would first have to convince Rutherford and Chadwick of their validity. For that reason, and because he was completely confident that the Vienna results would prevail, he made definite arrangements, at long last, to actually visit the Cavendish Laboratory to present them in person to Rutherford and Chadwick. As in the past, he left Vienna for Göteborg for about a month in the spring of 1927. On his return trip he stopped off in Cambridge, arriving there at 2:30 P.M. on May 16. That evening, in his room in the Lion Hotel, he wrote the first of three daily letters to Meyer, the first two of which were accompanied by long reports on his impressions and conversations.[172]

Chadwick impressed Pettersson as a "very serious somewhat dogmatic person," but "not unapproachable" and willing "to take me at my words." He was even "friendly," as was Rutherford, although at one point the latter "spoke excitedly several times," being quite upset over the "polemical tone" of the Vienna publications. In general, however, Pettersson experienced "the usual Anglo-Saxon hospitality": Chadwick invited Pettersson to dinner in his College (Gonville and Caius) the first evening, and Rutherford invited him to lunch at home with Lady Rutherford, F. W. Aston, and Kerr Grant the following day. Rutherford also invited Pettersson to a meeting of the Cambridge Philosophical Society that evening and to a meeting of the Royal Society in London on May 19, the day Pettersson left Cambridge for his return to Vienna.

Pettersson's two reports to Meyer (which ran to 13 typed pages) reveal that he had extensive discussions at the Cavendish Laboratory, especially with Chadwick, but also with Blackett, Aston, C. T. R. Wilson, and, of course, with Rutherford, who spent as much time with Pettersson as his numerous obligations (among them the presidency of the Royal Society) permitted. Chadwick and Pettersson discussed in detail a wide range of topics: the scintillation substances used, the methods employed for counting scintillations, the experimental equipment and techniques, and, to be sure, the detailed experimental results found in both laboratories. A number of these same topics came up for discussion with Rutherford as well, but in addition Rutherford held forth on his satellite model of the nucleus, particularly as he was now extending it to his interpretation of natural radioactivity.

In Pettersson's judgment, a major benefit of his visit was that a number of misunderstandings that had arisen from his publications had been cleared up, and as a result personal irritations had been largely swept away. One case in point was probably representative of the change in relations. L. F. Bates, whom Pettersson met in Cambridge, visited Vienna on vacation with his wife 3 months later. There, as Pettersson wrote to Meyer (who was in Bad Ischl), he had a wonderful time showing his "old foe" the city and the Institut für Radiumforschung, where they talked about their "old fights" in "the friendliest tone."[173]

Pettersson also gained the impression in Cambridge that Chadwick and Rutherford were opening their minds somewhat to the Vienna experimental results. Still, a clear residue of skepticism remained. The questions that Chadwick and Pettersson had addressed in print had now been aired privately, but—as Pettersson's reports to Meyer testify—none of the fundamental issues had been resolved. Pettersson's confidence in the Vienna results had not wavered, and since he could not bring his equipment to Cambridge, he urged anew that Chadwick visit Vienna. That was a suggestion that Chadwick himself now took much more seriously than before, and after a good deal of discussion the two finally agreed that the best mutually convenient time probably would be early in December, although Chadwick felt that the entire matter was so important that he did not like to postpone it so long. A cordial exchange of letters between Meyer and Rutherford soon after Pettersson had left Cambridge made this visit even more likely.[174] Chadwick's present position, Pettersson had told Meyer, was that he doubted that he was overlooking scintillations, hence that the Vienna counters actually were seeing so many, and he "just wants to convince himself of this."

That Chadwick and Rutherford earnestly sought to convince Pettersson of the correctness of their observations in Cambridge there can be no doubt. Thus, when Pettersson thanked Rutherford for the hospitality shown to him in Cambridge,[175] he told Rutherford that he was "especially indebted" to Rutherford and Chadwick for "the friendly spirit in which [they] discussed the questions regarding artificial disintegration," and that he "appreciated the kind advice and criticism [he] received in the course of these conversations." Continuing, Pettersson declared: "We all of us are looking forward to Dr Chadwick's coming visit to Vienna and hope then to have opportunities for discussing with him the details of the methods we are using and the possible objections which may be made against them more fully than what was possible during my stay in Cambridge. I am writing now to Dr Chadwick giving him supplementary information on certain points which interested him, but I am sure he will have to see our experiments himself in order to give a definite judgment on our results."[176]

Pettersson took it for granted that Chadwick's judgment would be favorable, for in the summer of 1927 he and his co-workers in Vienna pursued their researches as actively as ever. By the end of August 1927 he had completed a long monograph in Swedish, *Atomernas sprängning: En studie i modern alkemi*,[177] which Elisabeth Kirsch subsequently translated into German under the title *Künstliche Verwandlung der Elemente (Zertrümmerung der Atome)*.[178] This work, as Pettersson explained to Meyer in a letter of August 28,[179] was what the English call a "potboiler," a nontechnical, popular version of Pettersson and Kirsch's *Atomzertrümmerung*, which had been published the preceding year. In addition, Pettersson and Kirsch's co-workers Elisabeth Rona, Ewald A. W. Schmidt, Georg Stetter, Elisabeth Kara-Michailova, Marietta Blau, and Berta Karlik carried out a variety of studies elaborating or supporting the Vienna position in mid 1927.[180] In mid September, Georg Stetter presented a long review of the Vienna work at a conference of German physicists in Kissingen; this was subsequently published in the *Physikalische Zeitschrift*.[181] Stetter was particularly interested in trying to undercut certain recent experiments by W. Bothe and H. Fränz,[182] whose use of electrical counting techniques at the Physikalisch-Technische Reichsanstalt in Berlin-Charlottenburg had seemed to confirm the results of Rutherford and Chadwick.

Rutherford himself was not inactive. In early 1927 he lectured on nuclear structure,[183] and by August, as mentioned above, he had developed a quantitative theory of radioactivity based upon his satellite model of the nucleus.[184] In September 1927 he discussed his new theory at the Volta Conference in Como,[185] after which he had hoped to visit

Meyer at the latter's summer residence in Bad Ischl. (He was unable to do so because he was not feeling well at the time.[186]) That would have been a pleasant personal reunion, but it would not have resolved the scientific differences existing between their laboratories. By the fall of 1927 those differences were as sharp as ever, and neither side was prepared to bend. By the fall of 1927 a stalemate existed between the Cambridge and the Vienna researchers.

Private Exposé, December 1927

As the controversy ran its course, both teams became thoroughly enmeshed in a complex web of observational, experimental, and theoretical commitments, and each team recognized that the outcome of the controversy would seriously affect the reputation of the other, and of the laboratories, but in opposite senses. The only hope of resolution lay in visiting each other's laboratory to view directly the experimental procedures and evidence amassed there. Now it was Chadwick's turn. Early in the fall, his responsibilities in the Cavendish Laboratory, Rutherford's conference in Como, and another trip of Pettersson to Sweden, all prevented Chadwick from traveling to Vienna until the holiday season. He and his wife arrived there on Wednesday morning, December 7, 1927, where they stayed in the comfortable Hotel Regina, about ten minutes by foot from the Institut für Radiumforschung.[187]

Immediately after unpacking, Chadwick went to the institute, where he spent the rest of the day talking with Stefan Meyer, Egon von Schweidler, Karl Przibram, Adolf Smekal, and "Pettersson's people." To Meyer he brought the welcome news that Rutherford was just completing the final arrangements for the University of Cambridge to purchase the balance of the radium lent to him in 1908 by the Vienna Academy—for the grand sum of £3,000, payable in six annual installments beginning March 31, 1928. No progress, however, was made on the scientific questions at issue, nor was any progress made the following day, for, as Chadwick reported to Rutherford, it "was a holy day and no work could be done without danger to our future in the world to come."

Friday, December 9, was entirely different. Preliminary pleasantries had passed. Chadwick was determined to get to the bottom of things rapidly and directly, and Pettersson (now on the defensive) was adamant in standing his ground. The result was, as Chadwick wrote Rutherford from his hotel room in the evening, that he and Pettersson "ended up with a fierce and very loud discussion." Chadwick continued:

I am having the greatest difficulty in avoiding irrelevant matters. As far as I can see the only way in which we can hope to reach a definite conclusion is to repeat Schmidts experiments with Aluminum or rather extend them to carbon and it is essential that I should prepare the experiment. So far, I cannot get Pettersson to agree to this. He insists upon testing [?] my power of counting small range H particles and in showing me experiments with the Shimizu [cloud-chamber] apparatus. Tonight he proved to his satisfaction that [polonium] α-particles of range not greater than 2.5 cms. could disintegrate Al. The yield was less than 1 H for 10^7 α-particles according to their own calculations. . . . I naturally said that the experiment was to me not conclusive and further that the point was unimportant relative to the question of the disintegration of carbon which contains all the discrepancies in one result. This precipitated a most fiery outburst, and all kinds of small matters were magnified to great importance such as your use of Al foils for absorption purposes in the expts. described in your Bakerian Lecture. If this happens again I shall have to put things to Petterssen [sic] in their true perspective. I am afraid this visit will not improve our relations. Stefan Meyer Schweidler and all with no direct interest in the question are exceedingly pleasant and friendly but the younger ones stand around stifflegged and with bristling hair.[188]

The atmosphere was extremely tense. And Chadwick did nothing to make it less tense, as we know from the firsthand account of one of Pettersson's assistants, Elisabeth Rona, who was there at the time. Many years later Rona recalled: "The impression made on us by Chadwick in this short visit was not favorable. He seemed to us to be cold, unfriendly, and completely lacking in a sense of humor. Probably he was just as uncomfortable in the role of judge as we were in that of the judged."[189] Rona added that she later came to understand that Chadwick's "ordeal" during his wartime internment in Berlin "had much to do with his behavior" in Vienna. From Chadwick's point of view, of course, he was struggling to hold himself in check. He had just come across Stetter's long review article in the *Physikalische Zeitschrift*,[190] and Chadwick told Rutherford that if he had seen it too he would understand how patient Chadwick had been in not mentioning his having seen it to Pettersson. No matter; Chadwick's main goal was to "keep at this until the business is settled."

On Monday evening, December 12, Chadwick wrote to Rutherford a second time.[191] He reported that "Nothing very serious took place on Saturday." He had had to "explain things as far back as . . . [Rutherford's] 1919 papers," as well as the Bakerian lecture of 1920. He also had examined the Vienna apparatus for comparing α-particle and proton scintillations. "Quite nice," Chadwick commented, "but nothing to do with our argument." Finally, they had made "a short run" with Schmidt's apparatus, trying to see if polonium α particles could disintegrate carbon. Chadwick himself saw no disintegration protons beyond the range of the scattered α particles from the source. However, he told Rutherford: "Their counters, two girls, managed to find a few. Their methods of counting are quite different from ours and it was possible that I failed to see very weak scintillations."

Chadwick decided to examine the counting procedure as rigorously as possible on Monday. Elisabeth Rona, who was one of the counters, recalled the atmosphere in the laboratory: "All of us sat in a dark room for half an hour to adapt to the darkness. There was no conversation; the only noise was the rattling of Chadwick's keys. There was nothing in the situation to quiet our nerves or make us comfortable. Short spells of scintillation counting followed for each member of the group, and then the radiation source was exchanged with a blank, unknown to the persons who were doing the counting."[192]

Chadwick described his procedure to Rutherford in the following words: "Today . . . I arranged that the girls should count and that I should determine the order of the counts. I made no change whatever in the apparatus, but I ran them up and down the scale like a cat on a piano—but no more drastically than I would in our own experiments if I suspected any bias. The result was that there was no evidence of H particles."[193] Chadwick compiled the results of these experiments for Rutherford in the form shown in table 2. In each case, Chadwick had changed the amount of absorbing material present without informing the Vienna counters of the amount of the change. At the smallest absorption (1.2 mm of air, which was smaller than the calculated range of the scattered α particles), the counters had observed an average of 10.3 scintillations in 20 seconds; at increasing absorptions (all in excess of the calculated range of the scattered α particles), they had observed an average of 1.5, 1.8, 1.75, and 0 scintillations in the same period of time. These latter scintillations quite likely were not produced by disintegration protons, but perhaps by radioactive contaminants, because the counters also had observed an average of 2.25 scintillations in 20 seconds even when the polonium α-particle source was cut off entirely. Chadwick told Rutherford that he also had tested the Vienna counters for their ability to observe "natural" protons expelled from paraffin. The result was that their counts were "normal" at the low absorption to which they were accustomed, but "irregular at a greater

Table 2
The counts recorded by the Vienna observers with Chadwick as leader.

		Carbon exposed to α's		Range of α 3.5 cm Angle 140°
	Absorption	No. of Scintill[ation]s in 20 secs.	Average	
Scattered α's & H's	1.2 mm	13.7.11	10.3	
H's alone	8.3 mm	1.2.1.1.2.2.	1.5	
	14.7 mm	3.3.0.1.2	1.8	
	36.3 mm	1.5.1.0.	1.75	
	76.6 mm	0.0.0.	0	

α-Particles cut off	Average
1.0.3.3.0.6.2.3.	2.25

absorption." He added that as far as he could tell there "was no reason why the counters should be off colour." Chadwick concluded cautiously that these observations did not "prove that there is nothing from carbon" but that he thought they "make it doubtful that there is much."

Chadwick also explained to Rutherford that until now he had had no chance to discuss these results with Pettersson; he had seen Pettersson only for five minutes that day, as Pettersson's "whole family" had come to Vienna for the holidays. And Pettersson's co-workers "did not seem anxious to discuss them." These co-workers, in fact, came as a surprise to Chadwick. In Cambridge he himself participated regularly in the counting. In Vienna, however, as he wrote Rutherford: "Not one of the men does any counting. It is all done by 3 young women. Pettersson says the men get too bored with routine work and finally cannot see anything, while women can go on for ever." In his 1969 interview, Chadwick stated that Pettersson also said that he believed that women were more reliable than men as scintillation counters because they would not be thinking while observing, and that Pettersson preferred women of "Slavic descent" as counters because he believed that Slavs had superior eyesight.[194] Pettersson valued his co-workers highly, and they him, and hence it is not clear in what tone of voice he made these remarks. Chadwick, however, took them perfectly seriously.

Chadwick closed his letter to Rutherford with the comment that "I suppose I must now investigate the Shimizu [cloud-chamber] apparatus if [Pettersson and his co-workers] agree with me about the meaning of the above experiment." Whether Chadwick pursued this particular investigation is unknown; we only know (from his 1969 interview) that various excuses were made for the above results. The main excuse was that the counters' eyes were probably tired since the experiments had been carried out after 5:00 P.M.[195] Consequently, they decided to carry out a second series of similar experiments the following morning. The results were exactly the same. It was not, Chadwick emphasized, that the Vienna counters were dishonest. There was no question of cheating. Rather, they were deluding themselves. They had been informed of the nature of the experiments being conducted. Chadwick concluded that they saw what they were expected to see.

The Vienna counters, therefore, had fallen prey to a psychological effect, much as René Blondlot had in reporting evidence for N rays two decades earlier.[196] However, whereas R. W. Wood had published his exposé of Blondlot in the open literature for the world to read, a different solution was agreed upon in Vienna. Chadwick recalled that the situation was "extremely awkward."[197] Pettersson, when confronted with these experimental results and conclusions, became "very angry indeed." The two agreed, however, to meet the director, Stefan Meyer, in his office the following morning. When they did, Meyer became "very upset indeed." Little did Chadwick realize that on the preceding day Meyer himself had written to Rutherford, expressing his expectation of an amicable and easy resolution of the discrepancies.[198] Now, however, confronted with Chadwick's evidence, Meyer offered to do anything necessary to set the record straight, such as make a public retraction. Chadwick refused this suggestion, for without consulting Rutherford he knew what Rutherford would wish to do. Chadwick knew that Rutherford was adamantly opposed to public controversy. Chadwick also knew that under no circumstances would Rutherford do anything that might cause his friend Meyer pain, and a public exposé would certainly do that. Knowing Rutherford's feelings, then, Chadwick told Meyer and Pettersson that the Vienna experiments on artificial disintegrtion simply should be dropped, and nothing further should be said about them. It should be a private exposé, one known only to those directly involved in the controversy.

Epilogue

Thus the Cambridge-Vienna controversy was brought to a close in Stefan Meyer's office at the Institut für Radiumforschung, evidently on Wednesday, December 14, 1927. Chadwick had not misunderstood Rutherford's intentions. Rutherford himself expressed his views directly to Pettersson in a letter of January 9, 1928: "There are so few workers in this difficult subject that we must try and pull together and settle our differences as far as possible by private correspondence rather than by controversies in the scientific journals, which in my experience do nothing but cause irritation. . . . During the whole of my scientific career I have not had any serious controversy and always advise my students to be considerate where differences of opinions are involved."[199] The outcome of the controversy, in fact, was kept so secret that not even those close to it but outside the innermost circle were permitted to have knowledge of it. A half-century later, for example, Elisabeth Rona could write "As far as I know, the discrepancies between the two laboratories were never resolved."[200] And Otto Frisch, who was a student in Vienna between 1922 and 1927 working under Karl Przibram, remarked in 1967 "I still do not know how [Pettersson's scintillation counters] found these wrong results," although he suspected that they stemmed from psychological bias.[201]

Rutherford and Chadwick contributed directly to this nebulous state of affairs by their treatment of the controversy in their treatise *Radiations from Radioactive Substances*, written with C. D. Ellis and published in 1930.[202] In that work, Rutherford and Chadwick state that it "seems difficult to put forward any satisfactory explanation" of the "divergences" between the two laboratories, "and this is not the place to attempt it." Instead, "in order to avoid fruitless discussion," they "refer only to the experiments carried out in the Cavendish Laboratory and . . . deal as far as possible with those points on which there is little conflict of opinion." For a "full statement of the Vienna observations" the reader is referred to Pettersson and Kirsch's original papers and to their book *Atomzertrümmerung*. This approach, which stemmed directly from Rutherford's aversion to public controversy, inevitably left physicists of the period bewildered.

Pettersson, too, contributed to the uncertainty surrounding the end of the controversy. He had made an enormous investment in time, energy, and money by entering and pursuing the field of artificial disintegration after 1922, and together with Kirsch he had established the Institut für Radiumforschung as the only major competitor to the Cavendish Laboratory in that field during the following five years. He had become convinced, as Frisch put it, that he was "beating the English at their own game."[203] That entire investment collapsed to the ground in a few short days in December 1927. It was a severe shock. Pettersson left Vienna almost immediately, traveling with his wife and daughter first to Göteborg and then on to Nes, Norway, for a month's vacation. His letters to Meyer during January 1928 reveal his sensitivity toward Chadwick and Rutherford.[204] He was particularly concerned about how they might react to his publishing a manuscript that he had already drafted on the disintegration of carbon, and he left it to Meyer to decide if it should be revised, and if it should be sent to Chadwick and Rutherford for comment prior to publication. When Meyer did suggest some changes, Pettersson readily accepted them, not wishing "to impair a lasting *entente cordiale* between Cambridge and Vienna."[205]

That was a common wish. Chadwick, immediately after his return to Cambridge, wrote to Meyer, thanking him for his "kindness and hospitality" and expressing the hope that he might visit Vienna again "under more auspicious circumstances."[206] Rutherford, too, immediately wrote to Meyer, discussing, however, only the radium purchase, and thereby by inference conveying his desire to put the controversy in the past.[207] Pettersson, on his part, sent a spinthariscope of his own design to Rutherford, a gift the latter graciously acknowledged in the letter of January 9, 1928, noted above.[208] A couple of weeks later, Pettersson sent a copy of his revised manuscript about carbon to Chadwick, at the same time expressing his wish that Chadwick had had "an agreeable Christmas and New Year."[209] Soon Chadwick's own reserve began to diminish, and in later years he and his wife became good friends with the Petterssons.[210] Nor was that all. Younger researchers working both in Cambridge and in Vienna began to migrate in both directions. Blackett visited Vienna in January 1928,[211] and another Cavendish researcher, J. Chariton, followed in July.[212] From Vienna, E. A. W. Schmidt came to Cambridge also in July 1928, attending a conference on Chadwick's invitation.[213] Still later, in the 1930s, Elisabeth Kara-Michailova and Elisabeth Rona visited Cambridge, and Berta Karlik visited London and made side trips to Cambridge.[214]

In spite of these increasingly cordial personal relations, however, some of the scientific issues at stake in the controversy were not so easy to dispel; they had an inertia of their own. Thus, for at least 9 months after Chadwick's visit, Pettersson continued to maintain, in prominent jounals published in German and in English, that carbon and oxygen are disintegrable, and he held the same view much longer in private.[215] In addition, Pettersson, having completed his nontechnical monograph *Atomernas sprängning* in August 1927, he signed the preface to the German translation one year later, in August 1928, and he deleted none of the book's controversial assertions.[216] Pettersson's persistence was costly to him in an unanticipated way: In 1928 he competed seriously for the professorship of physics at the University of Stockholm left vacant by the death of Svante Arrhenius, and the committee of four empowered to recommend Arrhenius' successor was evenly split between Pettersson and Erik Hulthén. By November, as Pettersson's correspondence of the period with Meyer shows,[217] one of those who opposed Pettersson, Manne Siegbahn, had carried the day. Siegbahn was firmly of the opinion that Pettersson's Vienna results were seriously in error, since they conflicted with Rutherford and Chadwick's experiments in Cambridge and with Bothe and Fränz's experiments in Berlin. Not even warm letters of reference that Meyer had solicited on Pettersson's behalf from Marie Curie, Georg von Hevesy, and Kasimir Fajans undercut Siegbahn's opposition to Pettersson.[218] Hulthén was appointed, and Pettersson remained in Göteborg, increasingly directing his great energy, with outstanding success, to the field of oceanography.[219] Kirsch, too, turned to another field, geophysics, and in May 1928 he signed the preface to an extensive monograph entitled *Geologie und Radioaktivität*.[220] Pettersson and Kirsch's younger co-workers in Vienna continued to publish extensively in the field of artificial disintegration, but they confined themselves to experimental, instrumental, and observational techniques.

At the Cavendish Laboratory, Rutherford and Chadwick recognized that, even though they had prevailed, new steps had to be taken in two directions. First, a thorough study had to be made of the fundamental processes involved in the counting of scintillations—how the radiant energy of a scintillation is produced, and how this energy is detected by the eye. They turned this study over to two of their research students, J. Chariton and C. A. Lea, who by November 1928 had produced the period's most comprehensive investigation of these processes.[221] Second, and much more far-reaching, Rutherford and

Chadwick recognized that the days of the scintillation method were numbered. They saw that this method inevitably would have to be replaced by electrical techniques for counting individual charged particles, such as Greinacher was developing in Bern and Bothe and Fränz were using in Berlin. As Rutherford wrote Meyer on June 10, 1929: "I think, if much more progress is to be made, on artificial distintegration, it is essential to tackle it by electrical methods and count a large number of particles. The scintillation method is much quicker for a preliminary survey but is not ideal for quantitative investigations which are now necessary."[222] Rutherford and Chadwick therefore encouraged C. E. Wynn-Williams and others to develop and improve the electrical counting method at the Cavendish. By August 1929, Wynn-Williams, F. A. B. Ward, and M. H. Cave had demonstrated its fruitfulness and its potential;[223] one year later, Rutherford, Ward, and Wynn-Williams used it to "detect the presence of a long-sought-for group" of short-range α particles emitted by RaC.[224] This transition from the scintillation method to the electrical method of counting charged particles—from a human to an instrumental method of observation—was one of the most significant consequences of the Cambridge-Vienna controversy.[225]

Experimentally, Rutherford and Chadwick's conclusion that the Vienna observers lacked evidence for the disintegration of carbon carried the implication that oxygen also was not disintegrable by RaC or polonium α particles. This view was strengthened by F. W. Aston, who reported new isotopic-mass determinations in 1927 and concluded that both carbon and oxygen nuclei possess "a tightness of packing entirely in favour of the views of Rutherford and Chadwick as against those of Kirsch and Pettersson, who claim that protons can be detached."[226]

Pettersson attempted to rebut Aston's claim by showing that these reactions were energetically possible if the α particle were captured by carbon and oxygen nuclei. Pettersson's calculations were flawed,[227] but their intent was of central siginificance, as Chadwick too had recognized as early as 1925.[228]

Unfortunately, even in 1930, Rutherford, Chadwick, and Ellis were forced to state that the "information at present available is not sufficient to test accurately the balance between the energy and mass changes in the disintegrations."[229] Only seven years later, after many more mass-spectroscopic determinations had been made, M. S. Livingston and H. A. Bethe could conclude unambiguously that it is "known today that natural alpha-particles will disintegrate all of the lighter elements up to Ca . . . ., with the exception of H, He, C, and O."[230]

By 1928, however, the evidence that carbon is not disintegrable in-dicated that disintegration is not a general phenomenon, as Pettersson had believed. As a consequence, Pettersson's explosion hypothesis suf-fered a severe blow. Ironically, Rutherford's satellite model did not survive much longer either. In June 1928 it was challenged from a totally unanticipated direction. George Gamow, from Leningrad, arrived that summer in Göttingen, went into the physics library, came across Rutherford's most recent paper in the September 1927 issue of the *Philosophical Magazine*,[231] and before putting it down knew that Ruth-erford's satellite interpretation of radioactivity, as set forth quantitatively in that paper, was fundamentally incorrect—the α particles, in emerging from a radioactive nucleus, were actually tunneling through a quantum-mechanical potential barrier.[232] R. W. Gurney and E. U. Condon, at Princeton, came to the identical conclusion independently and virtually simultaneously.[233] The fundamental correctness of this new quantum-mechanical theory of α decay was rapidly and widely accepted. Ruth-erford, however, found it difficult to jettison his satellite model, which so beautifully united artificial and natural disintegration. He still included a discussion of it in the 1930 treatise *Radiations from Radioactive Sub-stances*,[234] over the strong objections of Chadwick.[235] Meanwhile, Ga-mow had applied his theory of α decay to the inverse problem of α disintegration and had concluded that natural α particles cannot dis-integrate iron and other heavy elements, as Pettersson and Kirsch had claimed.[236] Gamow's new theory demonstrated, above all, that classical models of the nucleus, whether Pettersson's or Rutherford's, would have to be replaced by quantum-mechanical ones.

The Cambridge-Vienna controversy had flourished on an intimate blend of observational, experimental, and theoretical issues, and in an atmosphere of intense competition. It had left the Institut für Radium-forschung, in Otto Frisch's words, as "a sort of *enfant terrible* of nuclear physics."[237] The Vienna team of Pettersson and Kirsch split up, as mentioned. The Cambridge team of Rutherford and Chadwick also dissolved to a degree—they published no joint papers after 1927, al-though they remained close colleagues in the Cavendish until Chadwick left for Liverpool in 1935.

The Cambridge-Vienna controversy was over, but it ended privately, with the principal protagonists leaving no trace of the reasons for its resolution in the open literature. Yet, the controversy ended definitively. The most authoritative judgment of the period was that of Livingston and Bethe in 1937. Referring in a footnote to the "long-standing controversy" between the Cambridge and Vienna researchers, Livingston and Bethe concluded: "Although the sincerity of these workers [in Vienna] cannot be questioned many subsequent experiments have proved their results almost entirely erroneous. . . . In the face of the existing evidence we are forced to eliminate these data from Vienna in the report to follow."[238] Many new challenges in nuclear physics already had supplanted the old ones, and it was time to put the Cambridge-Vienna controversy in the past.

Acknowledgments

I am grateful to P. H. Fowler for permission to quote from Rutherford's correspondence, to Lady Aileen Chadwick for permission to quote from her husband's correspondence, and to Berta Karlik, Roman Sexl, and Herbert Vonach for access to Stefan Meyer's papers. I also am grateful to The Royal Society for permission to reproduce figures 1 and 2 from its *Proceedings*, to Taylor and Francis Ltd. for permission to reproduce figures 3 and 5 from *The Philosophical Magazine*, and to The Institute of Physics for permission to reproduce figure 4 from the *Proceedings of the Physical Society of London*. This research has been supported in part by grants from the American Council of Learned Societies, the National Science Foundation, and the Bush Foundation, whose assistance I gratefully acknowledge.

Notes and References

1. E. Rutherford, "Collision of α Particles with Light Atoms. IV. An Anomalous Effect in Nitrogen," *Phil. Mag.* 37 (1919): 581–587; reprinted in *The Collected Papers of Lord Rutherford of Nelson, O.M., F.R.S.*, vol. 2, ed. J. Chadwick (London: Allen & Unwin, 1963) (hereafter referred to as *Collected Papers*).

2. E. Rutherford, J. Chadwick, and C. D. Ellis, *Radiations from Radioactive Substances* (New York: Macmillan, 1930), p. 281.

3. Karl K. Darrow, Contemporary Advances in Physics. XXII. Transmutation, Bell Telephone System Tech. Pub. Monograph B-596 (1931), p. 14. This report originally appeared in *Bell Sys. Tech. J.* 10 (1931): 628–655.

4. Rutherford, Chadwick, and Ellis, note 2, p. 545. See also pp. 54–55 and 544–548. For earlier discussions of the scintillation method see W. Makower and H. Geiger, *Practical Measurements in Radio-Activity* (London: Longmans, Green, 1912), pp. 48–50; E. Rutherford, *Radioactive Substances and Their Radiations* (Cambridge University Press, 1913), pp. 133–135. For a recent treatise see J. B. Birks, *The Theory and Practice of Scintillation Counting* (Oxford: Pergamon, 1964).

5. E. Rutherford and H. Geiger, "An Electrical Method of Counting the Number of α-Particles from Radio-Active Substances," *Proc. Roy. Soc. A* 81 (1908): 141–161; reprinted in *Collected Papers*, vol. 2, pp. 89–108. See also Thaddeus J. Trenn, "Rutherford's Electrical Method: Its Significance for Radioactivity and an Expression of His Metaphysics," in *Proceedings of the XIII International Congress on the History of Science* (Moscow: Editions "Nauka," 1974), section 6, pp. 112–118; "Die Erfindung des Geiger-Müller-Zählrohres," *Abh. Ber. Deutsch. Mus.* 44 (1976): 54–64.

6. Makower and Geiger (n. 4), p. 50.

7. E. Rutherford, "Collision of α Particles with Light Atoms. I. Hydrogen," *Phil. Mag.* 37 (1919): 537–561; reprinted in *Collected Papers*, vol. 2 (see p. 551).

8. Rutherford, Chadwick, and Ellis (n. 2), p. 550.

9. Rutherford, (n. 1), p. 590.

10. Ibid., p. 589.

11. For a discussion of the fundamental particles assumed to be present in nuclei during the period in question, see R. H. Stuewer, "The Nuclear Electron Hypothesis," in *Otto Hahn and the Rise of Nuclear Physics*, ed. W. R. Shea (Dordrecht: Reidel, 1983).

12. Rutherford (n. 1), p. 589.

13. See discussion in *Nuclear Physics in Retrospect: Proceedings of a Symposium on the 1930s*, ed. R. H. Stuewer (Minneapolis: University of Minnesota Press, 1979), p. 321.

14. For a biography see H. Massey and N. Feather, "James Chadwick 1891–1974," *Biog. Mem. Fellows Roy. Soc.* 22 (1976): 11–70. See also J. Chadwick, "The Rutherford Memorial Lecture, 1953," *Proc. Roy. Soc. A* 244 (1954): 435–447, especially pp. 443–445.

15. Massey and Feather, p. 65.

16. E. Rutherford, "Nuclear Constitution of Atoms," *Proc. Roy. Soc. A* 97 (1920): 374–400; reprinted in *Collected Papers*, vol. 3 (1965), pp. 14–38.

17. Ibid., p. 26.

18. Ibid., pp. 26–38. Thus, half his lecture related to these particles.

19. Ibid., p. 34.

20. J. Chadwick, "Some Personal Notes on the Search for the Neutron," in *Proceedings of the X International Congress on the History of Science* (Paris: Hermann, 1964), pp. 159–162; reprinted in *The Nucleus*, Harvard Project Physics Reader 6 (New York: Holt, Rinehart and Winston, 1968–69), pp. 25–31.

21. Rutherford, "Nuclear Constitution" (n. 16), pp. 35, 37.

22. Ibid., p. 37.

23. E. Rutherford and J. Chadwick, "The Artificial Disintegration of Light Elements," *Phil. Mag.* 42 (1921): 809–825 (reprinted in *Collected Papers*, vol. 3,

pp. 48–62); "The Disintegration of Elements by α Particles," *Phil. Mag.* 44 (1922): 417–432 (*Collected Papers*, vol. 3, pp. 67–80). For a preliminary report of the first paper see "The Disintegration of Elements by α Particles," *Nature* 107 (1921): 41 (*Collected Papers*, vol. 3, pp. 41–42).

24. This investigation formed the subject of the second paper listed in n. 23.

25. Rutherford and Chadwick, "Artificial Disintegration" (n. 23), p. 61.

26. S. Meyer, "Das erste Jahrzehnt des Wiener Instituts für Radiumforschung," *Jahrbuch Radioak. Elek.* 17 (1920): 1–29.

27. For biographies see the obituary notices by O. Hahn, *Z. Naturforsch.* 5a (1950): 407; F. A. Paneth and R. W. Lawson, *Nature* 165 (1950): 548–549; K. Przibram, *Akad. Wiss. Wien, Almanach* 100 (1950): 340–352; and H. Benndorf, *Acta Phys. Aust.* 5 (1952): 152–168.

28. Meyer describes this loan in his obituary notice of Rutherford, *Akad. Wiss. Wien, Almanach* 88 (1938): 251–262. See especially p. 255. See also A. S. Eve, *Rutherford* (New York: Macmillan, 1939), pp. 167–172.

29. S. Meyer and E. R. v. Schweidler, *Radioaktivität* (Leipzig and Berlin: Teubner, 1916). An enlarged second edition was published in 1927.

30. See n. 4.

31. Lawson, "Meyer" (n. 27), p. 549.

32. Meyer, "Rutherford" (n. 28), p. 255; Eve, *Rutherford* (n. 28), p. 325.

33. See for example Paneth, "Meyer" (n. 27), p. 548.

34. Pettersson to Meyer, November 28, 1921, in Meyer Papers, Institut für Radiumforschung und Kernphysik, Vienna (hereafter MP). A note by Meyer on this letter dated December 7, 1921, reads "willkommen." See also K. Przibram, "Festschrift des Instituts für Radiumforschung anlässlich seines 40jährigen Bestandes (1910–1950)," *Sitz. Osterr. Akad. Wiss.* 2a 159 (1950): 1.

35. For a biography and a bibliography see G. E. R. Deacon, "Hans Pettersson 1888–1966," *Biog. Mem. Fellows Roy. Soc.* 12 (1966): 405–421. See also *Svenska Män och Kvinnor* 6 (Stockholm: Bonniers, 1949), pp. 113–114.

36. Interview of Charles Weiner with James Chadwick, April 15–21, 1969 (Niels Bohr Library, American Institute of Physics), p. 62; Elizabeth Rona, *How It Came About: Radioactivity, Nuclear Physics, Atomic Energy* (Oak Ridge, Tenn.: Associated Universities, undated, ca. 1976), p. 63; Elizabeth Rona to author, private communication, February 16, 1979. In her publications of the period her first name was spelled "Elisabeth," and for the sake of consistency I have used this spelling throughout the text.

37. H. Pettersson and G. Kirsch, *Atomzertrümmerung: Verwandlung der Elemente durch Bestrahlung mit α-Teilchen* (Leipzig: Akademische, 1926), p. vii.

38. G. Kirsch and H. Pettersson, "Uber die Atomzertrümmerung durch α-Partikeln," *Sitz. Akad. Wiss. Wien* 2a 132 (1923): 299–307; "Long-range Particles

from Radium-active Deposit," *Nature* 112 (1923): 394–395; "Experiments on the Artificial Disintegration of Atoms," *Phil. Mag.* 47 (1924): 500–512. Pettersson mentions his double interruption in the second note on p. 500 of this last article. Also see "Long-range Particles from Radium-active Deposit," *Nature* 112 (1923): 540, where Pettersson corrects slightly his previous article in this journal and where the editor of *Nature* points out that the error was due partly to Pettersson and Kirsch.

39. H. Pettersson, "Zur Herstellung von Radium C," *Sitz. Akad. Wiss. Wien 2a* 132 (1923): 55–57.

40. Kirsch and Pettersson, "Experiments" (n. 38), p. 507.

41. Ibid.

42. Ibid., p. 512.

43. Kirsch and Pettersson, "Long-range Particles" (n. 38), p. 395.

44. L. F. Bates and J. S. Rogers, "Long-range Particles from Radium-active Deposit," *Nature* 112 (1923): 435–436.

45. G. Kirsch and H. Pettersson, "Long-range Particles from Radium-active Deposit," *Nature* 112 (1923): 687.

46. Ibid.

47. Rutherford to Meyer, November 24, 1923, MP.

48. L. F. Bates and J. S. Rogers, "Long-Range α-Particles," *Nature* 112 (1923): 938.

49. Ibid.

50. L. F. Bates and J. S. Rogers, "Particles of Long Range Emitted by the Active Deposits of Radium, Thorium and Actinium," *Proc. Roy. Soc. A* 105 (1924): 97–116. Also see "Particles of Long Range from Polonium," ibid., pp. 360–369. Bates published yet a third paper in 1924: "On the Range of α-Particles in Rare Gases," ibid., pp. 622–632.

51. D. Pettersson, "Long-range Particles from Radium-active Deposit," *Nature* 113 (1924): 641–642. See also "Uber die maximale Reichweite der von Radium C ausgeschleuderten Partikeln," *Sitz. Akad. Wiss. Wien 2a* 133 (1924): 149–162.

52. E. Kara-Michailova and H. Pettersson, "The Brightness of Scintillations from H-particles and from α-particles," *Nature* 113 (1924): 715. See also "Uber die Messung der relativen Helligheit von Szintillationen," *Sitz. Akad. Wiss. Wien 2a* 133 (1924): 163–168.

53. H. Pettersson, "On the Structure of the Atomic Nucleus and the Mechanism of Its Disintegration," *Proc. Phys. Soc. Lond.* 36 (1924): 194–202; "Discussion," ibid., pp. 202–204.

54. Ibid., p. 194.

55. Ibid., p. 196.

56. Ibid., p. 198.

57. Ibid., p. 202.

58. Ibid.

59. Pettersson, "Discussion" (n. 53), p. 203.

60. E. Rutherford, "Atomic Projectiles and Their Properties," *Engineering* 115 (1923): 798–800.

61. E. Rutherford, "The Electrical Structure of Matter," *Nature* Supplement 2811 (September 15, 1923), pp. 409–419.

62. E. Rutherford and J. Chadwick, "The Bombardment of Elements by α-Particles," *Nature* 113 (1924): 457 (*Collected Papers*, vol. 3, pp. 110–112). See also Rutherford to Meyer, March 26, 1924, MP, in which Meyer was informed of the imminent appearance of this paper.

63. E. Rutherford and J. Chadwick, "Further Experiments on the Artificial Disintegration of Elements," *Proc. Phys. Soc. Lond.* 36 (1924): 417–422 (*Collected Papers*, vol. 3, pp. 113–119).

64. Rutherford and Chadwick, "Bombardment" (n. 62), p. 111.

65. G. Kirsch and H. Pettersson, "The Artificial Disintegration of Atoms," *Nature* 113 (1924): 603.

66. Ibid.

67. Meyer to Rutherford, May 13, 1924, MP.

68. Rutherford and Chadwick, "Further Experiments" (n. 63), p. 115.

69. Ibid., p. 116.

70. Ibid., p. 119.

71. E. Rutherford and J. Chadwick, "On the Origin and Nature of the Long-Range Particles Observed with Sources of Radium C," *Phil. Mag.* 48 (1924): 509–526 (*Collected Papers*, vol. 3, pp. 120–135).

72. Ibid., p. 127.

73. Pettersson to Rutherford, July 13, 1924, in Rutherford correspondence, Cambridge University Library (hereafter RC).

74. Rutherford to Bohr, July 18, 1924, RC.

75. Rutherford to Pettersson, July 19, 1924, MP.

76. Rutherford to Meyer, July 19, 1924, RC; also quoted in Eve, *Rutherford* (n. 28), p. 299.

77. Meyer to Rutherford, July 24, 1924, MP.

78. G. Kirsch and H. Pettersson, "Uber die Verwandlung der Elemente durch Atomzertrümmerung. I.," *Naturwissenschaften* 12 (1924): 495–500.

79. G. Ortner and H. Pettersson, "Zur Herstellung von Radium C. II., *Sitz. Akad. Wiss. Wien 2a* 133 (1924): 229–234.

80. G. Kirsch and H. Pettersson, "Uber die Atomzertrümmerung durch α-Partikeln. II. Eine Methode zur Beobachtung der Atomtrümmer von kurzer Reichweite," *Sitz, Akad. Wiss. Wien 2a* 133 (1924): 235–241.

81. H. Pettersson, "Uber die Atomzertrümmerung durch α-Partikeln. III. Die Zertrümmerung von Kohlenstoff," *Sitz. Akad. Wiss. Wien 2a* 133 (1924): 445–461.

82. G. Kirsch, "Uber Atomzertrümmerung durch α-Strahlen. IV. Abbau von Stickstoff und Sauerstoff.—Helium als Abbauprodukt," *Sitz. Akad. Wiss. Wien 2a* 133 (1924): 462–476.

83. Pettersson to Rutherford, July 27, 1924 (draft, misdated 1923), MP; August 8, 1924, RC. All the quotations that follow come from this letter.

84. Pettersson to Meyer, July 27, 1924, MP.

85. Pettersson to Meyer, August 8, 1924, MP.

86. E. Rutherford, "The Nucleus of the Atom," *Engineering* 117 (1924): 458–459.

87. Chadwick to Rutherford, September 1924, RC; also quoted in Eve, *Rutherford* (n. 28), pp. 300–301.

88. Weiner-Chadwick interview (n. 36), pp. 35–36. Also see Chadwick, "Some personal notes" (n. 20).

89. Rutherford, "Electrical Structure" (n. 61), p. 419.

90. Rutherford, "Nucleus" (n. 86).

91. E. Rutherford, "The Natural and Artificial Disintegration of the Elements," *J. Franklin Inst.* 198 (1924): 725–744; see especially pp. 731–735. This important paper is not reprinted in Rutherford's *Collected Papers*.

92. E. Rutherford and J. Chadwick, "Scattering of α-particles by Atomic Nuclei and the Law of Force," *Phil. Mag.* 50 (1925): 889–913 (*Collected Papers*, vol. 3, pp. 143–163; see especially pp. 156–163). See also E. Rutherford, "Studies of Atomic Nuclei," *Engineering* 119 (1925): 437–438; "Atomic Nuclei and Their Transformations," *Proc. Phys. Soc. Lond.* 39 (1927): 359–372 (*Collected Papers*, vol. 3, pp. 164–180; see especially pp. 173–180); "Structure of the Radioactive Atom and Origin of the α-Rays," *Phil. Mag.* 4 (1927): 580–605 (*Collected Papers*, vol. 3, pp. 181–202).

93. Rutherford and Chadwick, "Scattering," p. 158; Rutherford, "Atomic Nuclei," pp. 177–178; "Structure," pp. 181–182.

94. P. Debye and W. Hardmeier, "Anomale Zerstreuung von α-Strahlen," *Phys. Z.* 27 (1926): 196–199. See also W. Hardmeier, "Anomale Zerstreuung von α-Strahlen," ibid., 28 (1927): 181–195.

95. See especially Rutherford, "Structure" (n. 92), pp. 197–202.

96. H. Pettersson and G. Kirsch, "Uber Atomzertrümmerung," *Phys. Z.* 25 (1924): "I. Methodik" (Pettersson), pp. 588–592; "II. Ergebnisse" (Kirsch), pp.

592–595. See also H. Pettersson, "Zur Methodik der Atomzertrümmerung," *Sitz. Akad. Wiss. Wien 2a* 134 (1925): 45–50.

97. H. Pettersson, "On the Forces at Nuclear Collisions and Coulomb's Law," *Ark. Mat., Astr. Fys.* 19B, no. 2 (1925–1927): 1–6; "The Reflexion of α-particles against Atomic Nuclei," ibid. 19A, no. 15 (1925–1927): 1–16.

98. H. Pettersson, "Uber das Kraftfeld des Atomkernes und Coulomb's Gesetz," *Sitz. Akad. Wiss. Wien 2a* 133 (1924): 509–515; "Uber die Reflexion von α-Teilchen an Atomkernen," ibid., pp. 573–588.

99. Pettersson, "Reflexion" (n. 97), p. 6.

100. Ibid., pp. 10 and 14.

101. Rutherford and Chadwick, "Further Experiments" (n. 63), p. 119.

102. Rutherford and Chadwick, "Scattering" (n. 92), pp. 143–163.

103. Ibid., p. 153.

104. Ibid., p. 162.

105. P. M. S. Blackett, "The Ejection of Protons from Nitrogen Nuclei, Photographed by the Wilson Method," *Proc. Roy. Soc. A* 107 (1925): 349–360. See also "The Rutherford Memorial Lecture, 1958," ibid. 251 (1959): 293–305, especially pp. 293–295.

106. E. Rutherford, "The Stability of Atoms," *J. Roy. Soc. Arts* 73 (1925): 389–402; "Discussion," ibid., pp. 402–403. This paper is not included in Rutherford's *Collected Papers*.

107. Ibid., p. 402.

108. This letter has not survived in the Rutherford correspondence; we know its date from Pettersson's reply to Rutherford.

109. Pettersson to Rutherford, February 27, 1925, RC.

110. A copy of Pettersson's withdrawn letter to *Nature* of July 1924 is included in the Rutherford Correspondence.

111. Rutherford to Pettersson, March 5, 1925, RC.

112. E. Rutherford, "Disintegration of Atomic Nuclei," *Nature* 115 (1925): 493–494 (*Collected Papers*, vol. 3, pp. 136–138).

113. Weiner-Chadwick interview (n. 36), pp. 40 and 52.

114. G. Kirsch, "Uber den Nachweis retrograder *H*-Partikeln aus zertrümmerten Atomen," *Phys. Z.* 26 (1925): 379–380.

115. G. Kirsch and H. Pettersson, "Uber die Reflexion von α-Teilchen an Atomkernen II," *Sitz. Akad. Wiss. Wien 2a* 134 (1925): 491–512.

116. Ibid., p. 512.

117. G. Kirsch, "Uber den Vorgang bei der 'Atomzertrümmerung' durch α-Strahlen," *Phys. Z.* 26 (1925): 457–465.

118. Ibid., p. 459.

119. Ibid., p. 464.

120. Chadwick to Rutherford, undated but probably July–August 1925, RC. The suggested date of October 10, 1925, in *Rutherford Correspondence Catalog*, ed. L. Badash (New York: AIP Center for History of Physics, 1974), is incorrect.

121. Lawson to Meyer, December 22, 1924, MP.

122. Carton 7 of the Tuve Papers (Library of Congress, Washington, D.C.) contains handwritten notes on the Cambridge and Vienna publications.

123. See n. 52 and n. 79.

124. E. A. W. Schmidt, "Uber die Zertrümmerung des Aluminiums durch α-Strahlen, I," *Sitz. Akad. Wiss. Wien 2A* 134 (1925): 385–404; G. Stetter, "Die Massenbestimmung von 'H'-Partikeln," *Z. Phys.* 34 (1925): 158–177 [see also "Die Bestimmung des Quotienten Ladung/Masse für natürliche H-Strahlen und Atomtrümmer aus Aluminium," *Sitz. Akad. Wiss. Wien 2A* 135 (1926): 61–69]; M. Blau, "Uber die photographische Wirkung natürlicher H-Strahlen," ibid. 134 (1925): 427–436 plus plate; "Die photographische Wirkung von H-Strahlen aus Paraffin und Aluminum," *Z. Phys.* 34 (1925): 285–295.

125. E. Rona, "Absorptions- und Reichweitenbestimmungen an 'natürlichen' H-Strahlen," *Sitz. Akad. Wiss. Wien 2a* 135 (1926): 117–126; M. Blau and E. Rona, "Ionisation durch H-Strahlen," ibid. 135 (1926): 573–585. Rona, in *How It Came About* (n. 36), explains that Stefan Meyer extended the invitation to her to come to Vienna while both were on vacation in Bad Ischl.

126. H. Pettersson and G. Kirsch, *Atomzerstrümmerung* (n. 37).

127. Ibid., p. 122.

128. Ibid., pp. 240–243.

129. R. W. L[awson], "Modern Alchemy," *Nature* 120 (1927): 178–179. See also Pettersson to Lawson, August 31, 1927, MP, in which Pettersson thanks Lawson warmly for his review.

130. H. Pettersson and G. Kirsch, "Atomzertrümmerung," in *Handbuch der Physik*, Band XXII, ed. H. Geiger and Karl Scheel (Berlin: Springer, 1926), pp. 146–178.

131. G. Kirsch, "Atomzertrümmerung," *Ergeb. exak. Naturwiss.* 5 (1926): 165–191.

132. Weiner-Chadwick interview (n. 36), p. 36. Here Chadwick points out that the paper published four years later as Chadwick and Rutherford, "Energy Relations in Artificial Disintegration," *Proc. Camb. Phil. Soc.* 25 (1929): 186–192 (*Collected Papers*, vol. 3, pp. 218–224) was actually written by him while Rutherford was away on his journey to Australia and New Zealand.

133. Rutherford and Chadwick, "Energy Relations," p. 223.

134. Chadwick to Rutherford, November 21, 1925, RC.

135. Rutherford to Bohr, February 8, 1926, RC.

136. Weiner-Chadwick interview (n. 36), p. 61.

137. J. Chadwick, "Observations Concerning the Artificial Disintegration of Elements," *Phil. Mag.* 2 (1926): 1056–1075.

138. Ibid., p. 1058.

139. Ibid.

140. Ibid., p. 1059.

141. Ibid., p. 1061.

142. Ibid., pp. 1061–1062.

143. Weiner-Chadwick interview (n. 36), p. 33.

144. H. Geiger and A. Werner, "Die Zahl der von Radium ausgesandten α-Teilchen. I. Teil. Szintillationszählungen," *Z. Phys.* 21 (1924): 187–203.

145. Chadwick, "Observations" (n. 137), p. 1063.

146. Ibid., pp. 1059 and 1071.

147. Ibid., p. 1072.

148. Ibid., pp. 1074–1075.

149. Ibid., p. 1075.

150. Meyer to Rutherford, December 17, 1926, RC.

151. Rutherford to Meyer, December 23, 1926, RC; reprinted in Eve, *Rutherford* (n. 28), p. 318.

152. G. Kirsch and H. Pettersson, "Die Zerlegung der Elemente durch Atomzertrümmerung," *Z. Phys.* 42 (1927): 641–678.

153. H. Pettersson and G. Kirsch, "The Artificial Disintegration of Elements," *Ark. Mat., Astr. Fys.* 20A, no. 16 (1927–28): 1–42. All quotations will be taken from this English version of the paper.

154. G. Kirsch and H. Pettersson, "Uber die Atomzertrümmerung durch α-Partikeln. V. Zur Frage der Existenz von Atomtrümmern kurzer Reichweite," *Sitz. Akad. Wiss. Wien 2A* 136 (1927): 195–224.

155. Pettersson and Kirsch, "Artificial Disintegration" (n. 141), pp. 5–8.

156. Ibid., p. 10.

157. Ibid., p. 11.

158. Ibid., p. 18.

159. Ibid., p. 19. Pettersson and Kirsch cite Geiger and Werner (n. 144), p. 196, to substantiate their claim.

160. Ibid., pp. 19–22.

161. Ibid., pp. 23–24.

162. Ibid., pp. 22–23.

163. Ibid., p. 26.

164. Ibid., p. 29.

165. Ibid., pp. 29–30.

166. Ibid., pp. 31–36. Pettersson and Kirsch refer to the papers of Blau, Grein-acher, and Ortner and Stetter. Pettersson had discussed these methods earlier as well; see "Zum Nachweis der bei Elementverwandlungen durch Atomzer-trümmerung entstehenden Produkte," *Mikrochemie* 4, no. 9/16 (1926): 177–184.

167. Pettersson and Kirsch, "Artificial Disintegration" (n. 153), p. 38.

168. Ibid., p. 40.

169. Ibid., p. 41.

170. H. Pettersson, "Die Zertrümmerung des Kohlenstoffatomes," *Z. Phys.* 42 (1927): 679–703.

171. H. Pettersson, "Uber die Atomzertrümmerung durch α-Partikeln. VI. Die Zertrümmerung von Kohlenstoff, II. Teil," *Sitz. Akad. Wiss. Wien 2a* 136 (1927): 225–242.

172. Pettersson to Meyer, May 16, 1927 (with enclosed "P.M.," 4 pp.), May 17, 1927 (with enclosed "P.M.," 9 pp.), and May 18, 1927, MP. The quotations that follow are drawn from the first letter and the two "P.M."'s.

173. Pettersson to Meyer, August 28, 1927, MP.

174. Meyer to Rutherford, May 25, 1927; Rutherford to Meyer, June 1, 1927, RC, MP.

175. Pettersson to Rutherford, May 31, 1927, RC.

176. Ibid.

177. H. Pettersson, *Atomernas sprängning. En studie i modern alkemi* (Stockholm: Gebers, 1927).

178. H. Pettersson, *Künstliche Verwandlung der Elemente (Zertrümmerung der Atome)* (Berlin and Leipzig: De Gruyter, 1929).

179. Pettersson to Meyer, August 28, 1927, MP.

180. E. Rona and E. A. W. Schmidt, "Untersuchungen über das Eindringen des Poloniums in Metalle," *Sitz. Akad. Wiss. Wien 2a* 136 (1927): 65–73 [see also "Die Herstellung von hochkonzentrierten Poloniumpräparaten durch Des-tillation," *Z. Phys.* 48 (1928): 784–789; G. Stetter, "Die Massenbestimmung von Atomtrümmern aus Aluminium, Kohlenstoff, Bor und Eisen," ibid. 42 (1927): 741–758; "Zur Umladung langsamer H-Partikeln," ibid. 42 (1927): 759–762; E. Kara-Michailova, "Helligkeit und Zählbarkeit der Szintillationen von magnetisch abgelenkten H-Strahlen verschiedener Geschwindigkeit," *Sitz. Akad. Wiss. Wien 2a* 136 (1927): 357–368; M. Blau, Uber die photographische Wirkung von H-Strahlen II," ibid. 136 (1927): 469–480; "Uber die photograph-ische Wirkung von H-Strahlen aus Paraffin und Atomfragmenten," *Z. Phys.*

48 (1928): 751–764; B. Karlik, "Uber die Abhängigkeit der Szintillationen von der Beschaffenheit des Zinksulfids und das Wesen des Szintillationsvorganges," *Sitz. Akad. Wiss. Wien 2a* 136 (1927): 531–561 plus plate; B. Karlik and E. Kara-Michailova, "Zur Kenntnis der Szintillationsmethode," *Z. Phys.* 48 (1928): 765–783.

181. G. Stetter, "Die neueren Untersuchungen über Atomzertrümmerung," *Phys. Z.* 28 (1927): 712–723.

182. W. Bothe and H. Fränz, "Atomzertrümmerung durch α-Strahlen von Polonium," *Z. Phys.* 43 (1927): 456–465.

183. E. Rutherford, "Atomic Nuclei and Their Transformations" (Twelfth Guthrie Lecture, delivered February 25, 1927), *Proc. Phys. Soc. Lond.* 39 (1926–27): 359–372 (*Collected Papers*, vol. 3, pp. 164–180); "Alpha-Rays and Atomic Structure.—IV," *Engineering* 123 (1927): 492–493 (delivered at Royal Institution, April 9, 1927).

184. Rutherford, "Structure" (n. 92), pp. 181–202.

185. E. Rutherford, "Structure of Radioactive Atoms and the Origin of the α Rays," in *Atti del Congresso Internazionale dei Fisici 11–20 Settembre 1927*, vol. 1 (Bologna: Zanichelli, 1928), pp. 55–64.

186. Rutherford to Meyer, September 22, 1927, RC, MP.

187. The information in this paragraph and the next one is drawn from the letter of Chadwick to Rutherford, December 9 [1927], RC.

188. Ibid.

189. Rona, *How It Came About* (n. 36), p. 20.

190. Stetter, "Die neueren Untersuchungen" (n. 181), pp. 712–723.

191. Chadwick to Rutherford, December 12, [1927], RC.

192. Rona, *How It Came About* (n. 36), p. 20. Pettersson described the normal counting procedure in Vienna as follows:

At least one-half hour before the experiment, the people who are to count the scintillations must sit in darkness so that their eyes achieve the greatest possible sensitivity. Only when everything is ready . . . are they led in and the counting begins, in which one after the other takes a position in front of the microscope. . . . On the command of the recorder the designated counter begins to count out loud the scintillations which he sees in the microscope. After a definite time, generally only 20–30 seconds, "Stop" is called out, and the total number of observed scintillations are written into the record. . . .

Although the counters change off among themselves, indeed, by inserting a rest period of several minutes after every second counting period, the strain on the observer for weak scintillations is so great that noticeable signs of fatigue set in if the counting lasts longer than 1 to 1 1/2 hours. Even with this restriction, the same counter may not be used more often than twice a week at most, since otherwise chronic symptoms of fatigue soon appear and the results are unreliable. [Pettersson, *Künstliche Verwandlung* (n. 178), pp. 85–86]

193. Chadwick to Rutherford (n. 191).

194. Weiner-Chadwick interview (n. 36), pp. 61–62.

195. Ibid., p. 62.

196. For a full account see Mary Jo Nye, "N-Rays: An Episode in the History and Psychology of Science," *Hist. Stud. Phys. Sci.* 11 (1980): 125–156.

197. Weiner-Chadwick interview (n. 36), p. 62.

198. Meyer to Rutherford, December 12, 1927, MP.

199. Rutherford to Pettersson, January 9, 1928, MP.

200. Rona, *How It Came About* (n. 36), p. 20.

201. Otto R. Frisch, "The Discovery of Fission: How It All Began," *Physics Today* 20 (November 1967): 44. See also Frisch's autobiography, *What Little I Remember* (Cambridge University Press, 1979), p. 64.

202. Rutherford, Chadwick, and Ellis, *Radiations* (n. 2), p. 286. We know that its composition began in 1926 from the Weiner-Chadwick interview (n. 36), p. 50.

203. Frisch, "Discovery of Fission" (n. 201), p. 44.

204. Pettersson to Meyer, January 3, 7, 14, 19, and 30, 1928, MP. See also Pettersson to Rutherford, January 30, 1928, MP.

205. Pettersson to Meyer, January 30, 1928, MP.

206. Chadwick to Meyer, December 21, 1927, MP.

207. Rutherford to Meyer, December 21, 1927, RC, MP.

208. Rutherford to Pettersson, January 9, 1928, MP. See also Pettersson to Rutherford, January 30, 1928, MP.

209. Pettersson to Chadwick, January 14, 1928 (draft), MP. This letter and the enclosed manuscript was actually sent in early February. See Pettersson to Meyer, January 30, 1928, MP.

210. Weiner-Chadwick interview (n. 36), pp. 62–63.

211. Blackett to Meyer, January 23, 1928, MP.

212. Chadwick to Meyer, June 23, 1928, MP, in which Chadwick solicits Meyer's help in securing an Austrian visa for Chariton, who is a Russian citizen.

213. Chadwick to Meyer, June 23, 1928; Schmidt (in Cambridge) to Meyer, July 26, 1928, MP.

214. Kara-Michailova (in Cambridge) to Meyer, October 26, 1935; June 7, 1936, MP. Rona, *How It Came About* (n. 36), p. 37. Karlik, personal conversation with the author, June 10, 1984.

215. H. Pettersson, "Die Zertrümmerung des Kohlenstoffs III," *Sitz. Akad. Wiss. Wien 2a* 137 (1928): 1–6; "The Artificial Disintegration of Atoms and Their

Packing Fractions," *Ark. Mat., Astr. Fys.* 21A, no. 1 (1928–29): 1–16, especially pp. 7–8 and 13; "Der Heliumkern als Baustein anderer Atomkerne," *Z. Phys.* 48 (1928): 799–804, especially pp. 800 and 802; "Artificial Disintegration by Means of α-Particles from Polonium," *Ark. Mat., Astr. Fys.* 21A, no. 2 (1928): 1–11, especially pp. 3 and 5; G. Kirsch and H. Pettersson, "Die Ausbeutefrage bei Atomzertrümmerungsversuchen," *Sitz. Akad. Wiss. Wien 2A* 137 (1928): 563–582, especially p. 575; "Uber die Ausbeute bei Atomzertrümmerungs-versuchen," *Z. Phys.* 51 (1928): 669–695, especially p. 684. Pettersson avoids the issue in "Die Sichtbarmachung von H-Strahlen" [ibid. 48 (1928): 795–798] and "Die Sichtbarkeit von β-Szintillationen," *Naturwissenschaften* 16 (1928): 463 (with Kirsch). For his private view, see Pettersson to Meyer, October 25, 1928; February 2, 1930, MP.

216. Pettersson, *Künstliche Verwandlung* (n. 178). See especially pp. 53, 61 (table includes carbon and oxygen as disintegrable elements), 65 (bar graph includes the same elements), and 112–114.

217. Pettersson to Meyer, February 27, October 10 and 28, November 3 and 18, 1928, MP.

218. Meyer to Curie, October 30, 1928 (with similar letters to Hevesy and Fajans); Hevesy to Meyer, November 3, 1928; Curie to Meyer, November 9, 1928; and Fajans to Meyer, November 25, 1928, MP.

219. Deacon, "Pettersson" (n. 35).

220. G. Kirsch, *Geologie und Radioaktivität: Die radioaktiven Vorgänge als geo-logische Uhren und geophysikalische Energiequellen* (Vienna and Berlin: Springer, 1928).

221. J. Chariton and C. A. Lea, "Some Experiments Concerning the Counting of Scintillations Produced by Alpha Particles.—Part I," *Proc. Roy. Soc. A* 122 (1929): 304–319; "Part II. The Determination of the Efficiency of Transformation of the Kinetic Energy of α-Particles into Radiant Energy," ibid.: 320–334; "Part III. Practical Applications, ibid.: 335–352. The authors specifically thank Chadwick "for suggesting the problem and for many helpful suggestions and criticisms" (p. 352).

222. Rutherford to Meyer, June 10, 1929, MP.

223. F. A. B. Ward, C. E. Wynn-Williams, and H. M. Cave, "The Rate of Emission of Alpha Particles from Radium," *Proc. Roy. Soc. A* 125 (1929): 713–730.

224. E. Rutherford, F. A. B. Ward, and C. E. Wynn-Williams, "A New Method of Analysis of Groups of Alpha-Rays.—(1) The Alpha-Rays from Radium C, Thorium C, and Actinium C," *Proc. Roy. Soc. A* 129 (1930): 211–234.

225. For a cogent discussion see Norman Feather, "The Experimental Discovery of the Neutron," in *Proceedings of the X International Congress on the History of Science* (Paris: Hermann, 1964), pp. 135–144, especially 136–137.

226. F. W. Aston, "A New Mass-Spectrograph and the Whole Number Rule (Bakerian Lecture)," *Proc. Roy. Soc. A* 115 (1927): 487–514 plus plate (quote

on p. 513). See also "Atoms and Their Packing Fractions," *Nature* 120 (1927): 956–959. Pettersson quotes Aston in "Packing Fractions" (n. 215), pp. 1–2.

227. Pettersson, "Packing Fractions" (n. 215,) pp. 1–16. Pettersson failed to calculate the threshold energy correctly (p. 5).

228. Rutherford and Chadwick, "Energy Relations" (n. 132), pp. 218–224.

229. Rutherford, Chadwick, and Ellis, *Radiations* (n. 2), p. 306.

230. M. Stanley Livingston and H. A. Bethe, "Nuclear Physics. C. Nuclear Dynamics, Experimental," *Rev. Mod. Phys.* 9 (1937): 295.

231. Rutherford, "Structure" (n. 92), pp. 580–605 (*Collected Papers*, vol. 3, pp. 181–202).

232. G. Gamow, "Zur Quantentheorie des Atomkernes," *Z. Phys.* 51 (1928): 204–212.

233. R. W. Gurney and E. U. Condon, "Quantum Mechanics and Radioactive Disintegration," *Phys. Rev.* 33 (1929): 127–140. See also R. H. Stuewer, "Gamow's Theory of Alpha Decay," in *The Kaleidoscope of Science: The Israel Colloquium Studies in History, Philosophy, and Sociology of Science*, vol. 1, ed. E. Ullmann-Margalit (Atlantic Highlands, N.J.: Humanities Press, 1985).

234. Rutherford, Chadwick, and Ellis, *Radiations* (n. 2), pp. 326–327.

235. For Chadwick's objections see Weiner-Chadwick interview (n. 36), p. 49.

236. G. Gamow, "Zur Quantentheorie der Atomzertrümmerung," *Z. Phys.* 52 (1928): 510–515, especially p. 515.

237. Frisch, "Discovery of Fission" (n. 201), pp. 43–44.

238. Livingston and Bethe, "Nuclear Dynamics, Experimental" (n. 230), p. 295, n. 1. Pettersson and Kirsch's work was also excluded from another review of the period, viz., R. Fleischmann and W. Bothe, "Künstliche Kernumwandlung," *Ergeb. exak. Naturwiss.* 14 (1935): 1–41.

10

Bubble Chambers and the Experimental Workplace

Peter Galison

The experimental physicist confronts nature through instruments, his daily work largely determined by the character of the apparatus. In high-energy experimental physics there was a striking transition in that character between 1950 and 1966. To appreciate the shift one need only think for a moment of the first bubble chambers, built in the 1950s, which were in many respects illustrative of the size and scale of earlier experiments. Typically the physicist worked by himself or with a few collaborators on a table-top device made in a machine shop. He designed, partially built, used, and modified the apparatus, and eventually he took the data, reduced it, and brought it to publication. His instruments cost on the order of thousands of dollars. Contrast this with the situation in the mid 1960s. The total number of people involved in running a large bubble-chamber experiment was now several hundred. They were divided into a wide assortment of semiautonomous subgroups. Specialists devised software and hardware for data reduction; engineers handled aspects of safety, design, and construction; lay scanners encoded raw data into Dalitz and effective-mass plots. Apparatus alone cost millions of dollars. Changes had occurred in almost every respect, from the kind of physics question being asked to the instruments and work structure that shaped routine tasks. Closely allied to these developments was a fundamental alteration in the criteria of experimental demonstration.

One way to explore these changes is through the comparison of detailed case studies of discovery. For example, one can refer to historical studies of the discovery of the positron, the muon, the neutron, the antiproton, parity violation, CP violation, and neutral currents.[1] Such case studies offer great detail on the interaction of many experimental and theoretical concerns all brought to bear on a single complex of physics questions. At the same time, they fail to convey a continuity in the physicists' interests and activities that transcends particular puzzles of physics.

That continuity may reside more in a tradition of physical apparatus than in specific physics problems. Indeed it is quite typical for an experimentalist to follow a given type of equipment through varied investigations within his specialty. In particle physics virtually all experiments are now attached to accelerators, and the move from cosmic-ray stations to accelerator laboratories marks one of most significant long-term trends of the discipline. Studies of individual accelerator laboratories therefore have a crucial role to play. They can depict the history of high-energy physics within a framework that complements case studies organized around the resolution of specific questions.[2] Histories such as those of the Lawrence Berkeley Laboratory and CERN will offer a view of the institutional and technical changes that have transformed the budget and the infrastructure of the postwar high-energy-physics community. Through these studies we can see how, along with a remarkable number of experimental discoveries, the move to accelerator laboratories was accompanied by changes in the experimentalist's laboratory life. As the accelerators grew in size, the experimenter became progressively more distant from direct control over parts of the experimental apparatus, and specialists were increasingly needed to supervise the scheduling, the maintenance, the construction, and the operation of the accelerators themselves.

While accelerator building was in full swing, during the late 1940s and the early 1950s, certain features of the experimentalist's work remained stable, or at least changed more gradually. In particular, despite the growth of the accelerator, the particle detector continued to fall almost completely under the experimentalist's control. Whether it was a cloud chamber, an emulsion stack, or a counter array, the detector continued as the proximate source of the data needed for experimental demonstrations. In addition, the detectors still offered the experimenter many of the same technical challenges. Could one make faster electronics, more distortion-free cloud chambers, or more sensitive emulsions? For many experimentalists, the skills, physics questions, and apparatus that they had acquired when building detectors for cosmic-ray physics could be applied (with some modifications, of course) to accelerator physics. These considerations suggest that we think of the transformation of the particle-physics laboratory as having occurred in two steps.

In the first step, the outer laboratory (constituted by the accelerator) was restructured while the inner laboratory environment (surrounding the detectors) remained roughly constant. This set the stage for a second step that took place during the 1950s and the 1960s. When physicists began to construct the first large-scale detectors, the effects of the growth in scale that had begun during the move to accelerators were repeated within the immediate experimental workplace around the detectors. The consequences of this second step for the physicists' laboratory life were even sharper than those of the first. Often the experimentalists could no longer build or operate the large devices—on which their experiments directly depended—without extensive collaborations among physicists and engineers. The new generation of detectors also altered the kind and quantity of data that had to be analyzed.

The removal of the physicist from the apparatus, the specialization of tasks, the increased role of computation, and the establishment of hierarchical collaborations have become hallmarks of high-energy physics experiments. These changes raise a host of questions. How has the role of the experimental physicist changed? What part of the laboratory activity counts as "the experiment"? In what ways have the criteria of experimental demonstration been altered? What parts do computer programming and engineering play in the experimental workplace? Since the 1950s many types of detectors have undergone the expansion that provokes these questions, but the first, the one whose pictures have become a symbol of particle physics, was the bubble chamber.

The bubble chamber has proved ideal for use with accelerators and useless for cosmic rays. It is therefore often described as if it had been invented for the big machines. However, I will argue here that, despite the bubble chamber's eventual use, its origin lay not with accelerators but squarely in the tradition of cosmic-ray research. This is true on several levels: The bubble chamber borrowed many of its technical features from the cloud chamber; both involve compressors, optics, and hydraulics. In the early stages of the bubble chamber's development the underlying processes of bubble formation and drop formation were thought to be very similar. Finally, the physics goals toward which Donald Glaser hoped to direct his bubble-chamber work grew out of cloud-chamber work. Beyond technical considerations, Glaser hoped to preserve a relation between experimenter and experiment that was present in his early cosmic-ray work and markedly absent in large detector groups based at accelerators. A major concern of this chapter is the relation of the organization of experimental work to the content of instrument use and design.

During the 1950s and the 1960s the bubble chamber served in an enormous number of advances. The η, the ω, the Ξ^0, the Y_1^* (1385), the K^* (890), the Y_0^* (1405), the Ξ^* (1530), and the Ω^- (1672) are but a few of the many particles and resonances discovered by bubble-chamber physicists. These discoveries played an essential role in the development and confirmation of Gell-Mann and Ne'eman's eightfold way.[3] Contributing to these successes were bubble-chamber groups from across the United States, Europe, and the Soviet Union, including those at Argonne National Laboratory, Brookhaven National Laboratory, CERN, ITEP, Rutherford Laboratory, DESY, SLAC, Rome, Bologna, Oxford, Dubna, and the Cambridge Electron Accelerator.

Though many groups eventually applied the bubble chamber to important matters of physics, it was Donald Glaser and his collaborators who first developed the instrument, and it was the group at Berkeley under the direction of Luis Alvarez that first constructed and productively used the large hydrogen bubble chambers. In a technical sense, Alvarez and his collaborators continued Glaser's work. The Berkeley team began by recreating, step by step, Glaser's invention. But the style of their work could not have been more different. Where Glaser sought to preserve his table-top, individually run experiment, Alvarez brought in a team of structural, cryogenic, and accelerator engineers to work with physicists, postdocs, and film scanners.

What counted as an experiment in 1966 was not the same as what counted as one in 1953. This chapter is an attempt to exhibit how that transformation took place and to examine some of its consequences for our very concept of an experimental demonstration.

Invention

Questions about the origin and the nature of cosmic rays played an unexpected role in physics at the California Institute of Technology for three scientific generations. In pursuit of an idiosyncratic theory of the origin of cosmic rays, Millikan stumbled across several important aspects of their character, such as the peak in their intensity in the upper atmosphere.[4] His student Carl Anderson discovered the positron while on an assignment to examine other experimental consequences of the same odd theory.[5]

The basic technique Anderson exploited was to examine the energy distribution of charged cosmic rays by measuring their curvature in a magnetic field as revealed by their tracks in a cloud chamber. Some twenty years after Anderson finished his thesis, he set his graduate student to the task of extending to high energies the cloud-chamber measurements of sea-level muon momenta. With great care the student, Donald Glaser, succeeded.[6] Very energetic muons were hardly bent by the magnetic field strengths then available. Worse, as one increased the magnetic fields the heat generated by the electromagnet caused convection currents in the chamber, ruining the track. The key to the device Glaser and his co-workers used lay in that by employing two cloud chambers separated by a strong magnetic field (Fig. 1) they could measure the angle of deflection, if not the elbow of the curve itself.[7]

Even though the basic device was partially assembled before Glaser's arrival,[8] his assignment forced him to become familiar with many details of the operation of the cloud chambers: high-pressure devices for expansion and recompression, photographic techniques, track-analysis methods, and the theory of droplet formation.[9] The project was, despite its success, "rather humdrum"; its main advantage, Glaser figured, was that it was likely to be over quickly.[10] Meanwhile, the apprentice to the cloud chamber and his graduate-student roommate Bud Cowan promised themselves that at least once each week they would try to think of something original to break the monotony of their routine thesis work.[11]

Several problems threatened the usefulness of the cloud chamber. Above all, the diffuse gas offered relatively few targets in which an incoming particle could interact. This meant that most particles simply passed through the chamber. If interactions did occur, they often did so in the dense walls of the chamber, where they could not be seen. Only a rare particle was sufficiently obliging to interact in a place where the vertex could be seen and the decay tracks studied. To compensate for this, another detection system, film emulsions, had been developed. Here the sensitive volume was a solid and therefore offered many more targets, but there were problems of a different sort. First, film collects tracks over its entire lifetime from its creation until its development. It was consequently difficult to disentangle tracks made at different times. Second, to make film sensitive to events not coplanar with the emulsion one must place the plates in expensive stacks; otherwise it is impossible to tell whether a track ended or simply left the emulsion's surface.

Figure 1
Double cloud chamber. A powerful electromagnet was set between two cloud chambers. By determining the angle of track deflection between the upper and lower chambers, Glaser could find the energy of very-high-energy muons. The bubble chamber grew out of an attempt to improve this apparatus for the study of these muons and the then recently discovered strange particles. From Glaser, n. 6.

Even if the cloud chamber had been capable of producing enough data, it was frustrating in its slowness to produce the most exciting events of the late 1940s. These came from Manchester, England, where in 1947 G. D. Rochester and C. C. Butler culled two inexplicable events from some 5,000 cloud-chamber pictures. Both shots seemed to depict the decay of a heavy particle into two other particles with a total mass on the order of 500 MeV.[12] For over a year no one, including Rochester and Butler, could repeat the filming of these elusive "V particles" or "pothooks" (named after the shape of the tracks into which the neutral particle decayed). Finally, by removing a cloud chamber to a mountaintop, Carl Anderson, R. B. Leighton, and collaborators were able to achieve a better view of the pothooks—they could photograph, on the average, one per day.[13]

Analysis of the events became the priority in Anderson's group. On his blackboard everyone could see Anderson's admonition: "What have we done about the pothooks today?"[14] When Glaser finished his thesis in 1950, he hoped finally to be able to do something original—with luck, something about the pothooks—by building a better detection device. Later Glaser remembered his desire to work on his own as his prime consideration in deciding among job offers at Columbia, MIT, Minnesota, and Michigan.[15] Although the University of Michigan had experimental groups centered around the synchrotron (the group of H. R. Crane), the cyclotron (the group of W. Charles Parkinson), and the cosmic rays (the group of Wayne Hazen),[16] Glaser "characteristically stated that he did not wish to join one of the existing research efforts . . . but preferred to pursue an independent course of research."[17] After all, this was the work pattern that Glaser had seen prosecuted with striking success by his advisor at Caltech. Anderson had found the positron working by himself, and later discovered the muon working with a single collaborator. Anderson successfully continued his cosmic-ray program, composed of small collaborations working around certain general questions, well into the 1950s. Michigan was willing to allow Glaser the freedom to continue this style of work. Upon his arrival in 1949 at Ann Arbor as an instructor, he began a two-year period of independent exploration, looking into a great variety of mechanisms in the hope of designing a new visual detector for use with cosmic rays.

H. R. Crane remembered the contemplative silence of Glaser's early period at Ann Arbor this way: "I took the credit for recruiting him, but for the next year or two I wondered whether I had been very smart, because there were few external signs of activity."[18] The focus of Glaser's attention remained cosmic rays; the large research teams associated with Van de Graaff machines had not appealed to him as a new graduate student in 1947, and the accelerator teams did not appeal to him in 1950. "There was," he has said, "a psychological side to this. I knew that large accelerators were going to be built and they were going to make gobs of strange particles. But I didn't want to join an army of people working at the big machines. . . . I decided that if I were clever enough I could invent something that could extract the information from cosmic rays and you could work in a nice peaceful environment rather than in the factory environment of big machines."[19]

Glaser's first research at Michigan was supported by the Michigan Memorial–Phoenix Project for the "Study of the High Energy Cosmic Rays" from June 1950 to January 1951 and involved $750.[20] By the time the short grant expired, Glaser had little to show. Nonetheless, in December 1950 the physics department extracted $1,500 from the School of Graduate Studies for Project R #250, "Investigation on Cosmic Ray Mesons"—$1,000 for a research assistant and $500 for supplies.[21] Fortunately for the historian, in May 1951 Glaser had to justify even these initial small grants.[22] There was complete continuity between the physics goals of his thesis project—exploring the upper end of the known energy spectrum of cosmic rays—and the new project. "The immediate goal of the research carried on under Phoenix Project No. 11 was to develop a method for measuring the momentum of cosmic ray particles of very high energies, perhaps up to a thousand billion volts. . . ." The measurements were of "interest for their bearing on the question of the nature and origin of cosmic rays."[23]

Studies of the origin and nature of cosmic rays represented a tradition not just of Anderson but of Anderson's teacher Millikan as well. Even Glaser's new instrument, with its two visual detectors separated by a powerful electromagnet,[24] was to function in much the same way as his old double cloud chamber. If the deflection was adequate, existing detectors were not. All extant detectors, Glaser figured, were based on the catalysis of an instability, be it chemical (emulsions), electrostatic (Geiger-Müller counter, amplified scintillation counter), or thermodynamic (cloud chamber).[25] In an effort to make a new track device, he set out a list of all instabilities he could think of and how they could be exploited in a fast, dense detector. One by one, he began to build new detectors, mostly without success.

One attempt involved letting the ion tracks polymerize a monomer. One can polymerize acryle nitrile; the monomer is soluble in certain fluids whereas the polymer is not. Glaser's hope was that, when a charged particle sped through the solute, the path of ionization would polymerize some of the molecules, leaving a solid precipitate. In Glaser's words, "the 'total fantasy' was to be able to lift out a 'solid Christmas tree of tracks' and measure them at leisure." These tracks then could be extracted, or at least photographed. This try was an utter failure, resulting in much brown liquid and no tracks.[26]

Glaser's second try was inspired by some work he had seen J. Warren Keuffel perform when they were both graduate students at Caltech.[27] One problem with ordinary Geiger-Müller counters was the rapid fall off of the field as one moved away from the center wire. This meant that an avalanche precipitated near the center wire developed much more quickly than one starting near the walls. To improve the resulting time resolution, Keuffel tried making a counter with a parallel instead of a cylindrical geometry.[28] As a passing observation on the the eighth page of his thesis, he noted that "the discharge is localized," posing "obvious possibilities as a means of determining the path of the particle." It would take at least nine more years before this "obvious" possibility could be realized as a track-following device, but meanwhile Glaser began an abortive attempt to do so.[29] When the glass electrode was coated with a conducting but transparent medium, the sparks could be localized to within a tenth of a centimeter. However, the conducting layer was chemically unstable and the device was useless after a day or two of operation.[30] Equally unpleasant was the tendency of the spark to wander around the plate, ruining attempts at sharp photography.[31]

By late February of 1951 the spark counter was certified as a lost cause.[32] Fortuitously, at about this time there was a resurgence of interest in another detector: the diffusion cloud chamber. In an expansion cloud chamber a sudden enlargement of the gas volume causes a temperature drop, supersaturating the gas, which precipitates around the charged tracks. By contrast, in a diffusion chamber the gas supersaturates as it approaches the cooled floor of the container. Recommending the device was its continuous sensitivity—since no expansion is needed, the device is never "dead." "It is hoped," Glaser wrote to the granting agency, "that such a chamber, which operated reliably for long periods of time, may allow a position determination with accuracy sufficient for the present experiment [of high-energy cosmic rays]."[34]

The indefatigable diffusion cloud chamber had intrigued Glaser for some time. In graduate school he and Cowan had speculated about its possible uses. Shortly after Glaser's departure from Caltech, Cowan had begun work on such a device; he published an often-cited review article on the subject in 1950.[35] With money from the Phoenix Project, Glaser hired David Rahm as a research assistant to assist in building one.[36] Designing the chamber was fairly straightforward: A 12–18-inch glass cylinder would be filled with ethanol, which would drift down toward a dry-ice-cooled plate at the bottom.

Once again disappointment struck. Glaser and Rahm soon convinced themselves that the instrument might make a fine demonstration device but could never be precise enough for cosmic-ray energy determination. After several other failures, Glaser abandoned chemical and electrostatic devices. His questions narrowed to one: Could a thermodynamic device be produced, analogous to the cloud chamber, in which bubbles instead of droplets formed at the site of an ion? In other words, instead of supersaturating a gas with a vapor one would superheat a liquid (heat it above its boiling point by holding it under pressure). Success this time depended on a prior question: Is the energy deposited by an ion in a liquid sufficient to cause a vapor bubble to grow? With this question, Glaser opened "Book 1" of two notebooks still surviving from his early work by writing "Study of the Formation of Bubbles in Liquids (Theory and Literature)."[37] Slowly and carefully, Glaser alternately translated, annotated, and summarized a book by Max Volmer, *The Kinetics of Phase Formation*.[38]

In 1888, J. J. Thomson had explained how water, for example, tended to coalesce around ions because the dielectric constant of liquid water was much larger than that of air.[39] He pointed out that the electrostatic effect added a term to the potential energy of the system equal to $\Delta V = (1/2)(e^2/r\epsilon)$ for a droplet of charge e and radius r. Thus, a region of high electric field will tend to pull in a dielectric in order to minimize the total energy of the system.[40] Thomson's result would seem to imply that a single charge on a bubble would collapse much faster.

In an undated list of queries preserved in one of the early notebooks, Glaser wonders "Is a electrically-modified fluctuation theory sensible as [Volmer] does it, or should we consider multiple charges as a way of getting at the same type of effect?"[41] This question sheet seems likely to have been written just before Glaser calculated the effect of multiple charges on a bubble's pressure. Though in retrospect we know that the electrostatic repulsion theory is wrong, it is essential historically for two reasons. First, it provided for Glaser a theoretical bridge from the principles of a cloud chamber to a rough plausibility argument for a bubble chamber. Second, the electrostatic theory gave a rough quantitative prediction for the conditions of temperature and pressure under which the device would work.[42] Here is the electrostatic argument:[43] The energy E of a bubble with a surface charge ne distributed over a surface of $4\pi r^2$ with a surface tension $\sigma(T)$ (T = temperature) is

$$E = 4\pi r^2 \sigma(T) + \frac{1}{2}\frac{(ne)^2}{\epsilon(T)r}, \tag{1}$$

where $\epsilon(T)$ is the dielectric constant of the liquid. From E we can deduce the outward force, F:

$$F = \frac{dE}{dr} = 8\pi r \,\sigma(T) - \frac{1}{2}\frac{(ne)^2}{\epsilon(T)r^2}. \tag{2}$$

Here it is assumed that the surface tension and the dielectric constant are roughly independent of r. Then the pressure P is given by

$$P = \frac{F}{\text{Area}} = \frac{2\,\sigma(T)}{r} - \frac{1}{8\pi}\frac{(ne)^2}{\epsilon(T)r^4}. \tag{3}$$

Setting $dP/dr = 0$ and solving for r yields r_0, the radius for which the pressure P is maximal. Inserting r_0 into equation 3, we deduce P_{max}, the maximum value of the combined electrostatic repulsion and surface-tension contraction. It follows that the condition for a bubble to grow to macroscopic size is

$$P_\infty - P > \frac{3}{2}\left(\frac{4\pi}{(ne)^2}\right)^{1/3}[\sigma(T)]^{4/3}[\epsilon(T)]^{1/3} = P_{max}, \tag{4}$$

where P_∞ is the vapor pressure of the liquid and P is the (mechanically adjustable) applied pressure. Many qualitative features are immediately discernible. Bubble formation is facilitated by increasing the expansion, decreasing the surface tension, decreasing the dielectric constant, or increasing the charge. In particular, the expression suggests operating at a high liquid temperature, which would augment P_∞ while reducing surface tension. (At the critical point the surface tension vanishes.)

Volmer's book offered Glaser a detailed review of the kinetic theory of droplet formation as well as a parallel development of bubble nucleation in liquids. The close analogy is stressed but not exaggerated, for there is one essential difference between the bubble and the droplet.

At the surface of the droplet the vapor pressure depends crucially on the curvature of the liquid-gas surface. But did this apply to a bubble as well? Glaser opened his notebook entries on Volmer by addressing just that section of Volmer's book where the question of vapor pressure occurs.[44] The crucial answer is that the pressure inside a bubble is not dependent on curvature and is just equal to the vapor pressure at the surface of the liquid. After skipping back and forth between chapters in Volmer to understand this surprising fact, Glaser looked up and translated W. Döring's article of 1937 in which the result was first proved.[45] Consider two bubbles (of insoluble gases) of different sizes in equilibrium in a liquid. If their partial vapor pressures were unequal, then a small capillary tube with a semipermeable membrane could transport vapor from one bubble to the other while blocking the gas. The liquid would evaporate from the high-vapor-pressure bubble and condense in the other. It would be a perpetual-motion machine. Similarly, the capillary could connect a bubble to the surface and the argument can be extended *mutatis mutandis*. Therefore, bubbles of any size must have vapor pressures equal to the saturated vapor pressure of the liquid. "The mistake" [of supposing the vapor pressure to depend on the radius] is often made," Glaser noted.[46] (Droplets are unlike bubbles because the droplet's radius-dependent surface tension determines the liquid pressure. The vapor pressure depends on this liquid pressure, not on the geometry *per se*.)

Immediately after discussing Döring's argument, Volmer turns to a comparison of the theory of bubble formation with experiment. However, he does so with a cautionary remark: "Tensile strength hasn't been done very well and for super-heating Wismer and Co. are principally quoted."[47] K. L. Wismer, a chemist at the University of Toronto, had with his collaborators studied how long a capillary tube of diethyl ether heated to 140°C could last before exploding. As Glaser reported from Volmer, this was typically about one second.[48] Volmer ascribed the inevitable explosion to local fluctuations,[49] but when Glaser arrived at the Michigan chemistry library to look up Wismer and Kenrick's paper he found something much more exciting than dirty glass: Not all of Wismer and Kenrick's capillary tubes had exploded almost instantaneously. They were puzzled by their results; Glaser was thrilled. What the physical chemists had found was that the superheated material sometimes burst out of its tube in a few seconds but at other times waited for well over a minute.[50] Wismer attributed the "capriciousness in the results" to inadequately polished glass walls. When Glaser plotted

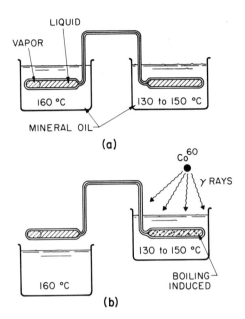

Figure 2
Radiation sensitivity apparatus. In Glaser's apparatus (a), a bubble of vapor diethyl ether in a capillary tube was heated to 160°C in an oil bath. The resulting pressure on the liquid diethyl ether was about 22 atmospheres, high enough to prevent the liquid from boiling even though it is above its normal boiling temperature (135°C). When Glaser removed the left side of the capillary tube from the oil bath (b), the pressure dropped to about 0.6 atmosphere. The liquid on the right side now became superheated, but would remain for some time in a metastable state. When a radioactive source was brought near, the liquid suddenly began to boil violently. This offered the possibility that a bubble chamber could be built. From Glaser, n. 53.

the results for a sample run with a temperature of 130.5°C, the points formed a Poisson distribution with an average waiting time of 60 seconds.[51] If nothing else, from his thesis Glaser knew the sea-level cosmic-ray flux. A brief calculation yielded the average time between cosmic-ray particles for a capillary tube of the size used by the chemists: 60 seconds.[52] The crucial test remained: Would a radioactive source induce boiling? Glaser connected two glass tubes by a glass capillary, partially filled the assembly with diethyl ether liquid, and dipped the contraption into two oil baths held at differing temperatures (figure 2a).[53] The higher temperature (and therefore the higher vapor pressure) of the left bulb forced the liquid to fill the right-hand tube. Boiling did not start because the pressure was too great. The ban on boiling was suddenly lifted when the hotter capillary tube (the left one in figure 2b) was removed from its bath, because the vapor pressure dropped, making the liquid superheated. Without a radioactive source, boiling would occasionally not begin for several minutes.[54] On the other hand, when late one night at the lab Glaser enlisted a graduate student to open a lead box of cobalt 60 that was thirty feet away the boiling began violently and suddenly.[55] In a brief letter to the editor of *Physical Review* received June 12, 1952, Glaser reported his radioactively induced boiling.[56]

Before prosecuting the work any further, Glaser had to leave on a previously arranged European trip.[57] His student, David Rahm, continued on the project by investigating the "temperature, pressure and ionization dependence of bubble formation in superheated liquids."[58] To create a high vapor pressure over the hot liquid diethyl ether, Rahm built a hand-cranked hydraulic piston-cylinder apparatus. The pressurized ether contained in the glass bulb was potentially explosive, so the whole device was covered by a large shield with clear plastic observation ports.

For two days after assembly, every trial fizzled. As soon as the pressure was released, boiling would set in. Eventually Rahm got the system to work properly and was able to record the waiting time as a function of temperature and pressure. The empirical data on the critical pressure change, together with Glaser's electrostatic theory, made it possible to calculate the number of charges on a bubble. Solving for n in equation 4 with standard values of $\sigma(T)$ and $\epsilon(T)$ gave $n = 37.5$.

Modifications of this experiment were possible until the glass broke. Throughout the summer the goal was clear, as Rahm emphasized in his report: "Bubble Chamber   A large bulbed system (3/4" diameter × 1-1/4") was used in the hope that tracks could be observed. The only thing that could be seen, however, was a single large bubble which almost always appeared to form on the walls of the bulb. . . . Further work on more carefully cleaned bulbs should be carried out."[59]

Scrupulous bulb cleaning became the order of the day when Glaser returned from Europe. Using Rahm's hand-cranked compressor, some new bulbs, and a high-speed movie camera, he resumed the search for tracks. From the notebooks we know that on October 14, 1952, Glaser shot 300 feet of super XX Panchromatic 16 mm film at 8,000 frames/second. Under "conclusions" he wrote the following:

1) bubbles can grow in ~1 msec from sizes invisible with the optics used to diameters ~1 mm. . . .

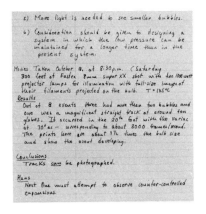
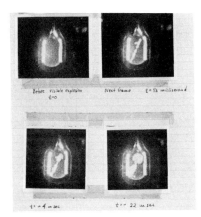

Figure 3
The first bubble-chamber tracks. Glaser first filmed distinct tracks on October 18, 1952, using a high-speed 16-mm camera. From GNB 2 (see n. 37).

4) Twice apparently very faint tracks of 4 or more bubbles were seen to precede a grown globe, but their faintness makes it possible to assume that these were dirt effects.
5) More light is needed to see smaller bubbles.[61]

Four days later, on October 18, with better illumination, Glaser recorded the following under the photographs reproduced here in figure 3:

<u>Conclusions</u>   Tracks <u>can</u> be photographed
<u>Plans</u>   Next one must attempt to observe counter controlled expansions[62]

On reflection, Glaser must have decided to take an intermediate step, for on October 20 he began to try a counter-controlled flash while maintaining a random expansion.[63] Some bubbles showed up on the photographs, but no tracks. Under "conclusions" he recorded that the sensitive time of the chamber was too short to produce useful results given that the counting rate of cosmic-ray particles was only 4 or 5 per hour. He therefore decided to press on with the attempt to make a counter-controlled expansion. Glaser enlisted an old family radio, a 1930 Colson sporting a loudspeaker with a very large magnet.[64] With a big vacuum tube to deliver a hefty current, the magnet could be made to release a valve very quickly. (That this was dubbed a "pop valve" needs no further explanation.) The work to this point had been done with $2,000. At the end of November 1952, the Phoenix Project doubled the yearly stipend to $3,000 for "New Methods of Detecting Ionizing Radiation by Its Effect on Phase Changes."[65] By December 18–19, 1952, a prototype counter-triggered bubble chamber was ready, pop valve and all. Glaser wrote: "Fifty-three pictures were taken of which 30 were at a sensitive time. Of these, 20 should have been good, but none were. . . ."[66] The failure was followed by a return to random expansions, amid much frustration.[67]

Above all, Glaser did not want to work on the "factory"-like big accelerators. "I wanted to help save cosmic-ray physics," he reminisced later.[68] To rescue his old specialty he needed to make the method adaptable to counter control—otherwise the bubble chamber was inevitably fated to be wed to the predictable pulses of the accelerator. Many attempts then followed. Glaser and Rahm tried injecting CO_2 into the diethyl ether to slow the formation of bubbles. No luck.[69] Another attempt began in April 1953 when Rahm and Glaser tried to exploit the "plink" heard when the violent eruption of boiling begins. To tap this signal the physicists attached a phonograph pickup to the side of the glass chamber.[70] The method worked, though not entirely satisfactorily; by the time the sound arrived at the pickup the bubbles had grown too large for precision measurement. By the fall they had a new method in which a photoelectric cell would trigger the flash when the optical disturbance of the bubbles indicated that a particle had passed through. Again, the resulting photographs were not spectacular.[71]

Even though Glaser was unable to rescue cosmic-ray physics with a new counter-controlled detector through this series of experiments, his basic bubble chamber was ready for presentation. On Saturday, May 2, 1953, in session XA of the Washington meeting of the American Physical Society, Glaser presented a ten-minute account of his results[72] to a somewhat depleted audience, it being the last day of the conference.[73] "My first paper," Glaser later wrote, ". . . was scheduled by the secretary of the American Physical Society, Karl Darrow, in the Saturday afternoon 'crackpot session'. . . ." Shortly after the conference, Glaser submitted a letter to *Physical Review*. It was promptly returned "on the ground that [he] had used the word 'bubblet' which is not in Webster."[74] When the letter was received without the offending word on May 20, 1953, it contained the outline of the charge-repulsion theory of bubble growth and a description of the counter-controlled flash–random expansion bubble chamber.[75] This published letter served as the point of departure for the many experimental groups that soon began to construct prototype chambers.

The surprising fact was that two very attentive listeners heard the bubblet producer's ideas. Luis Alvarez, from the Univerity of California at Berkeley, had met Glaser by chance at a reception the day before his talk; Darragh Nagle, from the University of Chicago, was one of the few people who had remained to hear the talk itself.[76] Unlike Glaser, neither had the slightest interest in saving cosmic-ray physics—indeed, both wanted nothing more than to use the bubble chamber to make full use of accelerator programs at their respective institutions. They had in mind a physics of a very different kind from that of the hand-cranked compressor and the loudspeaker pop valve.

Transition

It was Glaser's fantasy of a hydrogen-filled bubble chamber that intrigued Nagle and the Alvarez group. Nagle was just finishing a collaborative effort that had used liquid hydrogen as a target for the Chicago pion beam. If the target could double as a visual detector, a whole new realm of accurate interaction measurements would be possible.[77] The old collaboration had mastered the art of handling a liquid-hydrogen target 4 inches in diameter and 10 inches long, and had discovered some excited proton states (among other things).[78] Clearly, hydrogen, with its most simple nucleus, was the medium in which to do interaction experiments, even if high stopping power and decay questions were best addressed with heavier and more complicated nuclei. And elementary-particle accelerator physics was quintessentially the science of the simplest interactions.

Returning to Chicago, Nagle suggested to Roger Hildebrand that they work together on the hydrogen-bubble-chamber project. Since Glaser was also interested in making a hydrogen chamber, he began to collaborate with them, visiting them several times to participate in design sessions.

The switch to liquid hydrogen presented several problems, two of which stood out above the others. By weight, hydrogen gas is more explosive than dynamite, so safety features were needed far more than with diethyl ether. Second, whereas diethyl ether boils at atmospheric pressure at about 130°C, hydrogen does so at −246°C. Cryogenic expertise thus was required for the project to succeed. The low-temperature physicists Earl Long and Lothar Meyer, at Chicago, had worked with the hydrogen target group. Together, Nagle and Hildebrand, consulting Long and Meyer, began to repeat on nitrogen the radiation-sensitivity experiments Glaser had performed on diethyl ether. On August 4, 1953, the very first series of runs worked. No tracks appeared, but superheated nitrogen clearly was sensitive to a radioactive source. Two days later, hydrogen followed suit.[79]

Over the next two weeks, the Chicago physicists continued to explore the properties of their chamber. How should it be kept cold? How could the pressure be properly regulated? Above all, how could the bulb be kept exquisitely clean?[80] Late that summer, Hildebrand recalls trying (unsuccessfully) to contact Glaser in Europe, asking him about their collaboration.[81] In the event, the paper was submitted by Hildebrand and Nagle and received by *Physical Review* on August 21, 1953. When Glaser returned to the United States, he decided to continue his work separately.[82]

Despite their initial success in showing radiation sensitivity, the Chicago group still had no tracks. Toward this end, they turned in February 1954 to the building of a new chamber.[83] Most of the work centered on improving the expansion system, building bigger chambers, and improving the photography. Meanwhile, a simultaneous effort was made by Hildebrand's graduate students Irwin Pless and Richard Plano to obtain accelerator-generated meson tracks in a pentane chamber. This work succeeded.[84]

By the summer of 1953 certain physical properties of bubble chambers had become clear, above all that the chamber was sensitive to track formation for only several microseconds. This augured poorly for cosmic-ray applications, since by the time a counter control could effect an expansion the tracks could no longer be made visible. On the other hand, accelerators emitted particles at controllable times, so the chamber could be set up to expand at a propitious moment. Indeed, the short sensitivity time now became an advantage by excluding older (and therefore irrelevant) tracks. Glaser had hoped his invention would "allow physicists to work independently or at least in small groups." But now, he added, "I was trapped. It was just what the accelerators needed and it wasn't useful for cosmic rays."[85]

Thus, when Glaser came back to the United States from Europe in the fall of 1953, his attention turned to the Brookhaven National Laboratory cyclotron, where he and Rahm hoped to put the new detector through its paces. Arrangements were made in late 1953 for the Michigan physicists to travel to Brookhaven. Knowing all too well the difficulty of handling liquid hydrogen, Alan Thorndike from the Cloud Chamber Group wrote to Glaser the following: "We have no intention of getting into the liquid hydrogen business, but I think that a hydrogenous liquid such as ether would make a very valuable device for studying rare nuclear events, such as V-particle production, at the Cosmotron."[86] There is some irony in this remark, since Thorndike was to spend the better part of the next thirty years involved in the development of data analysis of hydrogen chambers. For the moment, building bigger versions of Glaser's device was the goal. Even here, Glaser cautioned Thorndike, there were difficulties: "My first reaction toward the possibility of making a successful ten-liter chamber is that it probably won't be feasible. One liter seems ambitious in view of my present knowledge of the problem."[87]

In the first experiments at Brookhaven, Glaser and Rahm propped the chamber on a table in front of a crack in the Cosmotron's radiation shielding. Photography was undertaken with Glaser's personal 35-mm camera. By capturing several π-μ-e events in the first roll of film, they impressed their sponsors sufficiently for them to pursue the accelerator–bubble chamber combination. When more funding was sought from the National Science Foundation, the Office of Naval Research, and the Atomic Energy Commission, three goals figured prominently. First, a much improved means of data analysis was needed to take advantage of the precise pictures now being copiously produced. Second, a heavy-liquid bubble chamber seemed promising, as it could offer high stopping power (making decays easier to study and photons more likely to convert into observable $e^+\,e^-$ pairs). Third, Glaser hoped that he could build a hydrogen chamber. Part-time salaries for Glaser, Martin Perl (a new faculty member), a secretary, and four research assistants added to the salaries for a full-time postdoc and a full-time machinist-technician came to about $25,000. Equipment, supplies, and machining ran about the same.[88]

The highest priority was accorded to the construction of a liquid-xenon chamber, a project that *inter alia* required most of the world's xenon supply. Here, Glaser recalled, was one last "unique niche that I could [fill] at Michigan without access to all this high technology and large engineering staffs."[89] With a working chamber now successfully demonstrated with an accelerator, funding (and xenon) flowed freely from the AEC. The first physics objectives centered around the high-stopping-power chambers, which promised results on the decay and production of strange particles without necessitating cryogenic engineering.

In the process of perfecting their new chamber, Perl, Glaser, and Brown succeeded in demolishng the old electrostatic theory of bubble formation. In this respect, success followed from failure. When the xenon chamber was completed, absolutely no tracks were to be seen. At about this time, Glaser and his collaborators learned from Los Alamos that gaseous xenon scintillated in the optical frequencies. This suggested that the energy that might have been exploited to form bubbles was escaping in the form of ordinary light. If it could be captured in the liquid, the chamber might be rescued. Happily, when tiny quantities of a quenching agent (ethylene) were mixed with xenon, tracks began to appear. (Apparently, by collisions of the second kind, the ethylene molecules were deexciting the xenon atoms before the latter could emit optical light. Instead of radiating away, energy was being converted into kinetic energy and internal molecular excitation of the ethylene.) Since even very small quantities of the absorber did the trick, the experiment suggested that bubble production was probably due to local heat trapping rather than electrostatic repulsion.[90] A more comprehensive treatment of the "heat spike theory" was worked out by Frederick Seitz, whose interest in bubble formation had begun in the days of the Manhattan Project, when it was feared that radiation-induced bubbles in reactor cooling water might lead to a meltdown. Inspired by Glaser's work, Seitz described in 1958 how a charged particle would cause local heating of the superheated liquid along its path, much as a hot needle would induce a long sharp line of nucleation.[91]

In short, during the transitional period of the mid 1950s both Glaser's group and the Chicago team succeeded in adapting a number of heavy-liquid bubble chambers for accelerator experiments. Though costs for both groups mounted into the mid five figures, the style of work did not shift radically. The switch to hydrogen chambers was much more difficult. Liquid hydrogen alone cost the Chicago group a sizable fraction of its 1954–55 budget.[92] Nonetheless, with funding from the ONR[93] and the AEC they were able, by late 1956, to produce tracks of pion-proton collisions in hydrogen.[94] They were not the first, however, to see tracks in liquid hydrogen. That honor went to the rapidly growing Berkeley collaboration, to which I now turn.

Physicist and Engineer

To a great extent, Glaser's experience in the early 1950s as an independent researcher using makeshift, low-budget equipment was out of step with the times. Too young to have been on the scene during the big war projects on radar and the atomic bomb, he had not been exposed to a very different style of research. The large teams at Berkeley,

Los Alamos, MIT, Oak Ridge, and the Metallurgical Laboratory in Chicago had established a new scale of physics research, lavishly funded by government and integrated horizontally among many physicists and vertically among engineers, administrators, students, and technicians. Many physicists expected the boom to end after the war; Lawrence, for example, foresaw a much reduced Radiation Laboratory at Berkeley.[95] By the time peace came in August 1945, however, the Radiation Laboratory, the government, and the military all agreed that laboratory activity should be continued at a high level.

Robert Seidel summarized the numerous projects undertaken in the postwar period in accelerator design and construction this way: "By accelerating science along the path of crash programs for large-scale technological development in science, the radiation laboratory helped set the tone of modern big science."[96] One physicist, Arthur Roberts, was moved to something like verse:

How nice to be a physicist in 1947,
To hold finance in less esteem than Molotov does Bevin,
To shun the importuning men with treasure who would lend it,
To think of money only when you wonder how to spend it,
Oh,
Research is long,
And time is short.
Fill the shelves with new equipment,
Order it by carload shipment,
Never give
A second thought,
You can have whatever can be bought. . . .[97]

Lawrence's style of physics was congenial to Luis Alvarez. Not only had Alvarez seen the building of the large, philanthropically supported accelerators, he had also participated in the two great scentific war efforts: radar and the atomic bomb. Each project was budgeted at around $2 billion; one learned in such an environment to plan on a large scale. It was in the midst of the war work of winter 1941 and spring 1942 that Alvarez made the great shift in his "laboratory lifestyle."[98] Before that time his work on cosmic rays and nuclear physics had been by himself, in a small group or with a few graduate students; afterward, for most of his career, he would hold line jobs in large research organizations.

The transition took place in the early stages of Alvarez's radar work at the MIT Radiation Laboratory. There, during the fall of 1941, he was made group leader of JBBL (Jamming Beacons and Blind Landing), and later head of Division F (Special Systems). In these capacities he and his group concentrated on three projects: the GCA (Ground Controlled Approach radar), which allowed planes to make blind landings, the MEW (Microwave Early Warning radar), and the Eagle Blind Bombing System, which was used over Europe. By 1943 most of the radar problems had been solved, and Alvarez transferred to the Manhattan Project, going first to Fermi's reactor at Chicago and then to Los Alamos, where he joined the team working on the implosion device. The implosion group grew quickly, and in June–July 1944 it underwent a reorganization that put Alvarez at the head of group E-11. This group's assignments were, first, to exploit the gamma-emitting radiolanthanide to follow the course of imploded material during tests, and, second, to investigate electric detonators for the imploding charge.

When the war ended, the budget was extended in new directions: Alvarez made use of $250,000 worth of surplus radar sets in his linear accelerator, $200,000 worth of surplus capacitors went to McMillan's synchrotron, and some $10 million was destined for the Bevatron.[99] One of the scientific benefits of the governmental largesse was the training and cultivation of an extraordinary group of engineers who enjoyed an unprecedented degree of autonomy. An example was the design of the Bevatron by William Brobeck.

Each of these three projects—the radar, the atomic bomb, and the linear accelerator—placed Alvarez in the nexus of scientific, engineering, and management concerns. Certainly together they would have been ample schooling in the execution of large-scale physics. But there was a fourth project that, if it had been pursued, would have been bigger than the three previous ones put together.

Late in the 1940s, the debate over whether to build the hydrogen bomb came to a head. At first the General Advisory Committee (GAC) recommended that the AEC not put priority on the thermonuclear weapon. Hesitation ended in the flash of the Russian atomic bomb explosion of August 1949. On January 31, 1950, President Truman gave his approval, and the crash program began. Lawrence surmised that the weapons program would soon require increased supplies of tritium (for the fusion bomb) and fissionable material (for a variety of atomic bombs). As he saw it, Alvarez's linear accelerator design could be scaled up to an enormous neutron foundry capable of producing both kinds of material. Had it been completed, this "Materials Testing Accelerator" (MTA) would have borne a price tag of $5 billion.[100] Despite a great deal of planning, only the front end of one of the projected machines was built; it alone cost over $20 million. By 1952, the entire enterprise was obviated by the discovery of rich natural sources of uranium in the Colorado plateau; the decision to postpone the project indefinitely was taken in August 1952.[101] Alvarez played a central role in much of the MTA work, an experience that (in addition to providing even more expertise in large-scale planning and management) solidified his ties with the AEC.

Before the MTA project ended formally, Alvarez began to reimmerse himself in fundamental physics. Samuel Allison of the University of Chicago wrote to Alvarez in September 1951 inviting him to summarize "your" recent work at an international conference.[102] Responding to the query, Alvarez wrote

I am slightly confused as to whether you mean "you" or "you all." I doubt that I would have anything that would interest the group, since I have been an engineer for the past year and a half and haven't spent much time in the laboratory itself. If you mean "you all" I could mention some of the highlights of the work at Berkeley. . . . I am looking forward to the conference as a chance to catch up on what is going on in the world of physics, since I feel a bit out of touch myself.[103]

As Alvarez "caught up" with particle physics, his experience in defense work would serve him well. Indeed, the engineering, the planning, and the management of the Berkeley bubble-chamber work were integral to its success.

Glaser's presentation thus caught Alvarez at a time when engineering was in his immediate past and accelerator physics on his present agenda. Immediately after the Washington meeting at which Glaser introduced the new detector, Alvarez resolved to set his lab the goal of building very large hydrogen chambers.

The work on the new instrument began at Berkeley much as it had in Michigan and Chicago: on a very small scale. And, as at Chicago, the hydrogen chamber had the highest priority. On May 5, 1953, Lynn Stevenson underscored the word *hydrogen* twice on the first page of the Berkeley group's notebook on bubble chambers. Then came data from the International Critical Table, along with plots of isotherms.[104] Alvarez later put it bluntly: "Who wanted nucleons tied together in a ball of junk? Still we thought we'd start from where Don [Glaser] had gotten to."[105] On these grounds, Stevenson began constructing a small diethyl ether chamber. On May 18, 1953, the prototype, Mark I, was ready. Stevenson noted: "Crappy chamber (not evacuated). At $t = 170$ bubbles formed. . . ."[106] The next machine, Mark II, produced bubbles soon after a radioactive source was inserted. On May 20, Mark III entered the world with a wimper: "glass slipped out of O rings and we evaporated bulb ether. good night. P.S. The reason this chamber might not have worked was the presence of minute glass particles from the filter."[107] Mark IV showed promise for a while on May 22, 1953— the source gave "a definite effect." The next step was to photograph the bubbles using pickup triggered strobes. Sadly, wrote Stevenson, "there was no success in viewing bubbles." The only thing Stevenson claimed to get was "a splitting headache ~2 atmospheres of ether vapor in the sinuses."[108] So it continued through June, July, and August of 1953.

While Stevenson's headaches with prototype ether chambers continued, a young Berkeley accelerator technician, John Wood, bypassed the Chicago group, Glaser, and the senior physicists at Berkeley. Working from the Hildebrand-Nagle design published in August, Wood produced a hydrogen chamber that, by October of 1953, clearly stood at the center of attention.[109] Then, more frustration: "Attempts were made to superheat the liquid and get bubble formation produced with a radium source. . . . There was no effect when radiation was incident on the chamber." Under "suggested improvements," the log book concludes that pressure and thermal gauges are needed and that the experimenters should "replace present chamber with a cleaner one."

Glaser's experiment on very smooth glass chambers had set a standard for everyone—without "cleanliness," it was universally assumed, the spontaneous bubbling initiating at the walls would ruin any attempt to produce tracks. Thus, Stevenson admonished himself in mid October to make a "cleaner" chamber. More work was done to get control of temperature. On January 25, 1954, the chamber was tried again. The experimentalists varied the delay between expansion and flash,[110] and fifty tracks were photographed. The first of these Polaroid photographs are preserved in Stevenson's first notebook.

While staring at these early photographs, Alvarez realized that his group had accidentally stumbled across the most essential feature of all.[111] True enough, boiling was beginning near the walls. Nonetheless, tracks near the center of the chamber were clearly visible. Thus, the failure to make a chamber as clean as Glaser's precipitated the Berkeley group's first great triumph. If accidentally "dirty" walls did not preclude track formation, neither would purposefully imperfect walls. A letter by Wood to *Physical Review* put it this way: "We were discouraged by our inability to attain the long times of superheat until the track photographs showed that it was not important in the successful operation of a large bubble chamber."[112]

Immediately, work began on larger bubble chambers in which the walls were not glass but metal.[113] Had it not been possible to use dirty walls, bubble chambers would forever have remained much too small to study highly energetic events. Even the 2 1/2-inch chamber designed by Parmentier and Schwemin[114] would have required such thick glass walls as to have made the device impractical to heat. Instead, the chamber was brass, 1 inch long, 2 1/2 inches in inside diameter, and capped with glass ends. It worked, and the pace quickened. On April 29, 1954, the first tracks showed up in the 2 1/2-inch chamber; by August, two 4-inch chambers were built, roughly on the same lines as the 2 1/2-inch one.[115] All the results were presented in December at the American Physical Society meeting in Berkeley.[116]

From the very beginning of bubble-chamber work at Berkeley, technicians and engineers played an essential role. None of the early papers were authored by physicists. Wood, Schwemin, and Parmentier were all technicians, and they worked quite autonomously, although they consulted with the physicists.[117] The establishment from the outset of close ties between engineering and scientific personnel certainly played a crucial part in accelerating Berkeley's dominance of bubble-chamber technology. The archival legacy of this coordination is a collection of thousands of "engineering notes," the first of which was filed in the engineering records of the lab and dated January 13, 1955.[118]

By late 1954 it was clear that Berkeley, at least for some time, was going to dominate bubble-chamber technology. The Chicago and Brookhaven groups looked west. Alan Thorndike wrote to Alvarez from the Brookhaven Cloud Chamber Group requesting a chance to work with the Alvarez group. Thorndike added that at Brookhaven "all the effort [was] going into the pentane-filled chamber approach." "But I think we all feel," he continued, "the liquid hydrogen filling, which you have been pioneering, is the thing that will really pay off. . . ."[119] For similar reasons, Darragh Nagle put in for some time in the Berkeley hills.[120] Roger Hildebrand spent the summer of 1954 with the Berkeley group and wrote back glowing reports of the growing project. On June 27, 1954, the remaining Chicago group (Darragh Nagle, Irwin Pless, and Richard Plano) got notice from their representative that the Berkeley group was designing a bubble chamber "4 inches diameter × 3 inches deep!"[121] Hildebrand suggested that his colleagues not "change any of our plans except to think hard about *big* chambers (for the future) (they work)."[122] Later tips that were passed along included the use of a liquid-hydrogen-level indicator and methods of sealing the glass-metal interface.[123]

Alvarez happily reported to Hildebrand in February 1955 that the 4-inch chamber was working in a reproducible fashion and that attention was moving to the physics of stopping positive and negative π mesons in hydrogen.[124] However, while the physics was extremely interesting, expansion continued. Shortly after the 4-inch chamber first gave tracks, construction began on a 10-inch chamber. Just two months later, in April 1955, Alvarez began drafting plans for a much larger instrument, $50 \times 20 \times 20$ inches in sensitive volume.[125] Building devices as big as the 10-inch bubble chamber required two important changes from the earlier, smaller machines. First, engineers joined the physicists and technicians in the design and construction of the chamber. Second, the complexity of the task forced the physicists to delegate some of the preparation of hardware and software for the processing of a vastly increased volume of data.

Thus, by this time the bubble chamber program had grown sufficiently large that its administration required both managerial expertise and extensive funding. Neither the one nor the other was acquired in the course of cosmic-ray research. For Alvarez, as for many of his contemporaries on the Manhattan and radar projects, the intervening war years had taught a great deal about how to run a large scientific-engineering project. For example, in structuring his staff Alvarez put Don Gow in a role that was "uncommon in physics laboratories, but . . . well known in military organizations"; he made him chief of staff.[126] Furthermore, some of the cryogenic hardware and much engineering expertise came through military connections, primarily via the Atomic Energy Commission–National Bureau of Standards cryogenic laboratory in Boulder, Colorado. This lab's first and primary task had been to provide large quantities of liquid hydrogen for the hydrogen-bomb project. Toward this end the laboratory produced two identical hydrogen compressors: one to be left in Boulder for cryogenic experimentation and one to be shipped to the Pacific atolls for the Eniwetok "wet" (i.e., liquid) hydrogen-bomb test.[127] With the equipment at Boulder, engineers and physicists explored the convenience and safety of storing and handling the low-temperature liquids.[128] For example, stainless steel, high-nickel steels, and aluminum alloys were tested down to 20°K for tensile and fatigue strength. Other programs sought to explore high-vacuum techniques and methods of transferring the liquefied gases.[129] Yet another group of workers began to study the thermodynamics of liquefied gases; enthalpy, entropy, and the Gibbs function all needed to be known to provide the large-scale refrigeration and production of the cryogenic liquids. None of these had been studied adequately over the wide range of conditions that were required.

After the development of the Teller-Ulam idea for building "dry" hydrogen bombs with Li^6D (instead of liquid deuterium), the lab's primary mission was finished, and the staff turned their attention to other applications.[130] Liquid oxygen was needed for propulsion in rocket testing, and portable cryogenic systems were required for various other Air Force airplane and missile uses, such as aerodynamic control surfaces, landing gear, communications equipment, cabin pressurization, and trajectory control.[131] Since none of these projects had the priority of the now defunct "wet" hydrogen-bomb effort, the physicists and engineers could turn to other tasks.

In the spring of 1955, the Berkeley bubble-chamber group began consulting with the experts in cryogenics at Boulder. Dudley Chelton, Bascom Birmingham, and Douglas Mann all traveled to Berkeley from Boulder to apply their engineering skills to the rapidly expanding needs of the particle phyicists.[132] Ordinarily, safety problems would arise in discussions among Alvarez, Don Gow, and Paul Hernandez; these would then be addressed in detail by Hernandez working with the National Bureau of Standards personnel. Some typical problems: How rapidly could the glass windows be cooled without fracturing? How often could they be recooled without breaking? Should excess hydrogen gas be vented or burned? Safety issues such as these became major problems in the construction of the 10-inch chamber and remained so for all larger devices.[133] Other joint testing projects included investigating the boiling, burning, and detonation of hydrogen.[134] To cope with these threats, Hernandez structured the engineering defense in three categories: blocking one failure from causing another, separating air from hydrogen, and designing equipment to withstand internal explosions.[135]

It is essential to understand just how dangerous—and frightening—these large quantities of liquid hydrogen were. The danger affected the decision of some laboratories not to build hydrogen chambers at all, and in those laboratories that did proceed the threat of explosion played a crucial role in the partition of work among engineers and physicists,

Galison

336

Figure 4
Engineering safety tests. Many of the components for the 72-inch hydrogen chamber had to be tested for behavior under conditions of high pressure and low temperature. Here, behind a concrete blast wall, an engineer at the National Bureau of Standards in Boulder, Colorado, tests a Joule-Thomson heat exchanger. Photo courtesy of D. Mann, NBS.

as well as in the daily routine of experimentation. Worries about safety were well founded. Many participants in the bubble-chamber work remembered the Hindenburg disaster,[136] and there was an accident in the operation of the 10-inch chamber, as Hernandez reported in 1956:

The 1809 psig rupture disc between the expansion line and the vent system failed causing a sudden drop of pressure in the bubble chamber. . . . about 4 liters of the 8 liters [of hydrogen] flashed almost instantaneously. The gas is believed to have reached supersonic speed for a part of the time as it travelled through the vent pipe. The gas literally screamed as it travelled down the vent line. When the hydrogen gas reached the flame it burned with a big "whoosh," estimated less than a second, with a flame estimated from 10 to 20 feet high.[137]

Fortunately, in this case the venting system worked and no one was hurt. Considerably less fortunate were those working on the 40-inch liquid-hydrogen chamber at the Cambridge Electron Accelerator in the early morning hours of Monday, July 5, 1965. One of the bubble-chamber windows failed. Hydrogen detonated and burned, killing one man, injuring six others, and causing over a million dollars' worth of damage.[138]

One result of the increasing complexity and danger inherent in the large bubble chambers was the initiation of a checklist procedure for the operating crew.[139] Such routinization "[eliminated] operating mistakes and improv[ed] the safety." It also "[gave] all crew members the same objectives and permit[ted] them to anticipate the next step of the operation." "Confusion," Hernandez concluded, was relegated to a lower level because "routine operating decisions [had] been eliminated."[140] Typical safety check-off entries were the following: "Barriers and warning tape installed definitely outlining danger area. Explosion-proof flashlights in area. Personnel wearing conductive shoes. Liquid nitrogen jacket filled well in advance to reduce thermal shock upon introduction of liquid hydrogen. Continuous explosimeter in proper operating condition."[144]

By the time of the 10-inch chamber, the scale, tone, and orientation of the Berkeley project could not have contrasted more sharply with those of the early Glaser work. Whereas Glaser sought to "save cosmic ray physics," Alvarez wanted to beat it. On the opening page of his April 1955 proposal for a new, very large bubble chamber (which was to culminate in the 72-inch chamber), Alvarez spelled out clearly that "the accelerator physicist is often like the hare, in competition with the tortoise-like cosmic ray physicist."[142] As long as the accelerator physicist was limited to the use of variants on the cosmic-ray physicist's detectors (cloud chambers, emulsions, and counters), the contest was more or less equal. Even though the accelerator produces a much higher flux density than the cosmos, the extra particles could not be detected. From Alvarez's perspective, only fast scintillation and Čerenkov counters could keep up with the accelerators, and they had very poor track resolution. (Not everyone agreed with Alvarez—important discoveries such as that of the antiproton led several Berkeley physicists to defend the counter tradition staunchly.) By pushing the cloud chamber to high pressures, the accelerator physicists could get about 50 times more tracks than the cosmic-ray physicists could achieve even on mountain peaks. Since the cosmic-ray particles have a higher energy maximum and thus interact more, the accelerator physicists actually held only a tenfold advantage. "This is an important increase," Alvarez asserted, "but not enough to justify much of the expenditure of ten million dollars, which went into the Bevatron."[143] Such a justification, Alvarez continues, could be based only on a fast-recycling high-density detection device such as the bubble chamber. Cloud chambers could, by 1955, be made to operate at pressures of over 30 atmospheres,[144] but the bubble chamber could perform at a density that would be the equivalent of 700–1,000 atmospheres. Since the number of events per length of track is proportional to the density, this represented an enormous improvement. Even better, instead of cycling every 15 minutes (the characteristic time for a high-pressure cloud chamber), the bubble chamber was ready in 5 seconds. In several ways, the Bevatron therefore called for a new detection device, and the bubble chamber fitted the bill. Physicists too played a role in shaping the plans for making a larger bubble chamber, primarily through Alvarez's consideration of the reaction[145]

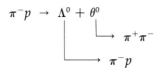

(The identification $\theta^0 \equiv \tau \equiv K^0$ would not be made until the downfall of parity in 1957.[146]) Since Λ^0 and θ^0 are neutral, they do not leave tracks in the chamber. In order for them to be studied, the active volume must be large enough for the neutral mesons to decay and for their decay products to leave measurable tracks. How large a chamber would one need to observe this double decay pattern? In the center-of-mass frame the π^- and the proton have equal and opposite momenta, so the outgoing Λ^0 and θ^0 must have equal values of $|\vec{p}|^2$. Solving for p is easy using energy conservation; knowing p, we immediately deduce that the speed of the θ^0 will be twice that of the Λ°, since the Λ^0 is almost twice as heavy as the θ^0. However, since the Λ^0 lives twice as long as the θ^0, the two particles travel approximately equal distances before decaying. For different π^- energies, one can plot the distance traveled before a typical particle will decay. In the center-of-mass frame, this will give a circle whose radius will grow for higher initial π^- energy. The inverse Lorentz transformation that takes us back to the laboratory frame from the center-of-mass frame expands the axis along which the pion is traveling so that the circle becomes an ellipse. Plots of these ellipses along which Λ^0 and θ^0 will decay (figure 5) were the prize exhibit of the Alvarez proposal. Adding some extra volume of hydrogen to serve as a target before the decay, some to observe the

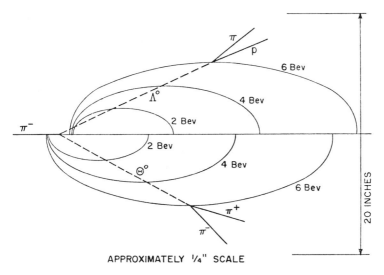

APPROXIMATELY ¼" SCALE

Figure 5
Physics motivation for "72-inch" chamber. Produced in a π^--proton collision, the Λ^0 and the Θ^0 decay by the weak interactions over a length that depends on their energy. The ellipses mark the typical point to which a Λ or a Θ of given energy (in billion electron volts) will travel; the dashed lines give the mean angle from the beam (horizontal axis). This diagram summarized the physics justification for the size of the big chamber. Soon, however, the physicists working with the chamber turned their attention much more to the strong than to the weak interactions. Diagram from Alvarez, n. 125.

decay products, and some to accommodate the finite size of the beam itself, one had (plus or minus a little hand waving) 50 × 20 × 20 inches. "This," Alvarez conceded, "is such a large extrapolation from the present 4 inch diameter chamber, that it would be quite foolhardy to make the jump in one stage. We therefore have well under construction a chamber 10 inches in diameter."[147] Without much ado, the length was soon expanded to 72 inches on the basis of the greater interaction lengths of other strange particles.[148]

When Alvarez voiced concern that the Bevatron was not being fully exploited, his words apparently fell on receptive ears. George Kolstad (chief of the Physics and Mathematics Branch of the AEC Division of Research) drafted a memorandum to T. H. Johnson (director of AEC Division of Research) arguing for more money: "With the increased application of atomic energy, the Russian publications and the great interest of U.S. physicists in this field, it would appear that a substantial increase in funds for high energy physics should be provided as soon as possible."[149] Kolstad had in mind $750,000 for bubble-chamber equipment at the Radiation Laboratory and an additional $2 million for associated facilities. Principal among the expenses were some $200,000 for a power supply, $370,000 for the chamber, and $150,000 for engineering. Some funds came directly through the Atomic Energy Commission, others by transfer to the NBS-AEC facility in Boulder. By the time a reexamination of these figures had been undertaken in late 1955, the cost of building the monster bubble chamber had increased to $1.25 million.[150] Safety features, data-processing equipment, and the magnet were to blame.

The Data Bottleneck

If the first great change in experimentation occurred with the integration of physics and engineering in the design and operation of the bubble chamber, the second took place in the treatment of data. Alvarez ended his 1955 prolegomena to future big machines with his views on the competition. The three visual detectors, he suggested, "all suffer from a common difficulty which is not present in counter experiments. Each event must be studied and measured individually. . . ."[151] In cloud chambers, the stereo pictures were typically "reprojected" through lenses set at the same angle from which the pictures were taken. By rotating a gimbaled frosted glass until the images came into focus, one could find the plane of interaction. Then the curvature could be measured with templates of known curvature. The process was slow and painful. With emulsions the situation was worse. Bubble chambers would provide in a single day enough data to "keep a group of 'cloud chamber analysts' busy for a year." "If this situation could not be improved," Alvarez continued, "the bubble chamber would be nothing but an expensive toy."[152] Instead of a physicist delicately adjusting a movable glass screen, Alvarez pictured a "relatively untrained person" following the tracks by turning a steering wheel. At selected intervals, the coordinates of the crosshairs would be automatically recorded. A computer could then interpolate the three-dimensional path the charged particle had followed. Summaries of events could be put out on IBM cards, and the fantasy was that future "experiments" would involve a physicist sitting happily behind a computer console, sifting automatically through boxloads of cards.

The fantasy was not very far from reality. Computers had been used for several essential tasks in the Manhattan Project, and after the war

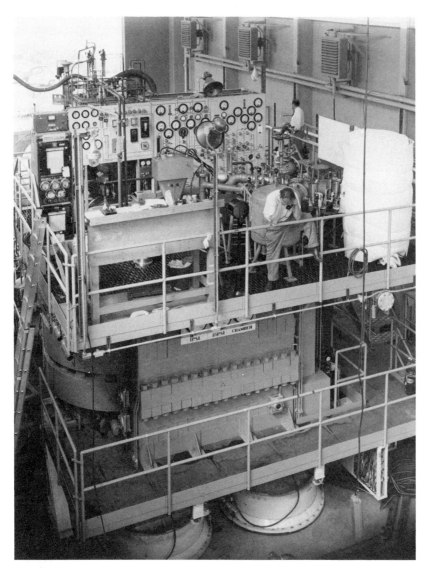

Figure 6
The 72-inch bubble chamber. Visible here are the hydraulic feet used for moving the device, the magnets (left side just above lower walkway), and the instrumentation for refrigeration (upper walkway). The actual chamber is obscured from view, and the compressors that pumped in pressurized hydrogen gas were located in a separate room. Operation began in March 1959. From UCRL Photo Archives.

Alvarez had seen the MANIAC at Los Alamos.[153] This, in conjunction with Alvarez's war work on automatic target tracking and automatic data readout (in the radar program), led to his optimism about data reduction. He wanted a device that could digitize individual stereo views and then reconstruct the spatial track.

Hugh Bradner and James Franck began work on such a machine. In 1957 the "Franckenstein" succeeded in automatically following and measuring tracks.[154] In 1959, after building an improved "Mark II Franckenstein," the measurement group was ready to take on the early output of the 72-inch bubble chamber. Despite their ambition, it was clear that even the grandest plans for extending the Franckensteins could not cope with the flood of pictures that were expected from the large bubble chamber.[155] A typical measurement proceeded in five steps: (1) Scanning. A nonphysicist would look for possible interesting events. (2) Sketching. A physicist would record rough data, including tentative identifications, and instructions for measurement. (3) Measuring. A nonphysicist would measure x-y coordinates of fiducial (calibration) marks and sample points on the tracks by "driving along the track" in the automatic following mode, using a steering wheel for rough guidance and pressing a pedal to produce an IBM punch card on which the coordinates would be encoded. (4) Computing. The computer would fit the tracks to a spatial curve. (5) Checking. A specialist would accept or reject the computer output (for example, offering a new particle hypothesis).[156]

As can be seen from the above outline, human intervention was considerable despite the advances in automation. Nonetheless, the die was cast. As the chairman of one session of a 1960 data-processing conference put it, "The evolution is towards the elimination of humans, function by function."[157] The first function to go was "sketching." The computer took over many of the decisions of the "sketchers," such as which two of three views to use, what number to assign each track, and whether a track stopped on the film.[158]

Programs that took over human labor bore such names as PANG, which reconstructed tracks, KICK, which performed a kinematical analysis by executing a least-squares fitting subroutine, and EXAMIN, which inserted hypotheses on center-of-mass variables into the data and decided whether there was a fit. SUMX, which summarized the data of many points,[159] elicited graphical displays so easily that it became evident that the physicist would "no longer [be] rewarded for his ability in deciding what histograms he should tediously plot and then examine."[160]

Further automation followed with the introduction of the Scanning and Measuring Projector (SMP).[161] With this device the operator pushed a cursor with a 1-cm^2 opening along the projected image of a track. An automatic electro-optical mechanism would locate the track precisely relative to a fixed array of dots of known position. Since the window of the cursor was not nearly so hard to "aim" as the Franckenstein crosshair, life was made considerably easier for the operators. Longer, easier shifts were possible. Furthermore, since the SMP measurements were made on line, the computer was able to check for consistency. If necessary, the machine could indicate to the operator that remeasurement was necessary. Behind this man-machine relationship was the philosophy that scanning for qualitative features is easy for humans whereas executing precise measurements is difficult. For the computer, the reverse is true. The SMP was designed to speed the process by following these strengths and keeping the two roles interactive.

While pursuing the "interactive" philosophy, the Berkeley group inverted the usual relationship of human operator to on-line computer. Before the introduction of the SMP, the operator "[determined] the sequence of operations, calling upon the computer to carry out arithmetic steps quickly and to aid in the human decision process." With the SMP, the machine set the sequence, "calling upon the human operator via the typewriter for those tasks such as pattern recognition, guidance along tracks, etc., at which a human is so adept."[162] Millions of times the computer would ask and the operator would respond in a dialogue dubbed a "script." Here are typical exchanges, with the operator responses underlined. (T2 means track 2, and π indicates measurement completed): "Roll 6748; Meas. Y; Reject N; T1 π; T2 π; T3 π." Not exactly Shavian, but every such exchange saved enormous amounts of time over the course of an experiment. Soon it was said that "the operators even learn to distinguish the patterns of sounds made by the typewriter and to recognize the messages which are transmitted without looking, as the vocabulary of these systems is rather limited."[163]

Further advances could be expected, one programmer wrote, when "a careful human engineering analysis of the entire measuring sequence" was undertaken.[164] Meanwhile, production was gradually speeding up for many reasons. Trivial mistakes by the operator were being corrected by the computer. Moreover, the operator knew that his mistakes would be caught, so he could "push his speed to the limit" since it was now "possible for him to determine what that limit is." Where a more automated device was placed near an older machine, the speed of the new device was "communicated by contagion and emulation to the operators of the non-automated Franckensteins of the same laboratory" in "an interesting example of psychology in man-machine relationships"[165] (an example, the reader might observe, of production management as well).

No matter how cleverly the scripts were rewritten, the exchanges between computer and operator seemed doomed to produce no more than an event every few minutes, or tens of thousands per year. Since the big bubble chambers would soon be churning out millions of events per year, a more completely automated device that could analyze pictures in seconds was needed. In labs across the United States, Europe, and the Soviet Union, groups began exploring the possibility that a fully automatic process could be perfected. At Berkeley, the Alvarez group continued to build by the "interactive" philosophy. Their work culminated in the Spiral Reader.[166] Using this device, the operator simply placed a cursor over a vertex; after that the machine took over, numbering, following, and measuring the tracks that emanated outward. As its name might suggest, the Spiral Reader (figure 7) was based on a photomultiplier tube that moved out from the vertex along a spiral beneath the projected bubble image. In this way both the radial and the angular position of points along the tracks could be determined. One of the most ambitious features of the system was the use of a filtering program to eliminate spots and scratches on the film. In 1967 the Spiral Reader was measuring over a million events per year, whereas in 1957 the Franckenstein had performed a mere 10,000.[167]

By 1966, data reduction had come of age and nostalgia for the student shop or even the departmental shop was irrelevant. "We operate a very large business," Alvarez noted, and he went on to outline just how large. In the United States alone the thriving bubble-chamber industry owned $15 million worth of scanning and measuring devices and spent about $13 million per year, of which $8 million was dedicated to technicians' salaries and $5 million to computer analysis. With the stakes this high, competition for the development of data-analysis techniques became fierce.

Principal among the competitors was a device known as the Flying Spot Digitizer (FSD). Its motivating philosophy was contrary to that of the SMP; instead of linking the human and computer interactively, the FSD sought to separate them.[168] Roughly speaking, the device works as follows: An operator measures a few points on the track and then turns over the operation to the computer. Next a flying spot of light

Figure 7
Data reduction with spiral reader. After the operator placed a spiral guide over the event vertex, an electro-optical reader labeled, measured, and computed the paths of tracks leading out from the vertex. At upper right and left is the computer that encoded the data onto tape. On the desk, one monitor showed the encoded image, the other an enlargement of the photographic image. At lower left is the interactive teletype that printed out inconsistency errors as the operator proceeded. From UCRL Photo Archives.

is projected onto the film so that it scans across the screen, skips to the next line, and scans again. Film from the bubble chamber is then placed between the screen and an array of phototubes that register when the dot has hit a dark spot (a bubble) on the film. The optics in the apparatus are constructed so that the dot's position at any time can be automatically recorded. When the phototubes give the signal, a registering device records the bubble's position. The FSD's enthusiasts hoped that eventually it would be able to recognize automatically the patterns of tracks without any intervention by humans. This precipitated a vigorous debate: Should one try to solve the general pattern-recognition problem, or should an active human element be maintained in data reduction?

In particular, some physicists felt that automation might preclude the possibility of unexpected discovery. Others considered the potential gain in statistics worth the sacrifice.[169] Alvarez argued first on the basis of engineering practicalities: Automatic scanning might be possible 25 years hence (from 1966), but that was far off. When the even more ambitious project of teaching computers to recognize patterns was raised, Alvarez demurred: "More important than [my] negative reaction to the versatile pattern recognition of digital computers is my strong positive feeling that human beings have remarkable inherent scanning abilities. I believe that these abilities should be used because they are better than anything that can be built into a computer."[170] For support, Alvarez turned to a historical example. Pluto was found by the use of a blink comparator, a device in which two images of the sky taken at different times are alternately flashed in front of an observer. It is remarkably easy for a person to pick out a "moving" object (in this case, the "wandering star," Pluto). It would be possible, Alvarez added, for a discovery like this one to be made automatically—but with enormous complications. For example, all the stars' positions would need to be computed in both plates. Comparisons would need to be made between each star's position in the two sets of data, and criteria would have to be introduced to compensate for dust and film distortions. Finally, all the moving objects would have to be checked against the proper motions of other planets, asteroids, and comets. Discussions of this type suggest that even the type of human intervention was under debate. It remains an open and terribly interesting question to ask how other modes of automation affect the relation of the experimentalist to the data.

Engineering affects attitudes as well as specific designs. Early in the 1960s it had seemed that measuring systems that "corrected" the tracks to ensure their conformity to conservation laws might prove worthwhile. Now it seemed that such systems were not cost-effective, given the distortions of turbulence, multiple scattering, and film distortion. A cheaper, practical analysis giving the most "events per dollar" was good enough. "The idea of 'good enough'," Alvarez remarked, "comes from engineering, and although it may upset instrument designers, I feel it is a concept we must embrace."[171] This was certainly the approach taken in the building of the Berkeley accelerators; now it could be applied to the detectors themselves. Engineering and industry offered helpful models in management as well as in instrumentation, as Alvarez noted: ". . . the head of our scanning and measuring group spends a large fraction of his time in a role that would be called, in industry, production management." For example, he would prepare daily, weekly, monthly, and yearly reports for himself and Alvarez, indicating such output data as the number of vertices scanned per operator-hour, the comparative outputs of operators, machines, and shifts, and the allocation of time to measuring, maintenance, instruction, and programming.[172]

The transition to corporate practice did not come painlessly. As Alvarez reminded an audience of physicists, it was well known among industrialists that little companies often go bankrupt trying to pass from an organization in which individuals handled many jobs to a situation in which high efficiency comes from "production line operation with very expensive production tooling."[173] The main difficulty arises with the initial inefficient use of supervisory personnel; once this integration is completed, the company operates at a vastly higher level of productivity. As in business, so in the analysis of physical data.

In these remarks, Alvarez identified a business-historical trend that Alfred Chandler has singled out as the essential feature of modern corporations. Chandler points to the introduction of middle management as the defining feature of the modern business enterprise.[174] Each box in the corporate flow chart stands for an organizational group that could function independently. By the mid 1960s, the Alvarez group clearly had such middle-level groups, such as bubble-chamber operation and development under James Gow and data-analysis development that at various times was under Hugh Bradner and Frank Solmitz.[175] Indicative of the relative autonomy of the subgroups are the many reports, conference talks, and even publications that issued from cryogenic, optical, and data-processing collaborations. Once the large structure became integrated, production did indeed speed up considerably. But Alvarez, knowing that his business methods for accelerating data analysis would startle his audience, reiterated the importance of a pragmatic approach: "For those of you who may be horrified to hear a scientist setting such unscientific goals, let me remind you that I have my 'engineering hat' on at the moment so I have no apologies."[176] Although the value of scientific goals could not be defined in production terms, clearly the preconditions for important work could.

Changes in the organization of work and in the evolving man-machine relationship deeply affected the day-to-day running of an experiment. Indeed, the nature of experimentation had reached a watershed. Kowarski prophesied in 1964 that "today's idea of a physical experiment is concerned mainly with the setting-up and operation (in the accelerator's real time) of the beam-defining and detecting apparatus; in the future, the 'runs' of such apparatus may become more comprehensive, and the individuality of an 'experiment' may shift to the dialogue between the physicist and the data-processing device."[177] That transformation had already begun.

The process by which tasks within a complex technological workplace are fragmented into smaller units has been studied for a number of professions (though more by labor historians than historians of science). For our purposes a particularly interesting example is that of computer programming. During and immediately after World War II, each task assigned to a computer had to be individually prepared at the level of machine language. Stored-program computers, introduced in the United States in 1950, revolutionized the relation of computer to user on every level, from the training of specialists to the types of problems the computer could solve.[178] Almost immediately a huge demand for programmers was created by the American government, which wanted to double the number of programmers from 2,000 to 4,000 in order to prepare the early-warning radar network code-named SAGE (Semi-Automatic Ground Environment). In addition to facilitating improvements in hardware, SAGE required a new staff structure. Early in the program's history the Systems Development Corporation was created to produce the programming workforce. This necessitated restructuring of software training and work organization from an apprenticeship system (largely within electrical engineering) to a much more specialized division of labor. Systems analysts, for example, were trained to supervise programmers, whose programming objectives and methods were highly specific. As this trend continued over the next decades, programming was further broken down into a hierarchy of skill levels. By the late 1970s a Chief Programming Team would make large-scale architectural decisions, less skilled programmers would fill in the modules outlined by the Chief Programming Team, and coders would complete the last details; finally, application programmers would employ the canned programs with minor changes for specialized uses.[179]

This shift in programming work organization is relevant on two levels. In general terms, the specialization of tasks and the isolation of less skilled from more skilled tasks parallels the evolution of laboratory life surrounding the bubble chamber. More specifically, we can see in the history of programming the outlines of the changing role of the computer at the Lawrence Berkeley Laboratory. The first important computer program written for the bubble-chamber group was PANG, the track-reconstruction software, which had to be prepared entirely in machine language. By the time KICK and SUMX were assembled (principally by Frank Solmitz and Arthur Rosenfeld) FORTRAN was available, enormously simplifying a still difficult job. Later, as this software was copied and distributed, its use became easier, with smaller changes needed for adaptation to specific experiments. Exemplary among such cases are the discovery at LBL of the strange resonances and the ω and η particles.

Physics Results

Programming, data reduction, engineering, and physics converged with the discovery of a wealth of new unstable particles in the late 1950s and the early 1960s. Already in 1957 the startling success of the bubble-chamber program led Alvarez to doubt the future of competing detectors. Writing to Edwin McMillan, he elaborated:

For one to appreciate such an assertion the way I feel it, he would have to look at tens of thousands of bubble chamber pictures, the way I have done. When one sees the way in which an interaction of charged particles gives rise to neutral particles which go some distance and then decay into charged secondaries which then may decay into other charged secondaries, or undergo charge exchange reactions, and reappear as neutral particles and so on, he can be as discouraged as I am about the future of counters.[180]

Just such a long string of decays would be needed to find the cascade zero, predicted to exist by Gell-Mann and Nishijima.[181]

From over 10,000 K mesons passing through the chamber, Alvarez hoped to find some of the elusive Ξ^0s. Writing to W. Libby, commissioner of the AEC, Alvarez announced in December of 1958 that his group had such a candidate.[182] With the help of the computer's kinematical analysis, the complicated production and decay of the cascade zero could be deduced:

$$K^-p \rightarrow K^0 \qquad + \; \Xi^0$$
$$ \;\; \rightarrow \pi^+\pi^- \qquad \rightarrow \Lambda^0 + \pi^0$$
$$ \rightarrow \pi^-p$$

Alvarez concluded as follows: "We, of course, hope to get several more of these particles before the run is finished, but even if we don't, I feel we will publish this event without reservations." No more cascade zeros showed up, so the singular event was reported, along with its characteristic photograph.[183]

Exhibition of such "golden events" was one way in which the bubble chamber sould be used to construct an experimental demonstration. Another, equally important use of the bubble chamber was to gather a statistical sample of many decays. During 1960–61, the bubble-chamber/data-reduction program at Berkeley used such statistical evidence to educe the existence of three particles, the strange resonances: $Y_1^*(1380)$, $K^*(885)$, and $Y_0^*(1405)$.[184] (Subscripts indicate the particle's isospin, the asterisk shows that it is an excited state, and the number in parentheses is the mass in MeV.) Illustrative of the mode of their discovery is the case of the $Y^*(1380)$. In this experiment, a monoenergetic beam of K^- particles was incident upon the bubble chamber in which the K particles interact. Forty-nine events of the type $Kp \rightarrow \Lambda^0 \pi^+ \pi^-$ were isolated using the computer to test various particle-identification hypotheses. Another 92 could have been either $\Lambda^0 \pi^+ \pi^-$ or $\Sigma^0 \pi^+ \pi^-$. The goal of the demonstration was to show that in fact the reaction does not suddenly produce all three final-state particles but instead occurs via the production of an intermediary, a new particle Y_1^* (now known as Σ^*):

$$K^- p \rightarrow Y_1^{*\pm} + \pi^\mp$$
$$\phantom{K^- p \rightarrow Y_1^{*\pm}} \longrightarrow \Lambda^0 \pi^\pm$$

Were this the actual production process, the pion produced with the Y_1^* should have peaked energy distribution around 285 MeV (some spread is expected, due to the uncertainty principle). Moreover, the pions made in the decay of the Y_1^* should peak in energy between 58 and 175 MeV. With the supposition that no intermediate Y_1^* existed, one would expect no such peaks in the pion energies. Two plots left the verdict unambiguous. In one, histograms were plotted for number of events versus energy of π^+ (and π^-). The peaks stood out unmistakably. In the other plot (a Dalitz plot), a point was drawn for each event, with the x coordinate representing the π^+ energy and the y coordinate standing for the π^- energy. If no Y_1^* had been present, the points would have been evenly sprinkled over the part of the plane allowed by energy conservation. Instead, and unmistakably, the points huddled together. Further analysis of the data indicated the spin and the isotopic spin of the particle, and similar analyses were used in the unveiling of the $K^*(885)$ and the $Y_0^*(1405)$. This work capped an eminently successful development effort, and for it Alvarez was awarded the 1968 Nobel Prize.

In addition to the success of the in-house physicists, one of the salient features of the new kind of experimentation was the distribution of the bubble-chamber film to outside groups. For example, in the summer of 1959, a visitor to the lab, Bogdan Maglic, began a search for the ω, an isospin-zero neutral meson.[185] Since charge is conserved when an ω decays into pions, it must go into neutral combination such as $\pi^+\pi^-\pi^0$. (The symmetry of the wave function discourages the ω from decaying into two pions.) Maglic then reasoned as follows: In a proton-antiproton collision the ω could be produced along with a $\pi^+\pi^-$ pair. The ω would then decay into three pions, leaving a total of five pions in the final state. Finding the event candidates was no mean feat. Maglic exploited the KICK program to identify the three-pion decays and to calculate the momentum of the unseen neutral one. With the energies and momenta of the pions, he could plot (by hand, since SUMX did not yet exist) the invariant mass of neutral three-pion combinations. A peak confirming the existence of the ω stood out in the region of 800 MeV. That no similar peak could be found for the charged three-pion combinations, such as $\pi^+\pi^+\pi^-$, indicated that the particle was indeed without isospin partners.

Soon afterward, another group led by Aihud Pevsner began examining data from pion-deuteron interactions.[186] Like Maglic, this Johns Hopkins–Northwestern collaboration was searching for a second neutral meson decaying to $\pi^+\pi^-\pi^0$. Again PANG reconstructed the tracks, KICK performed the kinematical analysis, and the results were used to plot the effective mass of the unseen neutral meson. The new particle, called η, weighed in at 550 MeV. As the η and ω discoveries indicate, by the early 1960s not only had the Bevatron been opened to collaborations with other laboratories; so had the bubble chamber and the data-reduction machinery.

Evidence for the new centrality of data reduction to the new definition of experimentation comes from another quarter as well. Emulsion groups all over the world had grown expert at the painstaking reconstruction of tracks and their corresponding kinematics. When the Alvarez group began to distribute some unprocessed film, many emulsion collaborations were eager to paricipate. Werner Heisenberg, on behalf of his colleagues at Göttingen, queried Alvarez about the possibility of obtaining film for the relatively large photographic-plate group of Gottstein.[187]

For bubble and emulsion groups the new machine offered the ability to produce hundreds of events where other methods could not even find one. Larger statistics in the ω and η discoveries aided in identifying the peak of the effective-mass histogram; equally importantly, the larger statistics allowed the determination of ΔE, the width or uncertainty in mass. Heisenberg's uncertainty principle tells us that $(\Delta T)(\Delta E) = \hbar$, where ΔT is the uncertainty in time. Therefore, good mass measurements also determine a particle's lifetime. Here was an unexpected benefit of the bubble chamber, since originally the 72-inch device was designed to measure lifetimes of the Λ^0 and Θ^0, which lived to the ripe old age of 10^{-10} second; the strongly decaying mesons survived but 10^{-23} second.

Over the early and mid 1960s, a great many more particles were discovered in bubble chambers at Berkeley, Argonne, Brookhaven, CERN, and elsewhere. Prominent among these finds were the discovery of the Ξ^* (1530) (independently by a collaboration using the 72-inch bubble chamber[188] and by a group working with the 20-inch chamber at the Brookhaven AGS[189]) and the Σ^*(1385) (by the Alvarez group). Indeed, when the Ξ^*(1530) was announced on July 5, 1962, at the CERN International Conference on High Energy Physics, it fitted neatly into the recently developed SU(3) particle classification scheme that Gell-Mann and Ne'eman had recently developed. Just five days later, at a plenary session of the CERN meeting, Gell-Mann had pointed out in a conference discussion that the SU(3) scheme could account for an equal mass spacing between the Δ(1238) and the Σ^*(1385), between the Σ^*(1385) and the Ξ^*(1530), and between the Ξ^*(1530) and a hitherto unseen particle. Perhaps, he concluded, "our speculation might have some value and we should look for the last particle, called, say, Ω^-." This particle would have strangeness 3, isospin 0, and mass 1,685 MeV. Its discovery at Brookhaven in 1964 dramatically confirmed the SU(3) symmetry.[190] Pursuing this symmetry would lead physicists toward the quarks and a new generation of physics.

By the time the Ω^- was found, plans were being set at laboratories around the world to exploit the bubble chamber for the study of neutrino interactions. It was a direction taken by many physicists who previously had been engaged with the strongly interacting resonance physics, including several from the Alvarez group, such as Lynn Stevenson. Of the many remarkable successes of neutrino bubble-chamber physics, the most influential was undoubtedly the discovery at CERN of weak neutral currents. As a confirmation of the electroweak theory it was remarkable; it also marked a great resurgence in European experimental particle physics.

As important as the physics was, the rapid growth of detectors and data reduction deeply affected the nature of laboratory life. Physicists were hopeful about the new possibilities for experiments, but worried as well. W. Jentschke (later to become director general of CERN) voiced concern that the growth of detectors and data-analysis equipment was driving more and more physicists and engineers into administration.[191] Similarly, at the Dubna conference Goldschmidt-Clermont regretted "the necessity to learn and use extensively the art of programming, or to enlist the help of professional programmers, at the price—amongst others—of a feeling of remoteness from the experiment."[192] For students, the danger was even greater, for while R. J. Spinrad took "the mechanization and acceleration of the more elemental tasks" to be "a great boon to the experienced worker," he added that this same automation might work to the detriment of the student for whom the task might not be "elemental" by virtue of his never having performed it. Students deprived of a chance to "get their hands dirty" would find themselves cut off from the "realities of the experiment."[193]

Concerns such as these about the routinization and specialization of laboratory life began to wear on Glaser and Alvarez as well. In 1960 Glaser moved to Berkeley to join the growing team of hydrogen-bubble-chamber workers. Shortly afterward, in large part because of his disaffection with the large team, he left physics for molecular biology.[194] Indeed, by February 1967 Alvarez too had begun to devote almost all his time to other projects, principally his balloon work on cosmic rays. Asked at that time about experimental nuclear physics with the bubble chamber, he responded that to him experimental nuclear physics was "just a little dull":

. . . so much work can be done by technicians. . . . You have technicians who run alpha particle spectrometers and beta ray spectrometers and gamma ray coincidence circuits. And the people working in the field are doing very much what our graduate students are doing; they are putting things into computers and analyzing the print-out, and they are pretty disconnected from the experimental side of it, in the same way that we are. I can't complain because our people don't go down and look at the bubble chamber very often or at the Bevatron. They ask the bubble chamber operators to expose a certain number of millions of frames of film, and then they ask somebody else to measure them, and then run them through computer programs, and then they start with computer program output and process the data.[195]

Summary and Conclusion

Instruments, Physics, and Work Organization

Interwoven in this essay are three levels of history. The first narrative attends to the specific physics questions that motivate the creation of a new instrument. Concerns such as the nature of strange particles, their decay properties, and their spectra were certainly on the minds of the Berkeley physicists and of Glaser as well. Interactions of strange particles were also of interest; later, hyperon decays and resonances became an essential part of the physics program. In the earliest stages of his exploratory work on new track detectors, Glaser was particularly concerned with the upper end of the cosmic-ray muon spectrum and the properties of the "pothooks." Summing up the contribution of bubble-chamber work in the 1950s and the early 1960s, one might well say that above all a classification scheme was established—the SU(3) scheme of Gell-Mann and Ne'eman. Later work, not discussed here, explored the weak interaction through neutrino experiments in even larger chambers.

Clearly, the history of results is essential. Results are used by theoreticians for inspiration, confirmation, and refutation of ideas, and by engineers and experimentalists in the design of equipment. Ought one to conclude that physics goals constitute the motor for instrumental innovation? Should one argue that the discovery of the bubble chamber was purely an outcome of the search for a better way of measuring the properties of muons and the new unstable particles? Did Alvarez's bubble-chamber program issue simply from the desire to explore K decays? There is some truth here, but clearly not the whole truth. After all, as Ian Hacking aptly put it, "experiments have a life of their own."[196]

To Hacking's well-chosen phrase one might add that instruments have a life of their own. It is their life stories that constitute the second layer of the history of experimentation. In Glaser's case, the cloud-chamber experiments carried over in a myriad of ways to the early bubble chamber. For example, the retrospectively wrong but historically essential electrostatic theory of bubble formation is of thoroughbred cloud-chamber extraction. The bubble chamber itself, with its creation of a metastable thermodynamic state precipitated by a sudden drop of pressure, holds a natural parallel with the cloud chamber. Finally, the product of both apparatuses is the same: detailed photographic tracks where one can see the interaction. Looking aside to a contrasting tradition of electronic "logic" detectors, we might also say of the two devices what they are not. Neither the cloud chamber nor the bubble chamber employs complicated electronics, neither depends on high-voltage technology, and neither produces a logically selected, high-volume output of statistical clicks. Where the image-type detectors produce fine-grained photographs, the logic detectors deliver coarse-grained but more plentiful tracks.

Before 1953 Berkeley detectors were a diverse lot of both image and logic detectors, with neither type overwhelmingly dominating the other. Some groups exploited emulsions, others applied cloud chambers or counters. With the sudden and massive research program of Alvarez, a new, much larger experimental program suddenly towered over the other groups. Even before the bubble chamber was invented, there was frustration with the clumsiness of counter, emulsion, and cloud-chamber work at the Bevatron. The bubble chamber offered more, but only on two conditions: The transition to liquid hydrogen had to be effected, and the pictures had to be processed quickly and effectively. Together these demands resulted in the integration of three new cadres of workers into the process of experimentation: structural and cryogenic engineers, computer programmers, and nonphysicist scanners. Each group brought its own methods and standards to bear on the construction and operation of the bubble chamber, fundamentally altering the nature of instrumentation. Cryogenic engineering, for example, drew on a tradition of safety testing established in industry and in the military. Computer programming likewise had its roots outside the evolution of fundamental physics.

We have seen in some detail how particular tasks were distributed. It has even been possible to observe the separation of skills within the scanning and measuring process. As Kowarski put it, the evolution is toward "the elimination of humans, function by function." This required a constant cycle of analyzing tasks, breaking them down to simple routines, and finally automating them.

These three historical categories—the history of physics questions and results, the history of instrumentation, and the history of work organization—afford us a particularly stark contrast between the early invention of the bubble chamber by Glaser and its subsequent exploitation by the Alvarez group. Moreover, we see how closely the three levels are bound together. For example, the heavy-nucleus bubble liquid is appropriate to a high stopping power and remains liquid at room temperature. Glaser hoped that he could use such a device in small cosmic-ray collaborations. Hydrogen is suited for studying particle interactions at the accelerator and demands a complex reorganization of the research team to fully exploit the apparatus. To ignore the experimental organization is to risk a grave misunderstanding not just of the day-to-day laboratory life but also of the very possibility of doing the physics, for example, of SU(3). Said one more time for emphasis: The organization of specialized subgroups and the integration of engineers, programmers, and scanners was as much a component of the changing experimental physics of the late 1950s and the early 1960s as the bubble chambers themselves.

Instruments and Arguments

In scale, purpose, and use, the bubble chamber had thus changed drastically from its invention in 1952 to its productive exploitation in the late 1950s and the early 1960s. Making use of this detector system, experimentalists began to develop new strategies for the demonstration of novel phenomena. This leads us to a difficult set of questions: What is the connection between the instruments we build and the arguments and entities we construct from them? What role, for example, did the presence of a visual track record of events play in the ascription of the term "particle" to short-lived resonances? What is it about photographic (or in more recent times electronic) records of certain particle interactions that make them so compelling as evidence for new particles or processes? What shapes our criteria for an "adequate" demonstration? When can a demonstration revolve around a "golden event" and when around high statistics? Here I explore only one facet of this complex of questions: How did the development of the large-scale bubble chamber affect its users' reasons for believing in the validity of its data? In other words, how did the new experimental procedures affect the manner in which real phenomena were extracted from the artifactual?

A particularly insightful approach to this distinction is provided by Hacking's analysis of our basis for belief in the images offered by various types of microscopes.[197] His grounds are three. First, Hacking points to the *similarity of results* obtained by apparatus of very different types based on different physical principles or applied to markedly different situations. For Glaser the bubble chamber's first promise of success came from the correspondence between the physical explosion of superheated diethyl ether and the known frequency of cosmic rays as measured by counter experiments. When the tracks were eventually seen, the correspondence with counters, cloud chambers, and emulsions became that much more vivid. Second, our beliefs are strengthened by the possibility of *intervening*; the predictable manipulation of a phenomenon gives added credence to its reality. In a microscope our success in observing a man-made calibration grid gives us faith in the verisimilitude of the enlarged images of hitherto unseen objects. This provides us with an appropriate way of describing the radiation sensitivity experiments in which Glaser could instantly and deliberately induce the violent boiling of superheated diethyl ether by exposure to Co^{60}. Finally, an understanding of the *physical principles* behind the apparatus inspires further confidence in the reports we glean from it. Often the description of the process can be provided in terms of older, well-understood theories, or even by low-level empirical generalizations and approximations. Glaser's work provides a fascinating twist, since his electrostatic repulsion theory (which aided in the construction of the detector) was later abandoned. Each of these factors certainly played a historical role in establishing the bubble chamber as a reliable detector of the paths of charged particles.

Thus far the discussion has centered around the certification of the first prototype chambers by Glaser. How did the problem of extracting a real effect change when the device was transformed by the Alvarez group?

First, new problems arise in the upscaling of the size of the machine— especially problems of *image distortion*. The most troublesome of these aberrations came from setting errors on the film, lens distortion, optical distortion from the windows, thermal turbulence in the chamber, and displacement of tracks by liquid motion.[198] Each of these distorting factors became the subject of working subgroups within various bubble-chamber collaborations. Only by understanding and controlling these effects could real phenomena be extracted from machine artifacts.[199] Second, equipment failures have their parallels in human faults. During bubble-chamber runs employing many scanners, it was common to calculate scanning efficiencies for each individual. From their coefficients, experimental errors and limits on certain processes could be adjusted. Though this sort of compensation for *personal errors* was hardly new with the advent of bubble chambers,[200] it became prominent when the scale of experimentation grew. More surprising, perhaps, were tests of human error at a higher, interpretive level. The Berkeley group developed a program, GAME, that faked histograms resembling the final reduced data of a particular experiment. With only random bumps as structure, a physicist would be asked to pick the real—putatively significant—signal of his experiment from a pile of histograms including the imposters. Alvarez wrote that "one can appreciate how many retractions of discovery claims have been avoided in our group by the liberal use of [this] program."[201] With apologies to the psychologists, we could label this tendency to project unwarranted structure on random information a "Rorschach effect." (As far as I know, the bubble-chamber project at Berkeley was the first to attempt systematically to compensate for our proclivity to see patterns.)

The third and most important innovation in the extraction of reliable signals comes through the extensive development of *data reduction*. As has been stressed above, the computer played a crucial role in this regard, taking over an increasing number of the intermediate steps between particle interactions and published histograms. One by one, scanning, measuring, track reconstruction, kinematic analysis, and experiment analysis were transformed through the exploitation of computers. First, these advances made possible experiments that would have been hopelessly time-consuming and expensive. Second, by statistically fitting a curve to the measured data points, the computer could produce a track more accurate than the photograph would naively reveal. A third task involved the resolution of ambiguities in particle identification. This was not an insignificant advance, since, as one physicist put it, "everyone has a favorite bump which he thinks can be explained by event misidentification."[202] Finally, the automated and routinized reduction of data to usable form (such as invariant-mass and Dalitz plots) increased the rigor of demonstration of real physical effects.

Together, the variety of automated tasks gave data reduction a far more prominent role in the construction of experimental demonstration than ever before in the history of physics. Indeed, its introduction permanently altered both the organization of work in particle-physics laboratories and the nature of experimental argumentation.

In sum: Our grounds for faith in our instruments' reports are manifold. In addition to tests by correlation among diverse instruments, our ability to intervene, and our understanding of underlying physical principles, we have seen many issues arise with the growing scale of particle physics. These include the development of subfields for the study and control of distortion, the understanding of personal errors, and the avoidance of spurious ascription of patterns. But, above all, the mark of the new large-scale physics is the creation of data reduction as an integral part of the experiment. All these techniques have figured in our vastly increased ability to extract real effects from the merely artifactual.

The path from instrument to argument is long. To capture something of the richness of even one device we have had to draw not just on a history of physics results but also on a history of machines and a history of work organization. On each of these descriptive levels Glaser's early development of the bubble chamber contrasts starkly with Alvarez's later exploitation of it. In the course of the evolution of the bubble chamber and other detectors, experimental physics may have crossed a divide. Our older concepts of exploratory experiments and experimental repetition no longer find their precise analogs in the new machines. Conversely, the computer-synthesized experimental runs and the multiplicity of internal checks that characterize the new apparatus have no precisely corresponding elements in earlier forms of experimentation.

The modern history of instruments and the patterns of their use may lack the glamour of the history of our overarching theoretical constructs. Nonetheless, these bubbling, sparking, clanking devices remain the weft and warp of physics.

Acknowledgments

I would like above all to thank Donald Glaser and Luis Alvarez for many helpful discussions and for making available the documents on which this chapter is in large part based. In addition, the following participants in early bubble-chamber work kindly conferred with me and shared notebooks, letters, and other documents: P. Hernandez, R. Hildebrand, D. Mann, D. Nagle, D. Rahm, P. Schwemin, R. Shutt, L. Stevenson, and A. Thorndike.

For making archival material available to me I would like to thank H. R. Crane of the University of Michigan Physics Department, V. Davis of the UCRL Archives, W. Kerr of the Michigan Memorial–Phoenix Project, H. Lowood of the Stanford University Physics Library, R. Seidel of the OHST at Berkeley, and J. Warnow of the AIP Niels Bohr Library.

For helpful conversations, I am particularly grateful to A. Davidson, J. Heilbron, L. Hoddeson, S. Schweber, R. Seidel, and B. Wheaton.

Support for this project has been provided in part by a grant-in-aid from the Friends of the Center for History of Physics, American Institute of Physics.

Notes

1. Some of the historical case studies on experimental physics after 1930 are the following: N. R. Hanson, *The Concept of the Positron* (Cambridge University Press, 1963); P. Galison, "The Discovery of the Muon and the Failed Revolution against Quantum Electrodynamics," *Centaurus* 26 (1983): 262–316; E. M. Purcell, "Nuclear Physics without the Neutron, Clues and Contradictions," in *Proceedings of the Tenth International Conference on the History of Science* (Ithaca, N.Y.: Cornell University Press); J. Heilbron, R. W. Seidel, and B. R. Wheaton, *Lawrence and His Laboratory: Nuclear Science at Berkeley 1931–1961* (Berkeley: OHST, 1981), pp. 81–85; A. Franklin, "The Discovery and Non-Discovery of Parity Nonconservation," *Stud. Hist. Phil. Sci.* 10 (1979): 201–257; P. Galison, "How the First Neutral Current Experiments Ended," *Rev. Mod. Phys.* 55 (1983): 477–509; A. Pickering, "Putting the Phenomena First? The Discovery of the Weak Neutral Current," *Stud. Hist. Phil. Sci.*, forthcoming; A. Franklin, "The Discovery and Acceptance of CP Violation," *Hist. Stud. Phys. Sci.* 13, no. 2 (1983): 207–239.

2. Some of the historical accounts of accelerator development are the following: M. S. Livingston, "Early History of Particle Accelerators," *Adv. Electronics Electron Phys.* 50 (1980): 1–88; M. Goldsmith and E. Shaw, *Europe's Giant Accelerator: The Story of the CERN 400-GeV Proton Synchrotron* (London: Taylor and Francis, 1977); *AGS 20th Anniversary Celebration*, ed. N. V. Baggett (Upton, N.Y.: Brookhaven National Laboratory, 1980); L. Hoddeson, "Establishing KEK in Japan and Fermilab in the U.S.: Internationalism and Nationalism in High Energy Accelerators," *Soc. Stud. Sci.* 13, no. 2 (1983): 1–48; R. W. Seidel, "Accelerating Science: The Postwar Transformation of the Lawrence Radiation Laboratory," *Hist. Stud. Phys. Sci.* 13, no. 2 (1983): 375–400; Heilbron, Seidel, and Wheaton, *Lawrence* (n. 1); D. Kevles, *The Physicists* (New York: Knopf, 1978). Two major accelerator-history projects are underway, one at CERN and one at LBL.

3. See A. Pickering, *Constructing Quarks: The Road to Unity in High Energy Physics* (forthcoming) for a discussion of the role of the bubble-chamber particle discoveries in the development of SU(3) as a classification scheme. More specifically, there is a fine overview of the Lawrence Radiation Laboratory in Heilbron, Seidel, and Wheaton, *Lawrence* (n. 1).

4. Galison, "Muon" (n. 1); R. Kargon, "Birth Cries of the Elements: Theory and Experiment along Millikan's Route to Cosmic Rays," in *The Analytic Spirit*, ed. H. Woolf (Ithaca, N.Y.: Cornell University Press, 1981); R. Seidel, Physics Research in California: The Rise of a Leading Sector in American Physics, Ph.D. Diss., University of California, Berkeley, 1978.

5. Galison, "Muon" (n. 1); Kargon, "Birth Cries" (n. 4); Seidel, Physics Research in California (n. 4).

6. D. Glaser, The Momentum Distribution of Charged Cosmic Ray Particles Near Sea Level, Ph.D. diss., California Institute of Technology, 1950, published as "Momentum Distribution of Charged Cosmic-Ray Particles at Sea Level," *Phys. Rev.* 80 (1950): 625–630.

7. Ibid., p. 4.

8. D. Glaser, interview with author, March 4, 1983.

9. Glaser, Ph.D. diss. (n. 6), pp. 8–11.

10. D. Glaser, interview with author, November 2, 1983.

11. Ibid.

12. G. D. Rochester and C. C. Butler, "Evidence for the Existence of New Unstable Elementary Particles," *Nature* 160 (1947): 855–857.

13. A. J. Seriff, R. B. Leighton, C. Hsiao, E. W. Cowan, and C. D. Anderson, "Cloud Chamber Observations of the New Unstable Cosmic Ray Particles," *Phys. Rev.* 78 (1950): 290–291.

14. D. Glaser, "Elementary Particles and Bubble Chambers" (Nobel Lecture, December 12, 1960), in *Nobel Lectures in Physics* (Amsterdam: Elsevier, 1964), p. 530.

15. Glaser interview, November 2, 1983.

16. For an idea of the kind of work Hazen, Crane, and Parkinson were doing, see W. C. Parkinson and H. R. Crane, Final Report, University of Michigan Cyclotron, Eng. Res. Inst., 1952; W. E. Hazen, C. A. Randall, and O. L. Tiffany, "The Vertical Intensity at 10,000 Feet of Ionizing Particles that Produce Penetrating Showers," *Phys. Rev.* 75 (1949): 694–695.

17. D. C. Rahm, "Donald A. Glaser," unpublished manuscript, 1969.

18. H. R. Crane, manuscript of unpublished colloquium talk to University of Michigan physics department, 1977. I thank H. R. Crane for making this available to me.

19. Glaser interview, March 4, 1983.

20. University of Michigan, Michigan Memorial–Phoenix Project Application for Grant from Research Funds, October 6, 1953. See under "Previous Grants for Research Purposes 1. Phoenix #11."

21. Ralph A. Sawyer to Donald A. Glaser, December 28, 1950. Copy is in files of E. F. Barker, Department of Physics, University of Michigan.

22. D. Glaser, University of Michigan Memorial–Phoenix Project Report on Research Project. Date: July 23, 1951. Project no. 11.

23. Ibid. Years later, after seeing these forgotten proposals, Glaser wrote: "It is embarrassing to admit it—but I really wasn't interested in high energy sea level muons after my thesis. What really interested me was the 'pothooks.' I must have asked for money based on high energy measurements because I thought that was the area in which I had demonstrated competence." (Glaser to author, March 6, 1984).

24. Glaser, Phoenix report (n. 20).

25. Glaser interview, March 4, 1983; Glaser to author, March 6, 1984.

26. Glaser interview, March 4, 1983.

27. J. W. Keuffel, Parallel-Plate Counters and the Measurement of Very Small Time Intervals, Ph.D. diss., California Institute of Technology, 1948, published in *Phys. Rev.* 73 (1948): 531 (abstract from APS meeting, UCLA, January 2–3, 1948), full account in "Parallel-Plate Counters," *Rev. Sci. Inst.* 20 (1949): 202–208; Glaser interviews, March 4 and November 2, 1983.

28. Keuffel, ibid.

29. Glaser, Phoenix report (n. 22).

30. Ibid.

31. Glaser interview, March 4, 1983; Rahm, interview with author, November 15, 1983.

32. Glaser, Phoenix report (n. 22).

33. The continuously sensitive cloud chamber was invented by A. Langsdorf, Jr. ["A Continuously Sensitive Diffusion Cloud Chamber," *Rev. Sci. Inst.* 10 (1939): 91–103] as part of his doctoral work at MIT. Work resumed on the device in 1950. See E. W. Cowan, "Continuously Sensitive Diffusion Cloud Chambers," *Rev. Sci. Inst.* 21 (1950): 991–996; T. S. Needels and E. E. Nielsen, "A Continuously Sensitive Cloud Chamber," *Rev. Sci. Inst.* 21 (1950): 976–977. Related work at Brookhaven began to be published in 1950; see D. H. Miller, E. C. Fowler, and R. P. Shutt, "Operation of a Diffusion Cloud Chamber with Hydrogen at Pressures up to 15 Atmospheres," *Rev. Sci. Inst.* 22 (1951): 280. The Brookhaven group went on to progressively larger and higher-pressure devices.

34. Glaser, Phoenix report (n. 22), p. 2.

35. Glaser interview, November 2, 1983; Cowan, "Continuously Sensitive Diffusion Cloud Chambers" (n. 33).

36. Glaser interview, November 2, 1983.

37. The following two conventions are introduced:
GP = Glaser papers, held by Donald Glaser at the University of California at Berkeley. These include notebooks and a few folders of letters, grant proposals, and miscellaneous documents. Among these papers are two bound notebooks in Glaser's handwriting, which will be referred to by their original numbers, GNB1 and GNB2. The first has no date; the second begins in October 1952. AP = Alvarez papers. These are divided into two parts: pre-1956 and 1956–present. The earlier papers are alphabetized and are stored in inverse chronological order in the archives of the Lawrence Berkeley Laboratory; the later ones, held by Luis Alvarez at LBL, are generally organized chronologically within separate files for each letter of the alphabet.

38. M. Volmer, *Kinetik der Phasenbildung* (Dresden and Leipzig: Steinkopff, 1939). Translated by Intelligence Department as *Kinetics of Phase Formation*, AMC ATI no. 81935 (unclassified), reproduced by Central Air Documents Office.

39. J. J. Thomson, *Applications of Dynamics to Physics and Chemistry* (London: Macmillan, 1888), pp. 164–166, especially p. 166: "We should expect an electrified drop of rain to be larger than an unelectrified one. . . ."

40. Ibid.

41. D. Glaser, Cloud Chamber Droplet Counts, undated one-page document, loose in GNB 2.

42. Actually, any mechanism that deposits between 10 and 1,000 eV will give roughly the correct conditions for bubble growth, so the fact that the electrostatic theory "worked" was not (in retrospect) a stringent test of its validity. As will be discussed below, the actual mechanism for energy transfer to the bubble is by the heat deposited by the penetrating ionizing particle. See C. Peyrou, "Bubble Chamber Principles," in *Bubble and Spark Chambers*, ed. R. P. Shutt (New York: Academic, 1967).

43. The first dated reference to the electrostatic theory is from June 1952: Glaser, "Some Effects of Ionizing Radiation on the Formation of Bubbles in Liquids," *Phys. Rev.* 87 (1952): 665. The actual inequality involving P, $\sigma(T)$, $\epsilon(T)$, and ne first appears in Rahm's progress report listed in n. 58. In many papers [e.g. D. Glaser and D. C. Rahm, "Characteristics of Bubble Chambers," *Phys. Rev.* 97 (1955): 474–479] Glaser indicates that the theory predated his first experiments on superheated liquids. These trials, he states, were "guided by a detailed physical model of the mechanism by which ionization could nucleate bubble formation."

44. GNB 1, pp. 15 ff.; Volmer, *Phasenbildung* (n. 38), pp. 156–165.

45. W. Döring, "Berichtigung zu der Arbeit: Die Uberhitzungsgrenze und Zerreissfestigkeit von Flüssigkeiten," *Z. Phys. Chem.* B36 (1937): 292–294, translated in GNB 1, p. 95.

46. GNB 1, p. 95.

47. GNB 1, p. 87. The works cited by Volmer (n. 38, p. 163) are K. L. Wismer, "The Pressure-Volume Relation of Superheated Liquids," *J. Phys. Chem.* 26 (1922): 301–315; F. B. Kenrick, C. S. Gilbert, and K. L. Wismer, "The Superheating of Liquids," ibid. 28 (1924): 1297–1308.

48. GNB 1, p. 87.

49. "Die Keimbildung, d.h. die durch lokale Schwankungen bedingte Blaschenbildung setzt stets vorher ein und lässt den überspannten Zustand zusammenbrechen." (Volmer, n. 38, p. 165.)

50. Kenrick, Gilbert, and Wismer, n. 47, p. 1298.

51. Data given in ibid., p. 1304.

52. A very useful account of the early work of Glaser and Rahm is found in D. C. Rahm, Development of Hydrocarbon Bubble Chambers for Use in Nuclear Physics, Ph.D. diss., University of Michigan, 1956. A reference to Glaser's cosmic-ray calculation appears on p. 4. There is a fascinating parallel between

these events and the development of the Geiger-Müller tube. For many years, Geiger, Rutherford, and Müller had despaired of making a useful counter because when the voltage was turned up to make the device sensitive, wild collisions ("wilde Stösse") were recorded which did not seem to be due to impurities in the air. Great effort was expended cleaning the walls of the tube before it was realized that cosmic rays were being seen. See T. Trenn, "Die Erfindung des Geiger-Müller Zahlrohres," Veröffentlichungen des Forschungsinstituts des Deutschen Museums für die Geschichte der Naturwissenschaften und der Technik Reihe A, Kleine Mitteilung Nr. 198 (1977).

53. D. A. Glaser, "Progress Report on the Development of the Bubble Chambers," *Nuovo Cimento* Suppl. XI (1954): 361–368.

54. Ibid. Compare D. Glaser, "The Bubble Chamber," *Sci. Am.*, February 1955, pp. 46–50.

55. Ibid. The graduate student was Noah Sherman.

56. D. Glaser, "Some Effects" (n. 43), p. 665.

57. Rahm to J. Warnow, September 22, 1976.

58. D. C. Rahm, progress report, August 1952. Typescript on deposit at AIP Niels Bohr Library. See cover letter to Warnow (n. 57). This work is summarized briefly in D. C. Rahm, Development of Hydrocarbon Bubble Chambers for Use in Nuclear Physics, Ph.D. diss., University of Michigan, 1956.

59. Rahm, Progress Report, section 5.

60. Rahm to Warnow, September 22, 1976.

61. GNB 2. The movies were first reported publicly in Glaser's "Progress Report" (n. 53), p. 366.

62. GNB 2, p. 7.

63. GNB 2, p. 11.

64. Glaser interview, March 4, 1983.

65. Ralph A. Sawyer to Donald A. Glaser, November 21, 1952. Copy is in files of E. F. Barker, Department of Physics, University of Michigan.

66. GNB 2, p. 17.

67. Glaser interview, March 4, 1983.

68. Ibid.

69. Glaser interview, March 4, 1983.

70. GNB 2, p. 37. See also Glaser and Rahm, "Characteristics" (n. 43), p. 367.

71. GNB 2, p. 55; Glaser, "Progress Report" (n. 53), p. 367; Glaser and Rahm, "Characteristics" (n. 43), p. 479.

72. D. A. Glaser, "A Possible 'Bubble Chamber' for the Study of Ionizing Events," *Phys. Rev.* 91 (1953): 496.

73. Glaser interview, March 4, 1983. Nagle remembers only himself and the other scheduled speakers in the audience (Nagle, interview with author, November 28, 1983).

74. Glaser to R. L. Weber, August 17, 1970. GP.

75. D. A. Glaser, "Bubble Chamber Tracks of Penetrating Cosmic-Ray Particles," *Phys. Rev.* 91 (1953): 762–763.

76. Alvarez, interview with author, March 7, 1983; Nagle interview, November 28, 1983.

77. Nagle interview, ibid.

78. H. L. Anderson, E. Fermi, E. A. Long, and D. E. Nagle, "Total Cross Sections of Negative Pions in Hydrogen," *Phys. Rev.* 85 (1952): 934–935; E. Fermi, H. L. Anderson, A. Lundby, D. E. Nagle, and G. B. Yodh, "Ordinary and Exchange Scattering of Negative Pions by Hydrogen," ibid., pp. 935–936; H. L. Anderson, E. Fermi, E. A. Long, and D. E. Nagle, "Total Cross Sections of Positive Pions in Hydrogen," ibid., p. 936.

79. R. Hildebrand and D. Nagle, Old Bubble Chamber Log (notebook, August 1953–November 1955). From personal papers of R. Hildebrand, University of Chicago, on p. 1.

80. Ibid.

81. R. Hildebrand, interview with author, February 17, 1983.

82. R. H. Hildebrand and D. E. Nagle, "Operation of a Glaser Bubble Chamber with Liquid Hydrogen," *Phys. Rev.* 92 (1953): 517–518.

83. Hildebrand and Nagle, Old Bubble Chamber Log (n. 79), p. 13.

84. Preliminary results: I. A. Pless and R. Plano, *Phys. Rev.* 99 (1955): 639(A); R. Plano and I. A. Pless, ibid. 99 (1955): 639(A). Pless's thesis is "Proton-Proton Scattering at 457 MeV in a Bubble Chamber," *Phys. Rev.* 104 (1956): 205–210.

85. Glaser interview, November 2, 1983.

86. Thorndike to Glaser, December 4, 1953. Thorndike files, BNL.

87. Glaser to Thorndike, December 14, 1953.

88. D. Glaser, Study of High Energy Nuclear Interactions Using Bubble Chambers and Further Development of the Bubble Chamber Technique, grant proposal to AEC Physics Division with cover letter to R. Trese. Copy in Glaser's files, University of California, Berkeley.

89. Glaser interview, November 2, 1983.

90. J. L. Brown, D. A. Glaser, and M. L. Perl, "Liquid Xenon Bubble Chamber," *Phys. Rev.* 102 (1956): 586–587.

91. F. Seitz, interview with author, November 29, 1983; "On the Theory of the Bubble Chamber," *Phys. Fluids* 1 (1958): 2–13.

92. Budget proposal sent May 28, 1954 to William E. Wright, Nuclear Physics Branch, Division of Physical Sciences, Office of Naval Research, and to George Kolstad, Physics Division, AEC. I thank Dr. Nagle for making these documents available to me.

93. H. L. Anderson, Proposal to Office of Naval Research for Supplemental Allocation for High Energy Accelerator Research 1 February 1955 to 31 January 1957, dated December 13, 1954. Also see letter, August A. Ebel LCDR, USN Nuclear Physics Branch, to H. L. Anderson, December 7, 1954 (Nagle papers).

94. D. E. Nagle, R. H. Hildebrand, and R. J. Plano, "Scattering of 10–30 MeV Negative Pions by Hydrogen," *Phys. Rev.* 105 (1957): 718–724.

95. Robert Seidel, "Accelerating Science: The Postwar Transformation of the Lawrence Radiation Laboratory," *Hist. Stud. Phys. Sci.* 13, no. 2 (1983): 375–400.

96. Ibid., pp. 399–400.

97. Arthur Roberts, "How Nice to Be a Physicist," mimeographed typescript dated April 27, 1947. AP.

98. For further discussion of Alvarez's work on the radar project see H. Guerlac's *History of Radar* (unpublished microfilm, Bancroft Library, U.C. Berkeley) and Alvarez's unpublished autobiography. Alvarez's work on the atomic bomb is summarized in his autobiography and in D. Hawkins, E. C. Truslow, and C. Smith, *Project Y: The Los Alamos Story* (Los Angeles: Tomash, 1983), pp. 112, 128, 203–206.

99. Seidel, n. 95.

100. See Atomic Energy Commission Objectives of the MTA Program. Report by the Director of Reactor Development. DC files, 1952. General Correspondence, document number AB-2153; DC 8990. For an excellent account of MTA see Heilbron et al., *Lawrence* (n. 1), pp. 62–75. For further documentation see DC files, LBL archives.

101. Heilbron et al., ibid.

102. Allison to Alvarez, September 6, 1951. AP.

103. Alvarez to Allison, September 11, 1951. AP.

104. The earliest Berkeley bubble-chamber notebook is located in the archives at LBL and will be referred to as BNB 1. Later notebooks were kept by Lynn Stevenson and remain among his private papers.

105. Alvarez interview, March 7, 1983.

106. BNB 1, p. 24.

107. Ibid., p. 25.

108. Ibid., p. 27.

109. Ibid., p. 30.

110. Ibid., p. 31.

111. Alvarez interview.

112. J. G. Wood, "Bubble Tracks in a Hydrogen-Filled Glaser Chamber," *Phys. Rev.* 94 (1954): 731.

113. A. J. Schwemin's chamber was first ready for testing on March 9, 1954. BNB 1, p. 39.

114. D. Parmentier and A. J. Schwemin, "Liquid Hydrogen Bubble Chambers," *Rev. Sci. Inst.* 26 (1955): 954–958.

115. L. W. Alvarez, "Recent Developments in Particle Physics," Nobel lecture, 1968, in *Nobel Lectures Physics 1963–1970* (Amsterdam: Elsevier, 1972).

116. D. Parmentier, A. J. Schwemin, L. W. Alvarez, F. S. Crawford, Jr., and M. L. Stevenson, "Four-Inch Diameter Liquid Hydrogen Bubble Chamber,"*Phys. Rev.* 98 (1955): 284.

117. Alvarez, Nobel lecture (n. 115), p. 250.

118. M. L. Stevenson, Bubble Chamber Development: Four-inch Bubble Chamber, Rad. Lab., U.C. Berkeley, Engineering Note Job Number 4311-14, file ms.

119. Alan Thorndike to Alvarez, November 4, 1954; Alvarez to Thorndike, November 16, 1954; Thorndike to Alvarez, November 23, 1954. AP.

120. Nagle to Alvarez, November 23, 1954; Alvarez to Nagle, December 13, 1954. AP.

121. R. Hildebrand to D. Nagle, I. Pless, and R. Plano, June 27, 1954. I thank D. Nagle for making this and the letters of August 20 and September 9 available to me.

122. Ibid.

123. Hildebrand to Nagle, September 9, 1954; Hildebrand to Nagle, Pless, and Plano, August 20, 1954.

124. Alvarez to Hildebrand, February 18, 1955. AP.

125. Luis Alvarez, The Bubble Chamber Program at UCRL. Stenciled typescript dated April 18, 1955, unpublished but widely circulated.

126. Alvarez, Nobel lecture (n. 115), p. 259.

127. Douglas Mann, interview with author, December 15, 1983.

128. F. G. Brickwedde, "A Few Remarks on the Beginnings of the NBS-AEC Cryogenic Laboratory," in *Proceedings of the 1954 Cryogenic Engineering Conference*, ed. K. D. Timmerhaus (New York: Plenum, 1960), p. 4.

129. Anonymous, "Research Facilities of the NBS-AEC Cryogenic Engineering Laboratory," in *Proceedings* (n. 128).

130. Mann interview, December 15, 1983.

131. P. Lieberman, "E.R.E.T.S. Lox Losses and Preventive Measures," in *Advances in Cryogenic Engineering 2* (New York: Plenum, 1960); G. Hohmann and

W. Patterson, "Cryogenic Systems as Auxiliary Power Sources for Aircraft and Missile Applications," in *Advances in Cryogenic Engineering* 4 (New York: Plenum, 1960).

132. Mann interview, December 15, 1983.

133. H. P. Hernandez, "Designing for Safety in Hydrogen Bubble Chambers," in *Advances in Cryogenic Engineering* 2.

134. Ibid. and D. B. Chelton, D. B. Mann, and H. P. Hernandez, UCRL, E.N. 4311–14 M, 33 (cited in ibid.).

135. Hernandez, "Safety" (n. 133), p. 338.

136. Mann interview, December 15, 1983; Alvarez interview, March 7, 1983.

137. Hernandez (n. 133).

138. M. Stanley Livingston, "Semi-Annual Report, July 1, 1965 through December 31, 1965," CEAL-1031, pp. 2–3.

139. UCRL EN 4311–17M6. Hernandez, n. 133, p. 348.

140. Hernandez, n. 133, p. 348.

141. RL-1353-5 (LBL check-off sheet used, e.g. November 4, 1955) for operation of 10-inch chamber).

142. Alvarez, n. 125.

143. Ibid., p. 4.

144. E.g., J. B. Elliott et al., "Thirty-Six-Atmosphere Diffusion Cloud Chamber," *Rev. Sci. Inst.* 26 (1955): 696–697. (Wilson Powell's group at Berkeley.)

145. The above and following discussions are based on pp. 10–12 of Alvarez's "Bubble Chamber Program at UCRL" (n. 125).

146. For an excellent history of this episode see A. Franklin, "The Discovery and Nondiscovery of Parity Nonconservation," *Stud. Hist. Phil. Sci.* 10 (1979): 201–257.

147. Alvarez, n. 125, p. 9.

148. Alvarez, Liquid Hydrogen Bubble Chambers, April 4, 1956, UCRL report 3367.

149. Memorandum: George A. Kolstad to T. H. Johnson, "Expanded Operating Costs for High Energy Physics." AP. Also see letters in DC files numbered DC 55-180, 183, 226, 227, 530, and 744.

150. Donald Cooksey to T. H. Johnson, November 21, 1955, in McMillan papers at Berkeley (document DC 53-131). My thanks to R. Seidel for pointing this item out to me.

151. Alvarez (n. 125), p. 18. At the time Alvarez's "Program" was written there was a heated debate over whether to build high-energy and low-flux machines or high-flux and low-energy machines. Bubble chambers were appropriate for

the former and counters for the latter; Alvarez's "Program" was part of this continuing controversy.

152. Ibid.

153. L. W. Alvarez, "Round Table Discussion on Bubble Chambers," in Round Table Discussion E Data Analysis, in Proceedings of the 1966 International Conference on Instrumentation for High Energy Physics, SLAC, September 1966, p. 271.

154. Alvarez, Nobel lecture (n. 115), pp. 262–263.

155. H. Bradner, "Capabilities and Limitations of Present Data-Reduction Systems," in *Proceedings of the 1960 International Conference on Instrumentation for High Energy Physics* (New York: Interscience, 1961), p. 225.

156. Ibid.

157. L. Kowarski, "V1a Introduction," in *Proceedings of 1960 Conference* (n. 155), p. 223.

158. A. H. Rosenfeld, "Current Performance of the Alvarez-Group Data Processing System," *Nucl. Inst. Meth.* 20 (1963): 422–434, on p. 422.

159. Ibid.

160. Alvarez, Nobel lecture (n. 115), p. 267.

161. See the preliminary reports: L. W. Alvarez, A Proposal for the Rapid Measurement of Bubble Chamber Film, UCRL Physics Notes, Memo 223, November 8, 1960; J. N. Snyder, "Some Remarks on a Data Analysis System Based upon the Scanning-Measuring Projector (SMP)," UCRL Physics Notes, Memo 326, August 25, 1961. A more complete report is L. W. Alvarez, P. Davey, R. Hulsizer, J. Snyder, A. J. Schwemin, and R. Zane, SMP-1 Scanning and Measuring Projector, UCRL-10109, UC-37, Instruments.

162. J. N. Snyder, R. J. Hulsizer, J. H. Munson, and H. Schneider, "Bubble Chamber Data Analysis Using a Scanning and Measuring Projector (SMP) On-Line to a Digital Computer," in *Conference on Automatic Acquisition and Reduction of Nuclear Data Karlsruhe 1964* (Karlsruhe: Ges. für Kernforschung m.b.H., 1964), p. 241.

163. R. I. Hulsizer, J. H. Munson, and H. N. Snyder, "A System for the Analysis of Bubble Chamber Film Based upon the Scanning and Measuring Projector (SMP)," *Meth. Comp. Phys.* 5 (1966): 157–211 (see p. 176) (this paper also includes a good chronology of SMP development); Y. Goldschmidt-Clermont, "Progress in Data Handling for High Energy Physics," in *XII International Conference on High Energy Physics, Dubna* (Moscow: Atomizdat, 1966) pp. 439–462 (see p. 445).

164. Snyder et al., "SMP," in *Karlsruhe 1964* (n. 162), p. 445.

165. H. D. Taft and P. J. Martin, "On-Line Monitoring of Bubble Chamber Measurements by Small Computers," in *XII International Conference* (n. 163),

p. 392; and Y. Goldschmidt-Clermont, "Progress in Data Handling for High Energy Physics," ibid., pp. 443–444.

166. See B. H. McCormick and D. Innes, "The Spiral Reader Measuring Projector and Associated Filter Program," in *Proceedings of 1960 Conference* (n. 155), pp. 246–248. Compare the results of the working version in Alvarez, n. 153.

167. Alvarez, n. 153, p. 288.

168. P. Hough and B. Powell, "A Method for Faster Analysis of Bubble Chamber Photographs," *Nuovo Cimento* 18 (1960): 1184–1191. There is a vast literature on the FSD and later PEPR (an even more fully automated system); however, this subject will have to await fuller treatment elsewhere.

169. See, e.g., the discussion after Macleod's talk in which the relative virtues of SMP and FSD devices were debated: *Proceedings of the 1962 Conference on Instrumentation for High Energy Physics* (Amsterdam: North-Holland, 1963), pp. 381–383. See also the discussion after L. Kowarski, "General Survey: Automatic Data Handling in High Energy Physics," in *Karlsruhe 1964* (n. 162), pp. 26–38.

170. Alvarez, n. 153, pp. 293–294.

171. Ibid., p. 272.

172. Ibid., pp. 277–288.

173. Ibid., p. 289.

174. Alfred D. Chandler, Jr., *The Visible Hand: The Managerial Revolution in American Business* (Cambridge, Mass.: Belknap Press of Harvard University Press, 1977), pp. 1–3.

175. Heilbron, Seidel, and Wheaton, *Lawrence* (n. 1), p. 95.

176. See Alvarez, n. 153, p. 289 and n. 115, pp. 262 ff. for discussions of who did what in the development of computer data analysis.

177. L. Kowarski, "Survey," in *Karlsruhe 1964* (n. 162), p. 36.

178. P. Kraft, *Programmers and Managers* (New York: Springer, 1977).

179. Ibid.

180. Alvarez to McMillan, September 17, 1957, widely circulated. W. A. Wenzel private papers, LBL.

181. M. Gell-Mann, "The Interpretation of the New Particles as Displaced Multiplets," *Nuovo Cimento* Suppl. 4 (1956): 848–866; K. Nishijima, "Charge Independence Theory of V Particles," *Prog. Theoret. Phys.* 13 (1955): 285–304.

182. Alvarez to Libby, December 12, 1958. AP.

183. L. W. Alvarez, P. Eberhard, M. Good, W. Graziano, H. Ticho, and S. Wojcicki, "Neutral Cascade Hyperon Event," *Phys. Rev. Lett.* 2 (1959): 215–219.

184. M. Alston, L. W. Alvarez, P. Eberhard, M. L. Good, W. Graziano, H. K. Ticho, and S. G. Wojcicki, "Resonance in the $\Lambda\pi$ System," *Phys. Rev. Lett.* 5 (1960): 520–524; Resonance in the K-π System," ibid. 6 (1961): 300–302; "Study of Resonances of the Σ-π System," ibid. 6 (1961): 698–705.

185. B. Maglic, L. W. Alvarez, A. H. Rosenfeld, and M. L. Stevenson, "Evidence for a $T = 0$ Three Pion Resonance," *Phys. Rev. Lett.* 7 (1961): 178–182. The actual data reduction was performed by Maglic alone, according to Alvarez's Nobel lecture (n. 115). The distribution of exposed photographs for analysis was not unique to bubble chambers—emulsion groups had done this also.

186. A. Pevsner, R. Kraemer, M. Nussbaum, C. Richardson, P. Schlein, R. Strand, T. Toohig, M. Block, A. Engler, R. Gessaroli, and C. Meltzer, "Evidence for a Three-Pion Resonance Near 550 MeV," *Phys. Rev. Lett.* 7 (1961): 421–423.

187. Heisenberg to Alvarez, September 21, 1956. AP. Gottstein had visited Berkeley to learn about bubble-chamber technique.

188. G. M. Pjerrou et al., "A Resonance at 1.53 GeV," in *1962 International Conference on High Energy Physics at CERN* (Geneva: CERN, 1962).

189. L. Bertanza et al., "Possible Resonance in the $\Xi\pi$ and $K\bar{K}$ Systems," *Phys. Rev. Lett.* 9 (1962): 180–183.

190. M. Gell-Mann, Comment, in *1962 International Conference* (n. 188), p. 805; V. E. Barnes et al., "Observation of a Hyperon with Strangeness Minus Three," *Phys. Rev. Lett.* 12 (1964): 204–206.

191. W. Jentschke, "Invited Summary and Closing Speech," *Nucl. Inst. Meth.* 20 (1963): 507–512, on p. 512.

192. Goldschmidt-Clermont, "Progress" (n. 163), p. 441.

193. R. J. Spinrad, "Digital Systems for Data Handling," *Prog. Nucl. Phys. Inst.* 1 (1965): 221–246, on p. 245.

194. Glaser interview, March 4, 1983.

195. Alvarez, transcript of interview by Charles Weiner and Barry Richman, February 14 and 15, 1967, American Institute of Physics. I thank L. Alvarez for permission to see the transcript.

196. I. Hacking, *Representing and Intervening* (Cambridge University Press, 1983).

197. Ibid., Chapter 11.

198. M. Derrick, "Bubble Chambers 1964–66," in *Proceedings of 1966 International Conference on Instrumentation for High Energy Physics*, SLAC, September 1966, pp. 450–452.

199. See Gargamelle Construction Group, "The Large Heavy Liquid Bubble Chamber 'Gargamelle': The Optics" or M. S. Dykes and G. Bachy, "Vibration of Bubble Chamber Liquid During Expansion," both in CERN Yellow Report 67-26.

200. Personal errors were often discussed in standard textbooks on experiments in the late nineteenth and early twentieth centuries. See for example A. de Forest Palmer, *The Theory of Measurements* (New York: McGraw-Hill, 1912).

201. Alvarez, Nobel lecture (n. 115), p. 267.

202. Derrick, "Bubble Chambers 1964–66, in Proceedings of the 1966 International Conference on Instrumentation for High Energy Physics, SLAC, September 1966, p. 452.

Contributors

Peter Achinstein is Professor of Philosophy at The Johns Hopkins University. He is the author of *Concepts of Science, Law and Explanation,* and *The Nature of Explanation.*

Richard Boyd , Professor of Philosophy at Cornell University, has written articles on the philosophy of science and on metaphysics.

Peter Galison is Assistant Professor in the Program in History of Science, the Department of Philosophy, and the Physics Department at Stanford University. His book *How Experiments End* will appear in 1985.

Owen Hannaway , Professor of the History of Science at The Johns Hopkins University, is author of *The Chemists and the Word: The Didactic Origins of Chemistry.*

Richard C. Jeffrey is Professor of Philosophy at Princeton University. He has written *The Logic of Decision, Formal Logic: Its Scope and Limits, Computability and Logic* (with George Boolos), and numerous papers on probability and induction.

Geoffrey Joseph , Associate Professor of Philosophy at the University of Southern California, writes on the philosophy of science and on metaphysics.

Ronald Laymon , Associate Professor of Philosophy at Ohio State University, specializes in the philosophy of science but is also working on projects in semantics and architectural aesthetics.

John Rigden is Professor of Physics at the University of Missouri–St. Louis and serves as editor of *The American Journal of Physics*. He is the author of *Physics and the Sound of Music*.

Dudley Shapere , Reynolds Professor of the Philosophy and History of Science at Wake Forest University, is the author of many articles and three books on the philosophy and the history of science. His most recent book is *Reason and the Search for Knowledge*. A new work, *The Concept of Observation in Science and Philosophy*, is to be published.

Lawrence Sklar is Professor of Philosophy at the University of Michigan. He has written *Space, Time, and Spacetime* and the forthcoming volume *Philosophy and Spacetime Physics*.

Roger H. Stuewer is Professor of the History of Science and Technology at the University of Minnesota. He is the author of *The Compton Effect: Turning Point in Physics* and editor of *Historical and Philosophical Perspectives of Science, Nuclear Physics in Retrospect*, and *Springs of Scientific Creativity* (with R. Aris and H. T. Davis).

Name Index

Rutherford, Sir Ernest, vii, x, 136,
239–246, 248–293, 294–307nn., 365n.

Salmon, W., 54n., 94
Schmidt, E. A. W., 267, 272, 280,
285–286, 290, 301n.
Schweidler, Egon R. v., 246, 284–285
Schwinger, Julian, 224, 237
Segrè, Emilio, 215, 220, 237
Seidel, Robert, 329, 359, 367n.
Sextus Empiricus, 98–99, 121n., 126
Shapere, Dudley, viii, 21 ff.
Sklar, Lawrence, viii, 1 ff.
Smart, J. J. C., 55n., 94
Sommerfeld, A., 209–210, 212, 237
Stern, O., 207, 210, 212–217, 220, 234n.
Stetter, Georg, 267, 283, 285, 301n.,
303–304nn.
Stevenson, Lynn, 331–332, 352, 359,
368n., 372
Stokes, G., 167–168, 171n., 173
Stuewer, Roger, vii, ix, 239 ff.
Synge, J. L., 153–154, 173

Tait, P. 170–171nn.
Teller, P., 178, 197, 203
Thirring, Hans, 272
Thomson, J. J., 241, 318, 364n.
Thomson, William (Lord Kelvin), 170n.,
171n., 173
Thorndike, Alan, 326–327, 333, 359,
366n., 368n.
Tuve, Merle, 267

van Fraassen, Bas, 50, 94, 120, 202n.,
203
Volmer, Max, 318–320, 363n.
von Kluber, 169n., 173
von Mises, Richard, 109
von Neumann, J., 122n., 197, 203

Ward, F. A. B., 292, 306n.
Weiner, Charles, 266, 269, 296n.,
300–302nn., 305n.
Whewell, William, 128–130, 132–133,
139, 143
Wigner, Eugene, 198, 203, 234n.
Wimsatt, W. C., 161, 170n., 173
Wismer, K. L., 320, 364n.
Wittgenstein, Ludwig, 2
Wood, John, 332–333, 368n.
Wynn-Williams, C. E., 292, 306n.

Zacharias, Jerrold, 219–220, 222–224,
227–228, 234, 237
Zeeman, Peter, 208, 237